Palgrave Studies in the History of Science and Technology

Designed to bridge the gap between the history of science and the history of technology, this series publishes the best new work by promising and accomplished authors in both areas. In particular, it offers historical perspectives on issues of current and ongoing concern, provides international and global perspectives on scientific issues, and encourages productive communication between historians and practicing scientists.

More information about this series at
http://www.palgrave.com/gp/series/14581

Richard J. Altenbaugh

Vaccination in America

Medical Science and Children's Welfare

palgrave
macmillan

Richard J. Altenbaugh
University of Pittsburgh
Pittsburgh, PA, USA

Palgrave Studies in the History of Science and Technology
ISBN 978-3-030-07179-0 ISBN 978-3-319-96349-5 (eBook)
https://doi.org/10.1007/978-3-319-96349-5

Cover credit: GL Archive/Alamy Stock Photo

This Palgrave Macmillan imprint is published by the registered company
Springer Nature Switzerland AG
The registered company address is: Gewerbestrasse 11, 6330 Cham, Switzerland

Acknowledgements

I am a fortunate individual. So many people provided incredible assistance and unflinching support for this project. Archivists everywhere: You are a noble profession! Whether at the Bentley Historical Library, College of Physicians, March of Dimes Archives, Rockefeller Archive Center, University of Pittsburgh Archives, or Watson Institute, among countless collections I visited, I thank you for your incredible knowledge and wonderful cooperation. Colleagues have been very special in my life. Philip W. Gardner, a Fellow at St. Edmunds College, University of Cambridge, facilitated my Visiting Fellowship, offered unconditional support, and proved an urbane and thoughtful companion through many meals and bottles of wine. Thomas O'Donoghue, at the University of West Australia, a friend for many years, has always provided intellectual support and good company, no matter where we were in the world. And William J. Reese, University of Wisconsin, as an editorial advisor for Palgrave Macmillan, endorsed this book for publication following a paper presentation at the British History of Education Society. Springer's editorial staff members, Christine Purdue and Megan Laddusaw, among others, proved extremely supportive and unbelievably patient during the publishing process. These qualities were only surpassed by their professional skills. Family is precious. Ian always thoughtfully enquired about my progress while Colin discretely asked about this book's status through his mother. Marianne, of course, remained steadfast through yet another scholarly project. Finally, Daisy, our sweet English Cocker Spaniel, kept me company in my study and eagerly shared my snacks as I worked on this book. Thank you all!

Contents

CHAPTER 1

Introduction: To Vaccinate, or Not to Vaccinate

The February 4, 2015, front page of the *Pittsburgh Post-Gazette*, the city where Jonas E. Salk developed and refined the polio vaccine, exclaimed, "Despite Objections, Vaccinations Urged."[1] Sixty years after his remarkable breakthrough, many Americans are following a troubling trend to delay or reject childhood immunizations, largely led by a generation that neither witnessed cases of measles, mumps, or rubella (i.e., whooping cough) nor trembled in fear each summer as polio epidemics raged, leaving illness, death, or disability in their wake. What prompted this headline? For several months, two dozen National Hockey League athletes, including Pittsburgh Penguin's player Sidney Crosby, contracted mumps. Nothing symbolized this situation more than this all-star center appearing before news cameras with swollen neck glands that prior December. A month earlier, a media storm had erupted over a measles outbreak at California's Disneyland affecting numerous unvaccinated children.[2]

Decisions to avoid, or at least postpone, immunizations can prove catastrophic, affecting more than a few high-profile, professional athletes and some children on holiday. On July 2, 2015, the *USA Today* reported that a hospitalized woman, in Clallam County, Washington, had contracted measles and died of pneumonia, a common complication from this highly contagious disease. She had become infected from another patient who had been diagnosed with it. Because she had been at this medical facility to receive "medications that suppressed her immune system," her illness and death were tragically coincidental. This article

R. J. Altenbaugh, *Vaccination in America*,
Palgrave Studies in the History of Science and Technology,
https://doi.org/10.1007/978-3-319-96349-5_1

attributed her death to a resurgence of measles.[3] Children's parents, for a variety of reasons, had chosen not to have them inoculated. In an era when such childhood diseases have been largely eliminated, such scenes appeared surreal. How do we explain these parents' actions?

Newspaper commentators at the time summoned the past in general and poliomyelitis in particular to shed light on this phenomenon. David Oshinsky, director of Medical Humanities at New York University and Pulitzer Prize winner for his 2005 book, *Polio: An American Story*, wrote "The Last Epidemic" in the October 18, 2014, edition of *Wall Street Journal*. The development of the polio vaccine signaled the pinnacle of the golden age of medicine. Kathleen Parker a *Washington Post* columnist, in early February of 2015, struck a sardonic tone, "Seeking a Vaccine for Ignorance," for her nationally syndicated article. Its subtitle read, "How Would Salk Be Treated If He Tried to Introduce His Polio Vaccine Today?"[4] Why did these writers invoke Jonas Salk? As medical historian James Colgrove states it, "The nationwide testing and subsequent licensing of Jonas Salk's polio vaccine were watershed events in the history of vaccination in America, bringing the value of the practice to the forefront of popular culture to an extent unequaled before or since."

This study reconsiders the "mythology that surrounds [that] vaccine's development." It reveals a complex experience. The discovery of the polio vaccine represented a remarkable breakthrough for medical science in its own right, marking the death of a dreaded childhood disease. But it was more than this. It also signaled the largest medical experiment using American schoolchildren. Furthermore, the public school system operated as an extension of the medical laboratory, supplying tens of thousands of young human subjects, thousands of trained educators to conduct it, and a medicalized environment. And it became the site of the first mass immunization of children. Finally, it supports the argument that "the fight against polio changed vaccination programs."[5]

Medical science certainly improved the lives of millions of human beings through discoveries that prevented diseases as well as treatments and cures for other illnesses. Nevertheless, it possessed a dark side: Test subjects faced risk. Advancements in vaccines and sera, through most of the twentieth century, relegated them to what scholars term "vulnerable populations;" that is, groups that have traditionally been exploited in scientific experiments. Race and social class often characterized them. African Americans have been exploited for medical experiments from the *antebellum* period through the 1970s; like them, poor and working-class subjects rarely knew they were even participants. Other historians have relied on the term "captive populations" to encompass military troops, prison inmates, and school students. Because of their institutionalization, they seldom, if ever, had an opportunity to *volunteer*. Countless examples exist.[6]

However, children have universally served as research subjects. For example, Albert Neisser, a German professor of dermatology and venereology, inoculated four children and three adolescents with syphilis in 1892. In the USA, physician M. Hines Roberts performed spinal punctures on "423 African-American infants from the newborn service of Atlanta's Grady Hospital," publishing his findings in 1925. Finally, the American medical profession in 1940 hailed children as "'little medical heroes' ... [who] had done so much to advance medical knowledge."[7]

This leads to Jonas Salk's trials with the polio vaccine. Historian Jane S. Smith points out that Salk, while at the University of Pittsburgh, wrote to Harry Weaver, research director at the National Foundation for Infantile Paralysis (the parent organization for the March of Dimes) speculating about human tests as early as June 1950, before the poliovirus typing program had been completed. Using ultraviolet light to inactivate the virus, Salk suggested trying it on "'Hydrocephalics and other similar unfortunates'" residing at local institutions by transferring them to Pittsburgh's Municipal Hospital to keep "'them under isolation and under very close supervision.'" He felt confident that he would easily "'obtain permission'" to use them. He also floated the idea of injecting "'inmates of prisons who might volunteer for such studies.'" Weaver balked at this audacious plan.

> Salk's dream was Weaver's nightmare–taking helpless children, wards of the state, isolating them in Municipal Hospital, injecting them with an unproved vaccine, and exposing them to live poliovirus, even in an attenuated form. Although Salk's proposal was not far from standard experimental procedures in 1950, Weaver knew better than to think the public would regard the test of any polio vaccine as a routine experiment.

Such premature testing could have resulted in a "medical atrocity." As Smith concludes, like other scientists, "[s]ecure in his conviction of his careful work, his good intentions, and the infallibility of his laboratory results, Salk never even imagined such a response."[8]

Medical researchers reflected the ethical context of their times. Although some general guidelines existed at different periods and in various countries, decisions remained at the sole discretion of the experimenter. Or, they chose to simply ignore them. No enforceable guidelines or regulatory bodies, public or private, domestic or international, existed.

Moreover, and completely overlooked, education played a significant role in the ultimate eradication of infantile paralysis. Fear of this disease certainly drove parents to submit their children to a nationwide trial, based at local elementary school buildings. However, the long-standing medicalization of American society as well as the concomitant development of school health programs acted as effective and influential instruments of persuasion. For decades, the media had touted breakthroughs, building a largely trusted medical culture by the 1950s. The public education system, meanwhile, developed a medical infrastructure to accommodate the largest experiment involving children ever attempted. Coupled with National Foundation director Basil O'Connor's keen use of mid-twentieth-century mass media, it assured public consensus and a high level of efficiency.[9]

Additionally, the public school system proved to be an indispensable instrument in the eradication of polio in the USA and as a means to facilitate mass immunizations, yet the histories of education, medicine, and public health have been treated as discrete entities or realities. The medical role and place of public education have been ignored, at worst, or relegated, at best, to the somewhat obscure fields of histories of health education and mental hygiene movement. Was the medicalization of schooling this simple? Within the history of medicine, children—often the actual objects of research—have been subordinated to the talents of famous scientists and the advances of their medical laboratories, ultimately disappearing from the historical scene altogether. And analyses of basic public health policies, which depend heavily on school vaccination regulations, seldom recognize that institution's central role in the larger picture.

This study also relies on the notion that education transcends schooling. It represents a social process, revealing what people learned, how they learned it, and what they did with it. Historian Lawrence A. Cremin broadly defines education as the "deliberate, systematic, and sustained effort to transmit, evoke, or acquire knowledge, values, attitudes, skills, or sensibilities, as well as any learning that results from the effort, direct or indirect, intended or unintended."[10] Education therefore is (and always has been) informal, relying on numerous sources of information, like printed matter, movies, and the Internet, and formal, compulsory and institutionalized. Both played key roles in the medicalization of American society, as well as spurring the most recent anti-vaccination movement.

Polio epidemics and the medicalization of American society, and its public education system, became intertwined during the first half of the twentieth century. This represented a rare confluence of grand human events. The distinction between the research institution and the school building likewise faded as researchers easily shifted their experiments from laboratories to classrooms. Once proved successful, school buildings served as the main dispensaries of the polio antigen. The broad-based, mass immunization campaign that followed changed American public health policy in profound ways.

A historical conundrum existed: Infectious diseases severely harmed children; vaccines eliminated many of them, alleviating suffering, pain, disfigurement, disability, and even death; but to develop these treatments, researchers, and sometimes parents, put children at risk as test subjects. These medical breakthroughs worked. Menacing epidemics abated and in some cases disappeared altogether. Everyone benefitted. However, now many parents and opponents claim that these very vaccines may be as harmful, if not more so, than the diseases themselves.

Anti-vaccination attitudes are not new. Medical historian Elena Conis sees three distinct iterations. Early opponents, during the decades straddling the nineteenth and twentieth centuries, "drew no distinctions between vaccines; they rejected them all out of an age-old belief in the wisdom and beneficence of nature" as well as religious freedom and individual rights. Those during the 1970s and 1980s appeared to be more discerning. They did not veto immunizations outright, seeing some, like polio and tetanus, as appropriate but hesitated about others, like diphtheria and pertussis, because the long-term effects had not been adequately charted. "In the later eighties and nineties, however, these generalized anxieties evolved into well-defined fears of specific harms, such as autism and learning disabilities, and specific chemical vaccine components, such as thimerosal and aluminum."[11]

This issue raises a fundamental question: Who owns the child? By this, I mean the oversight of a child's welfare. Is safety solely the will of a family? Or, does society see a child as an important part of a whole, a member too valuable to allow the whims of individuals to decide his or her fate? Thus, my focus here is on the tension between private and public responsibility.[12] This especially encompasses the concept and practice of health. When preventative measures exist to protect a child from a contagious disease, is this merely a private concern? Or, is this a public matter because it prevents him or her from pain, anguish, or even death? Or, worse yet, an infected child may serve as a vector, spreading a contagion, that is, a medical and social threat. Who ultimately should have the power to make these decisions about, or for, a child? This drives the contemporary American anti-vaccination phenomenon, as well as prodded its earlier counterparts. But the context is broader and more complex than this single issue. Child welfare transcends both anti-vaccination and pro-vaccination arguments: Children have been used as human subjects in vaccine development for hundreds of years. In the process of developing vaccines to protect them from dangerous illnesses, scientists and parents often placed children in harm's way, many became ill while some died.

This study employs an interdisciplinary approach, tapping the relationships between bioethics, education, medical research, and public health to reconstruct the dynamic relationship between medical science and children. Subtopics cover child and animal rights, judicial decisions, mass media, and the pharmaceutical industry. It uses the contemporary American vaccination phenomenon as a springboard to grasp its antecedents amidst early attempts at vaccine development. Why? Because this is a complex story. It does not follow steady and chronological progress; rather, it illustrates resistance to scientific advances, a counter-intuitive experience for many. As science scholar, Stuart Blume summarizes it: "When we … look at how vaccines have been made use of, and developed in the interest of public health, a notion of progress is less all-persuasive."[13] The social world introduces conflict. This study, therefore, not only notes earlier parents' concerns about the risks that vaccinations posed for their children but also examines the dangers children faced as frequent test subjects in vaccine development. In sum, it analyzes the legitimacy of contemporary vaccine refusals within a historical context, cutting through hyperbolic media coverage and polemical tomes supporting or opposing it—then and now—as well as describes the context of medical ethics through the twentieth century.

Notes and Sources

1. David Templeton, "Despite Objections, Vaccinations Urged," *Pittsburgh Post-Gazette*, February 4, 2015: 1–2.
2. Jeff Klein, "Crosby Has Mumps, So Does Brassard," *New York Times*, December 15, 2014: page D5. http://search.proquest.com/docview/1636086195?) (Accessed February 3, 2015); Lance Pugmire, "Mumps Is the Word on Two Ducks," *Los Angles Times*, November 13, 2014: page C3. http://search.proquest.com/docview/1623066791?) (Accessed February 3, 2015); Betsy McKay and Jeanne Whalen, "Vaccine Skeptics on the Rise–Latest Measles Outbreak Shines Light on New Trend Worrying Health Officials," *Wall Street Journal*, January 26, 2015: page A3. http://search.proquest.com/docview/1648007423?) (Accessed February 3, 2015). Another outbreak among hockey players occurred in 2017. See André Picard, "Mumps Outbreaks Underscore Need for Vaccinations," *The Globe and Mail*, February 24, 2017. http://www.theglobeandmail.com/news/national/mumps-outbreaks-underscore-need-for-vaccinations/article34134877 (Accessed February 24, 2017).
3. Meghan Hoyer, "Measles Kills First Patient in 12 Years," *USA Today*, July 3, 2015. www.usatoday.com/story/news/2015/07/02/measles-death-washington-state/29624385 (Accessed January 31, 2016).
4. David Oshinsky, "The Last Epidemic," *Wall Street Journal*, October 18, 2014: page C1. http://search.proquest.com/docview/1613381472?) (Accessed February 5, 2015); Kathleen Parker, "Seeking a Vaccine for Ignorance. How Would Salk Be Treated If He Tried to Introduce His Polio Vaccine Today?" *Pittsburgh Post-Gazette*, February 4, 2015: page B5.
5. James Colgrove, *State of Immunity: The Politics of Vaccination in Twentieth-Century America* (Berkeley, CA: University of California Press, 2006): 113 and 114, respectively.
6. Susan E. Lederer, *Subjected to Science: Human Experimentation in America Before the Second World War* (Baltimore: Johns Hopkins University Press, 1995): 142. Also see pages 115–16. Jane Smith, *Patenting the Sun: Polio and the Salk Vaccine* (New York: William Morrow and Company, 1990), uses the term "captive populations" to include prison inmates and school children (136–37). Michael A. Grodin, "Historical Origins of the Nuremberg Trial," in George J. Annas and Michael A. Grodin, eds., *The Nazi Doctors and the Nuremberg Code: Human Rights in Human Experimentation* (New York: Oxford University Press, 1992): 139.

7. Susan E. Lederer,"Children as Guinea Pigs: Historical Perspective," *Accountability in Research*, 10 (2003): 5 and 2, respectively. Further check College of Physicians, *History of Vaccines* (Philadelphia: College of Physicians, 2013): 20; Gerald L. Geison, "Pasteur's Work on Rabies: Reexamining the Ethical Issues, *Hastings Center Report*, 8 (April 1978): 26–33; Michael A. Grodin, "Historical Origins of the Nuremberg Trial," in Annas and Grodin, eds., *The Nazi Doctors and the Nuremberg Code*: 127, 129; Timothy R. Murphy, "The Ethics of Research with Children," *Virtual Mentor: Ethics Journal of the American Medical Association*, 5 (August 2003): 1–3. http://www.ama-assn.org/ama/pub/category/print/10817.html (Accessed April 29, 2004).
8. Smith, *Patenting the Sun*: 133–34.
9. Flurin Condrau, "The Patient's View Meets the Clinical Gaze." *Social History of Medicine* 20 (December 2007), notes "medicalization" has different meanings: One defines it as "how academic medicine claimed expert status for a range of social issues…." (531–32). Another is that it, at the same time, brought "a growing proportion of society under its control" (531–32). Yet another, but certainly related, interpretation is a "growing medical influence over the patient…." (532). For the purposes of this study, I am relying on a synthesis of these three, thus interpreting it in its broadest sense.
10. Lawrence A. Cremin, *American Education: The National Experience, 1783–1876* (New York: Harper and Row, 1980): ix. Further consult Cremin in *The Wonderful World of Ellwood Patterson Cubberley: An Essay on the Historiography of American Education* (New York: Teachers College Press, 1965) and *Traditions of American Education* (New York: Basic Books, 1977), as well as Bernard Bailyn, *Education in the Forming of American Society* (Chapel Hill: University of North Carolina Press, 1960).
11. Elena Conis, *Vaccination Nation: America's Changing Relationship with Immunization* (Chicago: University of Chicago Press, 2015): 132.
12. This represents a historical dichotomy that has been the subject of historic, judicial conflicts. Refer to Barbara Bennett Woodhouse, "'Who Owns the Child?:' *Meyer* and *Pierce* and the Child as Property," *William and Mary Law Review*, 33 (Summer 1992): 995.
13. Stuart Blume, *Immunization: How Vaccines Became Controversial* (London: Reaktion Books, 2017): 132.

PART I

Diseases, Death, and Disability

CHAPTER 2

Living on the Edge

Jacob A. Riis, a photojournalist and muckraker, published his famous exposé, *How the Other Half Lives* in 1890, portraying living conditions in New York City's shabby and overcrowded tenements. He especially focused on the tenuous existence of children. They lived, slept, and died in the streets, joined gangs, and often pursued a life of crime, that is, if they lived through the waves of epidemics that ravaged their neighborhoods. He wrote, "respiratory diseases, the common heritage of the grippe [i.e., influenza] and the measles, had caused death in most cases [along with] a few isolated deaths from diphtheria and scarlet fever." The mortality rate was such that "[t]here are houses in which as many as eight little children had died in five months."[1] Life appeared cheap.

These and other infectious diseases contributed to high levels of infant and child illnesses in the USA during the late nineteenth and early twentieth centuries. Average annual morbidity figures in the twentieth century, before vaccine development, stood as follows: measles 503,282; diphtheria 175,885; mumps 152,209; smallpox 48,164; rubella 47,745; polio 16,316; and tetanus 1314. Mortality rates proved equally serious, remaining steady at 20% during the late 1800s. Children younger than five years accounted for 48% of New York City's total annual deaths in 1875, still lingering at 40% by 1890. "The three leading causes … were pneumonia, tuberculosis, and diarrhea and enteritis, which (together with diphtheria) were responsible for one-third of all deaths. Of these … 40 percent occurred among children aged less than five years."[2] Tuberculosis, a

R. J. Altenbaugh, *Vaccination in America*,
Palgrave Studies in the History of Science and Technology,
https://doi.org/10.1007/978-3-319-96349-5_2

largely urban contagion, claimed the highest number of deaths among young children in that city. Therefore, "losing an infant and child was common to almost all families" throughout the country, but especially in ever expanding and denser urban areas.[3]

Children came in contact with bacteria and viruses in many ways, some of which overlapped. Impure water exposed everyone to diseases. Rats flourished on the garbage strewn in yards. Horse carcasses lay in streets where they died while dead cats and dogs graced numerous sidewalks. Swarms of flies hovered over decomposing bodies, carrying germs into unscreened doors and windows. Animal droppings fell and remained in streets and on walkways, likewise attracting flies. Humans though served as the most common vector as coughing and sneezing spread contagious water droplets, and dirty hands infected many individuals, directly and indirectly. However, milk, largely consumed by children, served as a notorious vehicle for the spread of disease.

Chicago represents a case in point, as medical historian Jacqueline H. Wolf illustrates. In 1897, 18% of that city's babies died during their first year, especially during summer months. Unpasteurized and unrefrigerated milk served as a petri dish for bacterial growth. Other variables contributed to this toll. The entire process, from cows to kitchen tables, proved risky. To increase profit margins, dairy farmers reduced their expenses by using distillery waste of grains and water to feed their cows to give them a larger amount of liquid. This produced, what many labeled, "swill milk," with little nutritional value. Farmers, shippers, and merchants frequently diluted milk with water up to 75%, contributing to malnourishment, compromised immune systems, illnesses, and deaths. Contaminants posed still other problems. Since dairy farms remained uninspected, dung-filled barns and filthy cows flourished, ensuring that, sooner or later, feces fell into milk pails. Uncovered eight-gallon containers proved vulnerable, allowing dirt and dust to drift into open cans, tainting them during transport. "In order to whiten milk," to make its appearance more appealing, sellers "customarily added handfuls of chalk dust to milk vats." Further, many used dangerous preservatives, like borax, boracic acid, and formaldehyde, to "delay milk spoilage." At local retailers, customers would sample (and test) it, using a common dipper or simply inserting one of their fingers. Consumers who preferred home delivery proved equally susceptible. Deliverymen often picked up used glass bottles from one household, refilled them in their wagons, and left them at the doorsteps of other homes. "In this way, tuberculosis, diphtheria, typhoid fever, scarlet fever, infant diarrhea, and a host of other illnesses became milk-borne diseases." Progress proved frustratingly slow, as Wolf enumerates: "Although Chicago's newspapers first called for pure milk in 1892, dairies in the Chicago milk shed–the vast area spread over seven states that supplied [that city]–did not have to seal milk vats until 1904, did not have to provide milk in bottles until 1912, did not have to pasteurize [it] until 1916, did not have to keep milk cold during shipping until 1920, and did not have to test cows for bovine tuberculosis until 1926." Meanwhile, many childhood diseases continued to exact a dreadful toll in this laissez-faire environment.[4]

Under Siege

The list of infectious diseases proved long and varied. Diphtheria, a bacterium (also known as bladders, canker, rattles, squinancy, throat ail, and distemper), causes swollen tonsils covered by a gray membrane that blocks breathing; it can sometimes lead to kidney failure. Like poliomyelitis, it rarely struck adults. The first serious, recorded North American outbreak occurred in 1735 and swept mercilessly through the colonies. Hampton Falls, New Hampshire, "suffered 210 deaths out of a population of 1200. About 95 percent of the victims were children, and in twenty families everyone below the age of twenty-one died."[5] Diphtheria became endemic in the nineteenth century. In 1875 alone, it accounted for 2329 deaths in New York City, rarely dipping below 1000 per year into the 1890s. It ravaged urban and rural areas alike. But the real figure may never be known, as medical historian John Duffy points out: "Since many deaths from diphtheria went unrecorded, the hundreds of infant deaths attributed to croup and other vague causes undoubtedly included some cases of diphtheria, the actual toll was probably larger than the statistics of mortality show." It proved so common that it "aroused … little concern," eliciting only "casual public reaction.… Doctors could do little about it, and the public attitude was one of resignation." Epidemics continued through the 1940s, numbering between 16,000 and 20,000 cases per year nationwide.[6]

Chickenpox a highly communicable virus, usually striking children aged two through eight, is easily contracted through the respiratory system. A distinct diagnosis for chickenpox remained elusive until the decades straddling the eighteenth and nineteenth centuries, since doctors often confused it with smallpox or scarlet fever; however, the rash caused "by varicella virus ... had nothing to do with poxviruses (or chickens)." These painful skin blisters opened the body to bacterial infections, such as streptococcus. Other serious conditions include respiratory problems, such as pneumonia. Finally, chickenpox never goes away, reawakening later in life as shingles.[7]

Measles, a highly infectious virus, has a ten-day incubation period. Its symptoms include fatigue, fever, cough, runny nose, and a full-body rash. This virus can lead, in extreme cases, to brain damage, deafness, and even death. According to science writer Seth Mnookin, it has "killed more children than any other disease in history."[8] Initial cases in North America date to 1635. It ravaged the colonies. The "most destructive measles epidemic of colonial times, the New England epidemic of 1713, claimed over 100 deaths"—many were adults.[9] Cotton Mather, a prominent Boston clergyman and prolific author, "lost his wife, maid, and three children to measles in a two-week period." Some sixty years later, between 800 and 900 children died of measles in Charleston, South Carolina.[10] "Late in the eighteenth century, however, the disease seems to have become endemic; by 1800 it was primarily a disease of childhood."[11] Infection at a young age conferred lifelong immunity, otherwise known as natural immunization. Measles spread westward as that century unfolded, devastating Native Americans who had no immunity to it. Nevertheless, seaport cities, like Baltimore, Boston, New York City, and Philadelphia, became reservoirs for diseases in general and measles in particular. In New York City, from 1804 to 1865, measles-related deaths increased annually from twelve to a high of 338. The mortality rate varied wildly, with 1.59 per 10,000 in 1804–1809 to 7.29 in 1835–1839. National mortality rates remained high at 43 per 10,000 between 1900 and 1910, but fell during the next twenty years. Virtually, every child contracted measles during the 1950s and early 1960s, annually numbering 3 to 4 million. Of these, 400 to 500 died, 48,000 became hospitalized, and 4000 suffered encephalitis. In 1966 alone, "one in three encephalitis cases died, and another third suffered permanent damage to the central nervous system."[12]

Mumps, another highly contagious virus, spreads through respiratory contact and causes fever, difficult breathing, and painfully swollen salivary glands, lasting between two and three weeks. In some extreme occasions, it infects the brain lining and spinal cord, "causing meningitis, seizures, paralysis, and deafness." When it strikes teens and adults, sterility can result as a rare complication.[13] Nevertheless, it never elicited "public dread."[14] Mumps became a "virtually universal disease in the pre-vaccine era" among children between six and ten years of age.[15]

Pertussis, otherwise known as whooping cough, represents a highly infectious bacterium, spread by the "inhalation of droplets aerolized by sneezing [and] coughing…." This illness strikes infants and causes protracted coughing that lasts three to four weeks, but sometimes can extend up to three months. Such persistent coughing can sometimes result in broken ribs. More seriously, it can hamper breathing, causing cyanosis. Unable to eat, because of coughing so long, many young children became malnourished. The first North American epidemic occurred in South Carolina in 1738. During much of the twentieth century, it ranged between 107,000 and 215,000 infections a year.[16]

Rubella (i.e., German measles), a virus, can cause a fever, rash, and runny nose, usually lasting only three days. Relatively harmless to children, by the "mid-1960s, rubella had joined polio and smallpox in the ranks of diseases actively instilling fear in parents, and particularly mothers." Pregnant women felt especially vulnerable because an infection threatened their fetuses through Congenital Rubella Syndrome, typically resulting in deafness but also could cause damage to the brain, endocrine system, eyes, and heart. Outbreaks ran in five- to seven-year cycles. "A nationwide rubella epidemic in 1963 and 1964 resulted in a reported 30,000 fetal deaths and the birth of more than 20,000 children with severe handicaps."[17] According to historian Elizabeth W. Etheridge's figures, "about three-quarters of them were deaf…." Six years later, when many of these children began their schooling, "[p]ublic education programs for children with impaired hearing were swamped."[18]

Smallpox, a disease that became known as the "greatest killer" of humans earned the sobriquet, "angel of death." While not strictly a children's disease, they suffered the highest mortality rates. This virus enters through the respiratory system and has an incubation period of about twelve days.[19] Symptoms begin with a fever, headache, lethargy, sore throat, and vomiting. The fever abates after a couple of days but resurges shortly thereafter. Sores first appear in all of the mucous membranes—the mouth, throat, and nasal passages. "For some, … drinking becomes difficult, and dehydration follows"; the patient continues to transmit the virus at this stage. At the end of two weeks, a rash spreads to the skin: arms, back, face, neck, the palms of hands, and soles of feet. With oozing pustules, a sickening odor, not unlike rotting flesh, emanates from the infected. This rash lasts about two weeks followed by the formation of scabs.[20] When death occurs, it usually happens ten to sixteen days after initial onset. Smallpox averages a 30% mortality rate, but actual figures can range from 10 to 50%; otherwise, it runs its course for the better part of a month. Pockmarks, permanent scarring with wide and deep pits, cover survivors, causing the most damage to the face. This virus can also result in blindness and loss of ear, lip, and nose tissue. Moreover, the patient stays highly contagious even beyond this period. "Dried blister fluid could reignite infection after three months while dried scabs remained viable for even longer."[21]

Imported to the colonies by English settlers, the first outbreak occurred in Boston in 1630. Additional "... smallpox epidemics ravaged Massachusetts and most of New England in 1648, 1666, 1667, 1689, 1702, 1721, 1731, 1751, 1764, and in the 1770s."[22] Fears of this disease, according to John Duffy, "inspired the majority of quarantine measures in the American colonies." Occurrences continued to ebb and flow, devastating previously unexposed First American populations. Public indifference, however, proved the biggest obstacle. "Members of the Boston Medical Association in 1811 agreed to administer vaccinations at a reduced fee and to vaccinate the poor free of charge." Only a few took advantage of this opportunity; that is, until a smallpox epidemic "swept through the United States from 1815 to 1817...." Then, that city's citizens rushed to exploit this offer. Vaccination proved so successful, sharply reducing infections, that Americans grew complacent again. When infections occurred during the 1840s, this outlook changed once more. General indifference returned when, by the late nineteenth century, the anti-vaccination movement discouraged participation. Smallpox, as a result, continued to exact a heavy toll. Between 1900 and 1904, about 48,000 cases occurred a year with 1500 deaths. Reported infections peaked at 100,000 in 1921, but fell to 10,000 annually by 1930s.[23] It was smallpox that ultimately "triggered the discovery of vaccination, one of the most valuable transferable technologies in the history of medicine and one that has saved hundreds of millions of lives."[24]

Tetanus, a neurotoxin, usually enters the body through a wound, often the result of an injury, attacks the spinal cord, and inflicts muscle convulsions resulting in the "fractures of bones and inability to eat or breathe" as well as rigidity. The backward arching of the trunk represents a common result. This noncontagious illness has an 80% mortality rate within the first four days, usually due to respiratory failure.

The highest number of recorded deaths, 1560, occurred in 1923. Tetanus has virtually disappeared, steadily declining from about 600 cases a year in 1947.[25]

Tuberculosis (TB), a bacterium also called consumption, became known as the white plague because of its usually slender victims' paleness. It proved ubiquitous, defying all attempts to contain it, invading the lungs by human contact, but can affect bones. When it attacks the spinal cord, it causes it to "collapse forward" producing what has often been referred to as "hunchback."[26] Robert Koch, the renowned German scientist who discovered the *tubercle bacillus* in 1882, claimed it represented the most dreaded infectious disease, worse than the bubonic plague and cholera, and estimated that one-seventh of all humans died of it. "Although twentieth-century observers have long blamed urbanization and industrialization for the high rate of tuberculosis in the nineteenth century, they have largely overlooked the growth of institutions as a likely contributor."[27] This includes, of course, the dense crowding of public school classrooms. Physicians, after decades of clinical work, determined that this disease most often struck during the childhood years. In response, beginning in the 1920s, tuberculosis campaigns focused on poor and working-class children; these efforts presumed that poverty contributed to it since tuberculosis had been declining among middle and affluent classes. The Pennsylvania Society for the Prevention of Tuberculosis in concert with local boards of education taught children "to keep their windows open, eat nourishing foods, wash properly, brush their teeth regularly, stand up straight, and get a good night's sleep." Sputum, they also learned, contained tuberculosis germs. Spitting in public, especially on sidewalks, eventually became outlawed.[28]

Mazyck P. Ravenel, in 1904, discovered that "bovine tuberculosis could be transmitted to humans, and in 1912 William H. Park and Charles Krumwiede found that contaminated milk was a significant source of various forms of the disease among children." Therefore, in addition to humans, cow's milk served as a vector. Purifying and distributing it became a key component in controlling infections. It began to decline in 1930s due to the Agriculture Department's "program to eliminate tuberculosis dairy cows…." Pasteurization represented another weapon.[29]

Contagions exerted a profound impact on late nineteenth- and early twentieth-century American society. Except for quarantine, a response to epidemics that dated back to feudal times, the medical arsenal remained reactive and, therefore, limited. As a result, childhood illnesses and deaths became routine: "... between 1866 and 1890 about 43,000 residents of New York [City] had died of diphtheria and croup and that more than 18,000 had succumbed to scarlet fever." It was not until the Progressive Era (1890–1920), when the concept of childhood underwent a profound shift, that society began to assume a proactive stance. However, one virus proved a wild card.[30]

A Medical Puzzle

Poliomyelitis, also known as infantile paralysis, but usually acknowledged simply as polio, represented the most elusive and deadly childhood disease. The earliest outbreaks occurred in rural or small villages: Norway in 1868, Sweden, 1881, and France, 1885. The first, major urban case took place in Stockholm in 1887. Attention shifted to the USA when Vermont recorded a sizeable epidemic in 1894. Still, most state and municipal health authorities did not require medical personnel to report infantile paralysis. When New York state officials instituted this in 1910, recorded figures, including New York City, numbered 112 in 1910, 139 in 1911, 1108 in 1912, 491 in 1913, 224 in 1914, and 257 in 1915. New York City suffered an epidemic in 1907 "when between 750 and 1200 children were afflicted." In 1911, polio struck both Cincinnati and New York City; Buffalo fell victim a year later. Yet, because of misdiagnosis, these figures remained unreliable and underreported. Finally, infantile paralysis erupted in 1916 with thirty-three states reporting 28,767 cases; New York state alone accounted for 13,223 of them. Baltimore, Boston, Philadelphia, and New York City hosted the most intense outbreaks. The latter absorbed the single worst onslaught with 7108 infections and 1654 deaths; 80% of patients were aged four or younger. While infection rates reached 7.9 per 100,000 between 1909 and 1915, this jumped to 28.5 per 100,000 in 1916. Mortality rates soared as well. During the 1907 New York City epidemic, infantile paralysis registered a 5% fatality rate while in 1916 that grew to 27%. In comparison, Vermont and Massachusetts recorded 17 and 23%, respectively.[31]

Onset took place over several days. Early symptoms mimicked a mild cold. Listlessness followed with children complaining about an "upset stomach." Within hours, they found one or more limbs to be "heavy," accompanied by a temperature spike. At this point, parents usually sought medical assistance. However, this did not always help because misdiagnoses proved common, ranging from 15 to 22%. Polio could be ignored as a result. Well into the 1950s polio remained "one of the most difficult of all diseases to recognize accurately." Many physicians initially identified it as appendicitis with spinal meningitis, influenza, kidney infection, meningitis, pleurisy, pneumonia, rheumatic fever, strep throat, or tonsillitis.[32]

Uncertainty continued with the ambulance ride. Not all hospitals welcomed potential polio patients. Others lacked the proper equipment and staff to accommodate young patients. Still others were already filled to capacity and refused to admit new ones. Once the staff confirmed polio, they isolated patients. This abrupt separation from parents was psychologically painful. Audrey, who contracted polio in 1943 at age five, found her hospitalization surreal: "I was taken 250 miles from home, admitted to the hospital, and didn't see my mother for two months. When I did … she was on one side of a glass window and I was on the other. She couldn't hold me, she couldn't comfort me. She couldn't even touch me." Likewise, parents became traumatized as they watched their child struggle to live.[33] No prescribed time frame had been formulated for hospital-based care and treatment. "And the release was rarely permanent. Many polio survivors spent years after their initial hospital stays undergoing more surgeries, post-surgical rehabilitation, and outpatient care."[34] Haven Emerson, New York City's public health director in 1916, reported the prognosis that grew out of this calamity: If the patient survived at all, care focused on the ability to "function."[35] This disease would become the leading cause of children with physical disabilities.

Paralysis usually, but not always, occurred between forty-eight and seventy-two hours of the first signs of infection. Chicago's Department of Health statistically pinned down the affected areas: 82% legs, 11% face, and 11% arms. If paralysis spread to children's diaphragms, medical personnel resorted to tracheotomies in an effort to save them. This did not always work. The invention of an artificial respirator in 1929, nicknamed the iron lung, reduced mortality rates. Until 1940, with the appearance of Australian nurse Elizabeth Kenny, rehabilitation remained crude. To this day, no treatment exists for this disease. "Poliomyelitis, like smallpox, sparked a popular terror far out of proportion to the number of deaths it caused." Two reasons explain this. First, while the infected person either recovered or died from a contagion, like the 1918 influenza pandemic, infantile paralysis primarily struck very young children, fueling public sympathy and parental fear. Second, no other disease evoked such intense emotions. Polio was highly visible: Children with physical disabilities wore braces or rode in wheelchairs.[36]

Polio thus created a "terrifying" reality: "the epidemics between 1930 and 1945 never reached the magnitude of the 1916 epidemic that devastated the northeastern USA, but they occurred with sufficient severity and regularity to keep the crippling of polio in the minds of anxious parents and physicians." Subsequent episodes only intensified matters, with 25,698 cases nationwide in 1946, 42,033 in 1949, and 57,879 in 1952. That year it paralyzed more than 21,000 individuals.[37]

How this virus spread remained a mystery for decades. Two competing theories vied to explain the transmission of polio. First, a "propagated" epidemic occurs when it is spread directly, that is, person-to-person, or other agent, like flies. The second involves a "common-vehicle" which disseminates the "causative agent," like milk.[38]

Milk as a potential source of this pathogen long dominated the attention of medical experts, even in light of inconclusive findings. In 1926, the American Medical Association released a pamphlet analyzing the 1925 polio epidemic that struck Cortland, New York, specifically an eleven-day period, beginning on December 14, when eight children, aged two through nine, fell ill. This painstakingly detailed study asserted that all of them had ingested unpasteurized milk from the same bottler. Data tables and vector maps peppered that report. Nevertheless, while pointing to milk as the culprit, the research team carefully couched its conclusion: "While this outbreak points to transmission through milk, we are not of the opinion that this is the usual mode of spread of poliomyelitis."[39] Nevertheless, milk, a mainstay of children's diets, remained under close scrutiny. Following the 1943 Texas outbreak, with 1271 reported cases, the largest to date in that state, scientists in their quest for answers focused on Fort Worth and methodically examined numerous case studies to discover how this disease spread from one person to another. They found sanitation procedures to be more than adequate, namely those involving the purity of food, milk, and water. They also included insects in their analysis and found few flies, mosquitoes, or roaches present in the areas with the highest number of infections. Finally, they tested chicken and rodent droppings and likewise found no signs of the virus.[40]

Nativists typically attributed a variety of epidemics during the 1800s and early 1900s to immigrants, blaming Irish newcomers for a cholera outbreak during the 1840s and East European Jews for cholera and typhus in 1892. New Yorkers, in particular, pinpointed Italians and their slum, known as "Pigtown," as the source of polio epidemics.[41] When a noted, medical authority like Simon Flexner, the Rockefeller Institute's director, faulted foreigners, because infections seemed prevalent in their communities, New York City's public health officials restricted their cultural events and "carefully" monitored those neighborhoods. The police proved especially vigilant toward Italians who dared to venture out of them. This attitude faulted the poor as well, regardless of ethnic or racial backgrounds. Health authorities argued these groups lived in unsanitary conditions, and their lack of hygiene posed a health threat to the entire population. However, not only did reality defy preconceived beliefs but contradictions existed. Infections remained virtually nil in the Barren Island community occupied by impoverished African Americans, Italians, and Poles.[42] Further, while observing the 1916 New York epidemic and attending the mayor's July 12 emergency meeting, Chicago's commissioner of health John D. Robertson, received reassurances from the health officers at Ellis Island and the Port of New York that the "disease was not imported from Italy."[43] Finally, higher rates of infection among native-born, middle-class children discredited this notion.

Polio outbreaks would remain seasonal, during the five hottest months: June, July, August, September, and October. Although a sharp discrepancy occurred in the 1916 geographical distribution of infections, this would not continue. Depending on the year, this disease struck some regions but skipped others. It ravaged some cities yet left others unscathed. New York, the most populated city in the country suffered greatly, especially in 1916, but Chicago, the second largest, seemed to be impervious to this virus. American children residing in all regions, whether urban or rural, eventually would be affected. Roland H. Berg, writing in *Poliomyelitis and Its Problems* in 1948, summarized the frustrating nature of this disease: "... *polio is an uncontrolled disease.* Nothing that medical science can yet do, can prevent one case or one epidemic from occurring; nor can health officials foretell where or when polio will make its visitation. It will strike where, when, and as severely as it pleases."[44]

Morbidity and mortality figures certainly paled when compared with other contemporary epidemics, causing many in the medical community to discount it. Haven Emerson articulated this point in 1916. Earlier cholera and yellow fever outbreaks claimed higher overall rates, dwarfing polio epidemics that chronologically overlapped it. This would prove especially true of the toll from the 1918 influenza pandemic.

The fledgling medical research community worked with relatively crude technology and instruments and drew on limited, and often ambiguous, epidemiological knowledge in its attempts to unlock the secrets of infantile paralysis. They grappled with clarifying symptoms, identifying origins, and finding cures, but they faced significant obstacles and seemingly unanswerable questions. "If polio mocked the dreams of middle-class culture, it mocked the gods of science even more."[45] Confusion reigned; false starts occurred; many desperate and, ultimately, futile efforts ensued. The fluid nature of medical science made the situation even more complex and policymaking difficult for public health officials.

Notes and Sources

1. Jacob A. Riis, *How the Other Half Lives* (1901; Rpt. New York: Dover Publications, 1971): 129. For grippe, refer to Kenneth F. Kiple, ed., *The Cambridge World History of Human Disease* (Cambridge: University of Cambridge Press, 1999).
2. Alexandra M. Levitt, D. Peter Drotman, and Stephen Ostroff, "Control of Infectious Diseases: A Twentieth-Century Public Health Achievement," in John W. Ward and Christian Warren, eds., *Silent Victories: The History and Practice of Public Health in Twentieth-Century America* (New York: Oxford University Press, 2007): 3. See also John Duffy, *The Sanitarians: A History of American Public Health* (Urbana: University of Illinois Press, 1992): 181; Alan R. Hinman and Walter A. Orenstein, "A Shot at Protection: Immunizations Against Infectious Disease," in *Silent Victories*: 63–65.
3. Duffy, *Sanitarians*: 180. Refer to pages 179, 207, as well as Elizabeth W. Etheridge, *Sentinel for Health: A History of the Centers for Disease Control* (Berkeley: University of California Press, 1992): 168; Levitt, Drotman, and Ostroff, "Control of Infectious Diseases": 7; Nancy Tomes, *The Gospel of Germs: Men, Women, and the Microbe in American Life* (Cambridge: Harvard University Press, 1998): 66.
4. Jacqueline H. Wolf, "Saving Babies and Mothers: Pioneering Efforts to Decrease Infant and Maternal Mortality," in *Silent Victories*: 138–39. For more on this topic, consult Philip J. Hilts, *Protecting America's Health: The FDA, Business, and One-Hundred Years of Regulation* (New York: Alfred A. Knopf, 2003): 23.

5. Duffy, *Sanitatians*: 11. Also see page 179, and Eric H. Christianson, "Medicine in New England," in Judith Walzer and Ronald L. Numbers, eds., *Sickness and Health in America: Readings in the History of Medicine and Public Health* (Madison: University of Wisconsin Press, 1997): 54; James Colgrove, *State of Immunity: The Politics of Vaccination in Twentieth-Century America* (Berkeley, CA: University of California Press, 2006): 82; Elena Conis, *Vaccination Nation: America's Changing Relationship with Immunization* (Chicago: University of Chicago Press, 2015): 39; Evelynn Maxine Hammonds, *Childhood's Deadly Scourge: The Campaign to Control Diphtheria in New York City, 1880–1930* (Baltimore: Johns Hopkins University Press, 1999): 178–79.
6. John Duffy, "Social Impact of Diseases in the Late 19th Century," in *Sickness and Health in America*: 418–19. Centers for Diseases Control and Prevention (CDC), "The Pink Book", https://www.cdc.gov/vaccines/pubs/pinkbook/dip.html (Accessed August 30, 2017).
7. Gareth Williams, *Angel of Death: The Story of Smallpox* (New York: Palgrave Macmillan, 2011): 2. Also consult R. H. Kampmeier, "Varicella Zoster," in *Cambridge World History of Human Disease*: 1092–93; Paul A. Offit, *Deadly Choices: How the Anti-Vaccine Movement Threatens Us All* (New York: Basic Books, 2011): 164; and Jonas Schmidt-Chanasit and Andreas Sauerbrei, "Evolution and World-Wide Distribution of Varicella–Zoster Virus Clades," *Infection, Genetics and Evolution*, 11 (2011): 2. http://www.sciencedirect.com.pitt.idm.oclc.org/science/article/pii/S1567134810002431 (Accessed March 8, 2017).
8. Seth Mnookin, *The Panic Virus: The True Story Behind the Vaccine-Autism Controversy* (New York: Simon and Schuster, 2011): 19.
9. Christianson, "Medicine in New England": 54. Refer also to Colgrove, *State of Immunity*: 151; Hinman and Orenstein, "A Shot at Protection": 69–70.
10. Hinman and Orenstein, "A Shot at Protection": 70.
11. Christianson, "Medicine in New England": 64.
12. Conis, *Vaccination Nation*: 52. CDC, "Measles History": https://www.cdc.gov/measles/about/history.html (Accessed August 30, 2017). See as well Andrew Cliff, Peter Haggett, and Matthew Smallman-Raynor, *Measles: An Historical Geography of a Major Human Viral Disease, from Global Expansion to Local Retreat, 1840–1990* (Cambridge, MA: Blackwell Publishers, 1993): 66–67, 98–99, 118, 162–63; Hinman and Orenstein, "A Shot at Protection": 70. Etheridge, *Sentinel for Health*, provides equally sobering figures: rare complications like encephalitis, "occurred in one case per thousand"; it left "one-third of these victims mentally retarded. One measles patient in every two or three thousand

died, one in two hundred had pneumonia, and one in twenty got ear infections" (168).

13. Paul A. Offit, *Vaccinated: One Man's Quest to Defeat the World's Deadliest Diseases* (New York: Smithsonian Books, 2007): 21–22.
14. Conis, *Vaccination Nation*: 65–66. Also consult College of Physicians, *History of Vaccines* (Philadelphia: College of Physicians, 2013): 33.
15. Hinman and Orenstein, "A Shot at Protection": 74. See also Albert E. Barskey, John W. Glaser, and Charles W. LeBaron, "Mumps Resurgences in the United States: A Historical Perspective on Unexpected Elements," *Vaccine*, 27 (2009): 6187; Robert J. Kim-Farley, "Mumps," in *Cambridge World History of Human Disease*: 887–88.
16. Hinman and Orenstein, "A Shot at Protection": 72. CDC, Surveillance and Reporting: https://www.cdc.gov/pertussis/surv-reporting/cases-by-year.html (Accessed August 30, 2017). Further check, Anne Hardy, "Whooping Cough," in Kiple, *Cambridge World History of Human Diseases*: 1095; Edward A. Mortimer, Jr., "Pertussis Vaccine," in Stanley A. Plotkin and Edward A. Mortimer, Jr., eds., *Vaccines* (Philadelphia: W.B. Saunders Company, 1994): 91, 99; Offit, *Deadly Choices*: xii–xiii. Mnookin, *Panic Virus*, chronicles the symptoms and sometimes fatal consequences of whooping cough in Chapter 23.
17. Conis, *Vaccination Nation*: 75. Refer as well to Colgrove, *State of Immunity*: 174–75; College of Physicians, *History of Vaccines*: 32; Etheridge, *Sentinel for Health*: 174; Hinman and Orenstein, "A Shot at Protection": 75; Robert Kim-Farley, "Rubella," in Kiple, *Cambridge World History of Human Diseases*: 987–88.
18. Etheridge, *Sentinel for Health*: 174. Also consult Offit, *Vaccinated*: 70–72.
19. Williams, *Angel of Death*: 1.
20. Elizabeth A. Fenn, *Pox Americana: The Great Smallpox Epidemic of 1775–82* (New York: Hill and Wang: 2001): 18. Moreover, refer to page 17 and Fenn's chart, "Communicability, Symptoms, and Pathogenesis," on page 19 that provides a daily breakdown of a smallpox infection; Offit, *Vaccinated*: 31; Williams, *Angel of Death*: 17–24.
21. Williams, *Angel of Death*: 9. See pages 26, 32, and College of Physicians, *History of Vaccines*: 15. Refer as well to page 11, and Fenn, *Pox Americana*: 19.
22. Christianson, "Medicine in New England": 54.
23. Duffy, *Sanitatians*: 28 and 54–55, respectively. Refer to pages 56, 200–201, and Hinman and Orenstein, "A Shot at Protection": 65; John Duffy, "Social Impact of Diseases in the Late 19th Century": 418.
24. Williams, *Angel of Death*: 179.
25. Hinman and Orenstein, "A Shot at Protection": 74. Refer, as well, to CDC, "Tetanus": https://www.cdc.gov/tetanus/surveillance.html

(Accessed September 2, 2017); Robert J. T. Joy, "Tetanus," *Cambridge World History of Human Diseases*: 1043–44.

26. Thomas M. Daniel, *Captain of Death: The Story of Tuberculosis* (Rochester, NY: University of Rochester Press, 1997): 10–11. Also see page 32.
27. Barbara Bates, *Bargaining for Life: A Social History of Tuberculosis, 1876–1938* (Philadelphia: University of Pennsylvania Press, 1992): 322. See as well Daniel, *Captain of Death*: 32; J. N. Hays, *The Burdens of Disease: Epidemics and Human Response in Western History* (New Brunswick: Rutgers University Press, 1998): 162.
28. Bates, *Bargaining for Life*: 274. Refer also to pages 273, 280, 291, 293, 296–300.
29. Duffy, *Sanitarians*: 199–200 and 263, respectively. Also consult Hays, *Burdens of Disease*: 175; Tomes, *Gospel of Germs*: 220.
30. John Duffy, "Social Impact of Diseases in the Late 19th Century": 418–19. Naomi Rogers, *Dirt and Disease: Polio Before FDR* (New Brunswick, NJ: Rutgers University Press, 1996): 31.
31. Hays, *Burdens of Disease*: 263. Also refer to Hinman and Orenstein, "A Shot at Protection": 66, and especially Haven Emerson, *The Epidemic of Poliomyelitis (Infantile Paralysis) in New York City in 1916: Based on the Official Reports of the Bureau of the Department of Health* (1917; Rpt. New York: Arno Press, 1977): 179.
32. Fred Davis, *Passage Through Crisis: Polio Victims and Their Families* (Indianapolis, IN: Bobbs-Merrill, 1963) 16 and 24, respectively. Further consult pages 17, 23, and Daniel J. Wilson, *Living with Polio: The Epidemic and Its Survivors* (Chicago: University of Chicago Press, 2005): 18–19, 25.
33. Audry in Richard L. Bruno, *The Polio Paradox: Understanding and Treating "Post-Polio Syndrome" and Chronic Fatigue* (New York: Warner Books, 2002): 62. See also Davis, *Passage Through Crisis*: 17, 52; Jane S. Smith, *Patenting the Sun: Polio and the Salk Vaccine* (William Morrow and Co., 1990): 40; Gareth Williams, *Paralyzed with Fear: The Story of Polio* (New York: Palgrave Macmillan, 2013): 141; Wilson, *Living with Polio:* 36–39.
34. Heather Green Wooten, *The Polio Years in Texas: Battling a Terrifying Unknown* (College Station: Texas A&M University Press, 2009): 159.
35. Emerson, *The Epidemic of Poliomyelitis*: 241–42. See page 179, and Richard J. Altenbaugh, *The Last Children's Plague: Poliomyelitis, Disability, and Twentieth-Century American Culture* (New York: Palgrave Macmillan: 2015): 5–7; Bruno, *The Polio Paradox*: 55; Robert W. Lovett, *Infantile Paralysis in Vermont, 1894–1922: A Memorial to Charles S. Caverly* (Burlington, VT: Vermont State Department of

Health, 1924): 149; Ralph K. Ghormley, "History of Treatment of Poliomyelitis," *Journal of the Iowa State Medical Society*, 37 (August 1947): 348; Hays, *Burdens of Disease*: 263; Aaron E. Klein, *Trial by Fury: The Polio Vaccine Controversy* (New York: Charles Scribner's Sons, 1972): 14; Jeffrey Kluger, *Splendid Solution: Jonas Salk and the Conquest of Polio* (New York: G. P. Putnum's, 2004): 14; John R. Paul, *A History of Poliomyelitis* (New Haven: Yale University Press, 1971): 148; Nina Gilden Seavey, Jane S. Smith, and Paul Wagner, *A Paralyzing Fear: The Triumph Over Polio in America* (New York: TV Books, 1998): 23. Statistics about the 1916 epidemic vary.

36. James Colgrove, *State of Immunity: The Politics of Vaccination in Twentieth-Century America* (Berkeley, CA: University of California Press, 2006): 114. Also check *Report of the Department of Health of the City of Chicago for the Years of 1907, 1908, 1909, 1910* (Chicago: Department of Health, 1911): 174, CPL; Altenbaugh, *The Last Children's Plague*: 31–45. For Kenny, consult Rogers, *Polio Wars: Sister Kenny and the Golden Age of American Medicine* (New York: Oxford University Press, 2014).
37. Daniel J. Wilson, "A Crippling Fear: Experiencing Polio in the Era of FDR," *Bulletin of the History of Medicine*, 72 (1998): 464. Further consult Colgrove, *State of Immunity*: 115; Kluger, *Splendid* Solution: 69, 130, 189.
38. Peter Haggett, *The Geographical Structure of Epidemics* (New York: Oxford University Press, 2000): 11–12. Hays, *The Burdens of Disease*, states on page 247, "The rapid rise of poliomyelitis in the early twentieth century baffled epidemiological opinion…."
39. A. C. Knapp, E. S. Godfrey, Jr., and W. L. Aycock, "An Outbreak of Poliomyelitis: Apparently Milk Borne" (Chicago: American Medical Association, 1926): 14. This pamphlet reprinted an article of the same title and by the same authors: *Journal of the American Medical Association* 87 (August 28, 1926): 635–39.
40. Harold E. Pearson, Gordon C. Brown, Robert C. Rendtorff, Gerald M. Ridenour, and Thomas Francis, Jr., "Studies of the Distribution of Poliomyelitis Virus in an Urban Area Duringan Epidemic," *American Journal of Hygiene*, 41 (March 1945): 208–10.
41. Tony Gould, *A Summer Plague: Polio and Its Survivors* (New Haven: Yale University Press, 1995): 4–5. See p. 21, as well to Saul Benison, "Poliomyelitis and the Rockefeller Institute: Social Effects and Institutional Response," *Journal of the History of Medicine and Allied Sciences*, 29 (January 1974): 80–81; Howard Markel, *Quarantine! East European Jewish Immigrants and the New York City Epidemics of 1892* (Baltimore: Johns Hopkins University Press, 1997): 4–5, 11; David M.

Oshinsky, *Polio: An American Story* (New York: Oxford University Press, 2005): 2, 19–20; Rogers, *Dirt and Disease*: 46–47.

42. Hays, *The Burdens of Disease*: 263. Also refer to "Infantile Paralysis a Scourge and Puzzle," *New York Times*, July 9, 1916, Magazine Section: 14; "Noted Scientists Organize to Curb Infant Paralysis," *New York Times*, July 13, 1916: 3; Gould, *A Summer Plague*: 26–27.
43. *Report of the Department of Health of the City of Chicago for the years of 1911 to 1918 Inclusive*: 222, CPL. Rogers, *Dirt and Disease*: 58–59.
44. Roland H. Berg, *Polio and Its Problems* (Philadelphia: J. B. Lippincott, 1948): ix. Italics in original.
45. Oshinsky, *Polio*: 90. Moreover, check pages 93–94.

CHAPTER 3

Bad Odors, Nasty Dust, and Dangerous Bugs

In 1922, Margery Williams published what would become a classic, American children's book, *Velveteen Rabbit or How Toys Became Real.* In it, a plush bunny, a surprise Christmas gift for the young main character, simply named "Boy," became a cherished and constant companion. Constructed of a cheap cotton fabric with a velvety pile and stuffed with sawdust, it became dirty and threadbare over time as Boy played with it in the garden and cuddled it every night. He treated it like a live pet. The Rabbit remained with Boy as he succumbed to a severe bout of scarlet fever. He eventually recovered but the doctor, believing that diseases thrived in inanimate objects, ordered Boy's bedroom disinfected, including burning all of his toys. The timely intervention of a fairy saved his beloved, stuffed animal from its fiery fate and magically made it real. The incarnated Rabbit visited his former companion in the garden soon thereafter, and Boy found something strangely familiar about this creature, ending the story on a happy note.

This children's tale gives us a glimpse of the state of medical science at that time, one that encompassed fomites. Both the medical and public health communities believed that inanimate "objects could harbor the dried microbes of disease for months and even years…." These typically included clothing, dust, and playthings. As a result, it became common during epidemics to decontaminate entire buildings and incinerate all of the patients' possessions.[1]

R. J. Altenbaugh, *Vaccination in America*,
Palgrave Studies in the History of Science and Technology,
https://doi.org/10.1007/978-3-319-96349-5_3

Scarlet fever, of course, has nothing to do with fomites; it is a highly infectious disease caused by group *A hemolytic streptococci*, spread through intimate contact with infected humans, "such as occurs in overcrowded homes and classrooms," as well as through contaminated milk. A sore and swollen throat combined with a high fever marks this disease, along with nausea and vomiting. Before antibiotics, mortality rates could reach 30%.[2]

Fomite-based beliefs also applied to viruses. During the 1916 polio epidemic, a New York City attorney sent his family to Adams, Massachusetts, to shield them from infection, later forwarding toys through the postal system. Three of his daughters contracted poliomyelitis: the eight-year-old died. The *New York Times* reported this tragedy casually, and without qualification, observing those playthings "carried germs of infantile paralysis."[3] Even in the early twentieth century, medical science embraced some antiquated ideas about the source of contagions as well as how people contracted them. Medical science had achieved some significant breakthroughs, but largely maintained a "muddled" history.[4]

In addition to fomites, the American medical community, like its European counterpart, embraced zymotic (or miasma) theories to explain the source of diseases and how they proliferated, asserting that noxious odors from cesspools, open sewerage, and slaughterhouses, "disrupted the body's balance.... [P]hysicians often attempted to restore its balance through purgatives, diuretics, and emetics."[5]

Health and Sanitation

Sanitary science, according to medical historian Nancy Tomes, "stressed the ubiquity of airborne infection and the disease-causing properties of human waste and organic decay." This followed two stages. The first, during the 1880s and 1890s, built on existing miasma theory. Contagions found portals through toilets, sink drains, and windows. Modern plumbing, namely flush toilets and running water, and simple house ventilation, i.e., fresh air, served as concrete solutions to these often lethal threats. The second period, in the early 1900s, emphasized casual contact through "[f]omite infection." Domestic science addressed these vectors by eliminating dirt and dust, stressing personal hygiene by covering the nose and mouth during sneezes and coughs, and promoting screened windows to prevent flies from coming into contact with humans.[6]

A sense of urgency arose with ever denser populations assembling in increasingly smaller spaces. "[C]ompulsory school attendance, becoming more common, created new opportunities for such diseases as polio and diphtheria." Therefore, dynamic urban centers, as brokers of trade and destinations for foreign travelers, and their ever-growing institutions, where large crowds congregated, assumed greater ecological and concomitant medical roles in human lives.[7] "In the decades after the Civil War, health departments, that had been established to deal with epidemic emergencies, assumed sanitary responsibilities such as garbage disposal, street cleaning, and food regulation."[8]

The Pittsburgh Study, a sociological survey conducted by the Russell Sage Foundation in 1907, saw this as a social imperative. Researcher Margaret Byington found highly congested living conditions in Homestead, a steel mill town along the Monongahela River. Tenements occupied entire city blocks, each accommodating up to 239 families, many of whom rented sleeping space to lodgers to supplement their meager household income; these additional inhabitants only intensified already overtaxed spaces. To make matters worse, few apartments had running water or indoor toilets. A "court" in the middle of each of these complexes contained both outhouses and water pumps. With such close proximity, the wastewater often leeched into the drinking water. Local health officials did not "inspect the wells nor analyze the water from them.... One outbreak of typhoid was traced directly to a well which had been used by a number of families...."[9]

"Sanitarians" ultimately succeeded in their quest for safe drinking water. As early as 1893, the Massachusetts State Board of Health became the first to employ a sand filtration system to remove bacteria. "By 1911 about 20 percent of America's urban population was using filtered water." The death rate, as a result, fell from 31.3 per 100,000 in 1900 to 7.6 in 1920.[10]

Public health measures, like clean drinking water and safe milk, as well as parents educated in the use of hygienic practices in their homes, such as isolating sick family members, protecting food from fly contamination, sterilizing baby bottles, and washing hands, caused child mortality rates to decline. In 1910, Chicago's health department leaders boldly pronounced that in the "light of the sanitary achievements of the past ten years, it may be safely postulated that when sanitary administration reaches its highest standard of efficiency there will be no more epidemics of the preventable diseases...."[11]

What they did not know was that the sanitation movement created perfect conditions for the proliferation of poliovirus. This disease, John R. Paul of the Yale Poliomyelitis Commission (later the

Yale Poliomyelitis Study Unit) and the first grantee of the National Foundation for Infantile Paralysis/March of Dimes writes, always "existed as a worldwide *endemic* infection of infancy and such a situation resulted in an absence of epidemics." Infant immunity, though "short lived, provided passive antibodies, derived from their mothers." They lost these at "about six or seven months of age" with a certain number beginning to "acquire active antibodies and immunity through unapparent infection. These percentages increased ... until generally by the age of fifteen between 80 and 100 percent of children are antibody positive and presumably immune." Indeed, as Paul continues, "urban population's immunization was going on at a surprising rate in infancy and early childhood without accompanying disease." While clean water, pure food, waste management, and hygienic practices and education sharply reduced the causes of bacterial infections,

> ... the effect on enteric viral infections such as poliomyelitis was less favorable. In fact it was the reverse and had the effect of transforming a situation which favored endemic infection into one that encouraged epidemics. What happened was that the circulation of poliovirus became more spotty and intermittent as the twentieth century progressed; children arrived at school age and even adolescence without having been exposed or infected, i.e., they remained as susceptibles. Accordingly, they became increasingly vulnerable, and when, inevitably, the virus was introduced after an interval of some years, it spread rapidly through an awaiting susceptible population....

The true impact of polio emerged during World War II. In 1940, morbidity rates stood at 4:100,000 Americans; that rate doubled during the next four years and doubled once more between 1945 and 1949. It jumped to 25 per 100,000 from 1950 to 1954. In 1952 alone, polio incidence peaked at 37 per 100,000.[12]

The Science of Diseases

Germ theory arose from the cumulative scientific efforts and findings of a disparate group of researchers. From England, this included physician John Burdon Sanderson and surgeon Joseph Lister; from France, physician Casimir Davaine and chemist Louis Pasteur; and from Germany, physician Robert Koch. By the early 1900s, remarkable laboratory insights began to pour forth. Researchers identified organisms that caused anthrax, bubonic plague, cholera, diphtheria, dysentery, erysipelas gangrene, gonorrhea, leprosy, pneumonia, scarlet fever, syphilis, tetanus, typhoid, and whooping cough. "[I]n the twenty-one golden years between 1879 and 1900 the microorganisms responsible for major diseases were being discovered at the phenomenal rate of one a year."[13]

No one characterized this phenomenon more than Pasteur. His work during the 1850s and 1860s led to formal presentation of the "germ theory of infection" to the French Academy of Medicine in 1878. He successfully applied this principle to anthrax, cholera, and rabies immunizations on animals. His most controversial test case involved Joseph Meister, a nine-year-old Alsatian youth who had sustained fourteen bites from an allegedly rabid dog. At the behest of his parents, Pasteur, assisted by two colleagues from Académe de Sciences, injected the attenuated experimental rabies serum over the course of ten days; the boy remained unaffected, persuading Pasteur that his cure had worked. To ensure this was the case, a few weeks later he injected Meister with active rabies; nothing happened. Pasteur's antitoxin appeared to work. American magazines and newspapers lionized him and celebrated his discovery.[14]

The New Public Health

These breakthroughs, first bacterial and later viral, profoundly altered public health polices, but only slowly. According to Nancy Tomes, germ theory became easily absorbed into existing public health policies based on sanitation. Since microorganisms caused illnesses, and since disinfection killed these germs, diseases could be better controlled. Microbes now drove the penchant for cleanliness and hygiene. White china toilets concretely symbolized the earlier movement but during the early 1900s "the gospel of germs began to take on more far-reaching forms" that would increase the power and scope of public health. "Medical officers could compel quarantines that sealed borders, order disinfection of private property, and forcibly isolate individuals in hospitals."[15] More importantly, this began the search for the "magic bullet." Medical scientists embarked on vaccine development from "concoctions of tamed germs that would confer ... protection against deadly diseases."[16]

The New York City Department of Health, a pioneer of municipal health, wielded significant power, featuring a staff of forty, including twenty-one medical inspectors and disinfectors, as well as its own police force to compel residents to follow its dictates. In 1889, based on Robert Koch's 1882 discovery of tuberculosis bacterium, it banned spitting, warned against cohabiting with an infected individual, urged the sanitization of eating and cooking utensils used by consumptives, ordered the separation of clothes for washing and bowel discharges, recommended the destruction of household furniture and pets suspected of being exposed to the TB bacterium, and condemned infected mothers who nursed their infants. Hermann M. Biggs, director of the newly created Division of Pathology, Bacteriology and Disinfection, earned accolades from politicians and the press alike for controlling that city's 1892 cholera outbreak. He felt fully justified in using the "fullest expression of state power with respect to public health," focusing on that city's tenements, where immigrants, the poor, and working class resided, during the late nineteenth century. Health authorities, beginning in 1894, posted placards at "tenement houses where cases of diphtheria, scarlet fever, or measles were found." This action alerted the general public and warned potential visitors, but especially protected the "middle- and upper-class population."[17]

In 1906, Chicago's health department stipulated clear guidelines for the funerals of those who succumbed to contagious diseases. Their bodies would be immediately sealed in coffins with private funerals held within thirty-six hours of death. It further forbade flowers and attendance by crowds, especially children. Health officials also implemented education activities to inform the public about hygienic practices. These included lecturers who spoke at church congregations and settlement houses, press bulletins published in multiple languages, and movies such as "The Fly Pest." One leaflet, titled "Some Sanitary Suggestions," consisted of a list of dictums to prevent tuberculosis:

> Consumption is a House Disease.
> It is Caused by Germs.
> Do Not Spit on the Floors.
> No Spit, No Consumption....

Although this represented an oversimplified notion of the cause of tuberculosis, public health officials at the time felt supremely confident about their education efforts.[18]

Uncovering the role of microorganisms in the spread of diseases facilitated the development of vaccines, feeding naive scientific exuberance at the dawn of the new century. "The germ theory of disease, when taken together with the extension of vaccinations to a general preventive principle, decisively altered the relations of humans and infectious disease." The prevention of all contagions therefore appeared quite realistic.[19]

Such research expanded the use of humans for scientific purposes. Experimenters felt comfortable exploiting "infants, dying patients, and mentally impaired individuals to demonstrate," medical historian Susan E. Lederer explains, "to demonstrate the pathogenic effect of microorganisms.... In 1895 New York pediatricians, Henry Heiman, for example described the successful gonorrheal infections of a four-year-old boy ('an idiot with chronic epilepsy') and a twenty-six-year-old man in the final stages of tuberculosis."[20]

Vaccination

Artificial immunization boosted public health initiatives. It began with variolation, exposing healthy individuals to "smallpox in the hope of preventing natural attacks of the disease." As medical historian Gareth Williams stresses, this represented "the first successful attempt to manipulate the body's immune system so as to throw up defences [sic] against a specific infection. Two traditions existed. Chinese accounts, dating back to 1550, relied on "'nasal insufflation;'" that is, blowing powdered scabs or dried blister fluid into nasal passages. African and Arab traditions involved "'cutaneous inoculation.'" In Sudan, this took the form of applying pus from an infected person to the arm of a healthy individual." Therefore, by the "mid-seventeenth century, variolation was commonplace in China and parts of Africa and cutaneous inoculation was also being practiced in isolated foci in various European countries...." For the 97 to 98% who survived, it imparted permanent immunity.

Western adoption of this procedure is well known. Lady Mary Wortley Montagu, the wife of the British ambassador to Turkey, noticed a paucity of scarred faces and discovered that this was due to inoculation, or as Turks termed it, "ingraftiing." It transpired every September during special social events. "The whole process," Williams continues, "was a casual seasonal ritual, so straightforward that it was entrusted to old women." They scratched children's arms with a large needle and inserted smallpox pus into it and bandaged the wound. They became slightly ill for several days, but recovered with immunity. In letters home, in 1717, Montagu painted the entire process in bright, almost cheerful hues, seeing it as completely safe. So much so, she had her three-year-old son inoculated a year later in Constantinople.

Montagu and her children returned to England when her husband was recalled to London. When a smallpox outbreak occurred there, she insisted that her three-year-old daughter be immunized under the close scrutiny of three doctors representing the Royal College of Physicians. Although skeptical of this "bizarre heathen custom," the outcome surprised them. The next step unfolded when the King ordered a public trial, using six condemned prisoners at Newgate Prison. For their service as *volunteers* they would receive a royal pardon. One of them, Elizabeth Harrison, agreed to take this experiment a step further by subsequently exposing herself to smallpox. She remained free from infection, "received her pardon and passed safely back into obscurity." The next, and final, stage of this trial involved twelve children. "Ethical approval," Williams points out, "would not even have been thought of, but would anyway have been impossible to obtain in six cases, who were all orphans from the Parish of St. James, Westminster." Whitehall announced the successful outcome with "pomp." The inoculation of royalty and "upper classes" soon followed, coopting it as a "respectable [western] medical innovation."[21]

The laboratory of an epidemic demonstrated its effectiveness on a large scale, according to Stephen Coss's historical account, but not without controversy. Boston, with a population of 11,000, had become the largest town in the North American colonies by 1721. Smallpox struck in April of that year and by early June many homes flew red quarantine flags over their doorways. October represented the worst month: "Boston had come to resemble a ghost town." People died at the "rate of thirteen a day." The wheels of a cart collecting dead bodies echoed nightly through desolate streets.

Cotton Mather, a well-known theologian, stumbled across an explanation of smallpox inoculation in *Philosophical Transactions*, a Royal Society of London publication. Onesimus, one of his slaves, who had been born and raised in North Africa, reinforced this account by sharing his own experience and showing him the scar as evidence. Mather corroborated this by interviewing several other members of that "town's African community." However, he was rebuffed when he attempted to share this information: "[t]he notion of giving someone a disease in order to save him from dying of that disease was so preposterous that every other doctor in Boston had refused to discuss it," except one, Zabidiel Boylston.

Boylston proceeded with "the experiment" on the morning of June 26, 1721, withdrawing clear fluid from a smallpox patient's blister and inserting it into cuts on his slave Jack, Jack's eighteen-month-old son, as well as his own six-year-old son. After nine days of a mild fever and rash, all survived and appeared immune. On July 17, the *Boston Gazette* printed Boylston's letter describing this breakthrough. When Boylston performed it on townspeople, civic and medical authorities ordered him to cease and desist. On July 21, at a public hearing, Boylston "refused repeated calls to disavow the procedure. Instead he reiterated its safety and efficacy as he had experienced it to that point." Although formally condemned, Boylston remained undeterred, continuing to treat individuals begging him for protection, including Mather's son, Samuel. When Boston's medical community intensified its attack on Boylston in late September, many Bostonians, emotionally drained by the epidemic, became terrified that Boylston's procedures would exacerbate it. Fearing for his life, Boylston hid in his house, only treating a few patients in October. However, during his "semi-seclusion," Coss adds, the "*Boston News-Letter* delivered stunning news that the King had authorized an inoculation experiment at Newgate Prison in London." With validation of his work, Bostonians suddenly viewed Boylston in a new light and "began flooding to [him] in large numbers." Boylston concluded his procedures in May 1722. By then, "... the worst epidemic of that disease in Boston was over."

The final tally proved sobering. An estimated 6000 individuals suffered illness with 844 deaths, a 14% mortality rate, jumping to 50% among children and the elderly. "Untold hundreds more had been left blinded, mentally disabled, debilitated with severe arthritis, or grotesquely disfigured." In contrast, of the 280 Bostonians inoculated only 6 died, amounting to only 2%. "The Boston inoculation experiment," Coss concludes, "was a victory for reason over ... unquestioning obeisance to accepted scientific notions." The method Boylston "employed and validated would have profound ramifications not only for the battle against smallpox, but also for the fight against polio, measles, tetanus, rabies, and other deadly illnesses."[22]

By the American Revolution, "it was widely accepted" in North America. Lower morbidity and mortality rates outweighed the risk.[23] So much so that General George Washington ordered the Continental Army immunized against smallpox in 1777–1778. This represented "the first large-scale, state-sponsored immunization campaign in American history."[24] Meanwhile, English physicians carried variolation to Germany and Russia. Other European countries soon followed suit. "Variolation," for Gareth Williams, "was one of the greatest successes of medicine at that time," leading directly to vaccination.

Edward Jenner is renowned for pioneering smallpox vaccination. Born in 1749 and raised in Gloucestershire, in the English countryside, Jenner became apprenticed to a surgeon at age thirteen, where he initially learned about variolation. After pursuing additional training at St. George's Hospital in London in 1770, Williams adds, he returned home to practice medicine. Jenner did not focus on cowpox until age forty-nine. He found a "couple of people" who had contracted cowpox and exposed them, through variolation, to smallpox; they appeared immune. Jenner turned next to "unprotected subjects." He planned to treat them with cowpox and then expose them to smallpox. He obtained fluid from a milkmaid's ulcerated hands and "scratched it into the arm of James Phipps, the eight-year-old son of Jenner's gardener…." He closely observed young Phipps for several days who developed a "transient fever but otherwise remained well." To ensure that the boy had indeed been immunized, Jenner infected James with smallpox. Jenner repeated this process twenty times, from different smallpox donors, on Phipps. He appeared immune. Jenner expanded his sample to "another dozen subjects, mostly children … including his own one-year-old son Robert." Again, Jenner appeared successful. Jenner privately published his findings as *An Inquiry into the Causes and Effects of the Variolae Vaccinate* in 1798. It generated "excitement with its promise of something safer and better than variolation."[25]

However, all of these smallpox episodes, from London to Boston to Gloucestershire, bequeathed a dubious legacy. The cowpox experiments using children set "both the [medical research] style and technique that would predominate for the next 150 years." Generally speaking, vaccine research ranks among the oldest of medical experiments.[26]

Furthermore, many on both sides of the Atlantic greeted vaccination with mixed emotions. Some resisted. The most extreme critics attacked the use of cowpox for immunization, stemming from "primitive beliefs that animals could pass on their traits to humans."[27] They feared that humans would start sprouting hooves and horns from their arms and heads following immunization, or resort to animal behavior, like chewing cud. The medical community thought otherwise, and "vaccination spread far and wide with remarkable speed."[28]

By the late nineteenth and early twentieth centuries, it became an important component of public health policy in urban America despite the fact that "[n]o public health measure inspired more ill will than compulsory vaccination."[29] This forced New York City's health commissioner Ernest J. Lederle to deploy "paramilitary vaccination squads" during the 1901–1903 smallpox epidemic. He even sent them, with police escort, to forcibly inoculate people residing near infected individuals, anytime of the day or night, scratching the "arms of the poor at the rate of fifteen hundred per day." Immigrant parents loudly and vociferously protested "having their ... children's arms scraped by the vaccinators...." Such opposition ebbed as fear of infection took hold. "But with each new outbreak in another of the island's crowded tenement districts, the dispatched vaccination corps met fresh resistance. Over time, the corps would ever more closely resemble a military outfit." By the end of 1902, that department had succeeded in giving 810,000 inoculations. Boston's and Chicago's health departments employed a similar approach.[30]

Vaccine research for other children's diseases continued to make strides "Much of the ... testing," according to Susan Lederer, "was performed on children, because they lacked previous exposure to infectious diseases." Relying on "benevolent deception," investigators escaped scrutiny and condemnation because "dangerous human experimentation, therapeutic or nontherapeutic, has been more easily forgiven if the clinical gains were great and the experimenter turned out to be right." Moreover, institutionalized children always precluded any need for formal consent.[31]

Diphtheria represents a case in point, and serum development in many ways foreshadowed mid-century polio vaccine experiments. The initial breakthrough occurred in Europe. In 1894, Emil von Behring and Carl Fraenkel, along with Shibasaburo Kitasato at the University of Berlin, building on the experiments of Emile Roux and his colleagues at the Pasteur Institute in Paris, created an antitoxin for the prevention and treatment of diphtheria. Von Behring won the Nobel Prize in 1901 for his work. However, it acted solely as a prophylactic. Hermann Biggs and William H. Park, who pioneered New York City's Bureau of Laboratories, introduced this antitoxin to the USA in 1895. Eleven years later Ernst Lederle, that city's former commissioner of health, founded Lederle Laboratories to produce it. As doctors administered it, the child mortality rate declined.[32]

Three additional discoveries followed in relatively quick succession. In 1907, Von Behring mixed diphtheria toxin and antitoxin, thus cultivating a "long-lasting immunity agent against the disease" while in 1913 Béla Schick, a Viennese doctor, devised an easy and effective skin test to determine diphtheria immunity. "With this pair of developments, scientists had both the means of inducing immunity as well as a simple and relatively straightforward way of measuring the efficacy of the procedure." Park, director of the Bureau of Laboratories, and Abraham Zingher, his colleague and one of Schick's former students, embarked on a twofold experiment: First, they tested children; second, if they found them at risk, then they gave them von Behring's toxin-antitoxin (TAT).[33] "Within three years," according to medical historian Evelyn Maxine Hammonds, "Park and Zingher had used the Schick test and TAT on some 12,000 healthy children in different New York City institutions and in some 1,500 convalescing cases of diphtheria and scarlet fever at the Willard Parker Hospital." They also employed this procedure on children housed at asylums, foundling homes, and orphanages. Working through these institutions they did not need permission to conduct their tests, nor did they bother to make an effort. "In essence, these children and adults were captive subjects for the studies being conducted by the health department."[34] World War I interrupted this effort, mobilizing health personnel and diverting medical supplies to the military. Finally, in 1923 Gaston Ramon, at the Pasteur Institute, used formalin to inactivate the toxin molecule, attaining 100% effectiveness. [35]

Researchers isolated the causative agent for pertussis (i.e., whooping cough) in 1906. Numerous attempts to develop an antigen followed. The first received a license in 1926. "By 1931, at least fourteen different vaccine preparations against pertussis had been developed, but there was little uniformity of opinion about which was the best formulation, or when or how the vaccine should be administered." Louis Sauer, "an Illinois pediatrician and researcher … had tested his formulation widely in public schools in Evanston…." Following the publication of his "promising results in medical journals," the press made it public knowledge.[36] Meanwhile, Pearl Kendrick, a microbiologist at the Michigan Department of Health, and Grace Eldering created another version by "simply killing pertussis bacteria with carbolic acid, an antiseptic." In a 1939 field trial involving 4000 children, they injected half with it; the remaining 2000 served as an observed-control group. "The results were clear, 348 unvaccinated children got whooping cough, only

52 vaccinated children suffered the disease."[37] Although the "American Academy of Pediatrics (AAP) ultimately gave its stamp of approval to both the Sauer and Kendrick vaccines in 1943," support remained uneven while debate over efficacy continued.[38]

In 1896, New York City's health department began to produce a tetanus antitoxin. A commercial version became available in 1938, but failed to become widely used until World War II when doctors employed it as a prophylaxis on wounded soldiers. It sharply reduced tetanus rates compared to troops wounded during World War I. Based on this success, the AAP recommended it as part of routine childhood protection in 1944. In 1948, it was combined with diphtheria toxoid and pertussis vaccine in a single injection. The AAP endorsed DTP for infant immunization in 1951.[39]

Until the introduction and widespread use of antibiotics, a vaccine to combat tuberculosis stirred controversy. In 1882, Robert Koch developed an experimental serum from the TB bacterium and injected himself with it. Koch believed he had found a cure, sparking "[a] circus atmosphere.... Koch's colleagues conducted public demonstrations at which patients were injected with tuberculin." However, the rush to find the "magic bullet" for this contagion clashed with the results of a sober and thorough German commission's 1891 report that Koch's serum provided marginal improvement at best for a minority of patients.

In 1900, Léon Charles, Albert Calmette, and Jean Marie Camille Guérin, microbiologists working at the Pasteur Institute, conducted their own experiments to find a TB vaccine, apparently achieving it with the Bacili Bilié Calmette-Guérin (BCG) antigen. They continued to work through the conflagration of World War I and in 1921 they tested it on a three-day-old infected infant whose mother had died of consumption immediately after giving birth; the newborn's prognosis was not hopeful. They administered doses of BCG to the baby during the first week of its life. Without any ill effects, they deemed the experiment a success. They injected 600 additional children with BCG three years later. It once again appeared safe and effective. It went into full production, becoming widely used throughout Europe and Canada, with 100,000 doses given to infants between 1924 and 1928. "The League of Nations certified BCG as safe for human use in 1928."

However, something went terribly array in 1930. After the inoculation of 251 children in Lübeck, Germany, 67 died and another 108 became infected, while 5 died of other causes. According to historian

Thomas M. Daniel, postmortem examinations demonstrated that the children had received a contaminated batch of serum. "Both French and German commissions investigating the tragedy exonerated BCG, but the reputation of the vaccine had been sorely damaged." In spite of this evidence, that disaster solidified American opposition. Further, the US Public Health Service (USPHS) conducted its own field trials of BCG between 1935 and 1938 on 3000 Native Americans, injecting 1500 with the serum and another 1500 with a placebo. Researchers continued to follow their progress with annual examinations for the next ten years, then intermittently for another ten years. "Tuberculosis occurred in sixty-eight of the control subjects but in only thirteen of the BCG-vaccinated subjects. There could be little doubt that BCG was protective." Nevertheless, USPHS officials refused to endorse BCG. In 1947, they field tested BCG again in Muscogee County, Georgia, using 11,000 school-aged children between the ages of five and nine. Public Health Service administrators again ruled BCG ineffective, yet it remained in use throughout Europe, and the World Health Organization endorsed it in 1973; by 1989, three billion doses had been given worldwide.[40] The US medical establishment opted to attack tuberculosis with antibiotics.

Remedy or Hazard?

Philadelphia's child mortality rates, indicative of national trends, declined sharply between 1870 and 1930, from 174 out of 1000 infants before the age of one to 75. Infectious diseases "accounted for 45 percent of all deaths occurring in 1870, about 40 percent in 1900, and about 24 percent of those in 1930." Sanitation improvements, like purified water and pasteurized milk, certainly helped. Immunization not only contributed to this decrease but held the promise of further reductions.[41]

Vaccines proved a key weapon in modern medicine's arsenal. Even before any firm knowledge of germ theory, smallpox immunization saved lives. With the emergence of virology in the 1930s, the picture clarified, facilitating additional breakthroughs. "Vaccines were developed against several diseases, including cholera, plague, and typhoid, and by the 1910s the term 'vaccine,' which originally had meant only the preparation of cowpox that provided immunity against smallpox, began to be applied more broadly to any product designed to produce active immunity."[42]

With knowledge about hygiene and the ability to develop vaccines, medical science held the promise of not only controlling childhood diseases but perhaps eradicating them altogether. Public health agencies attempted to translate these breakthroughs into policies through sanitation regulations and immunization campaigns. It would seem natural that parents would overwhelmingly embrace these life-saving opportunities, but they did not.

Notes and Sources

1. Nancy Tomes, *Gospel of Germs: Men, Women, and the Microbe in American Life* (Cambridge: Harvard University Press, 1998): 9. Consult page 237, as well as Thomas M. Daniel, *Captain of Death: The Story of Tuberculosis* (Rochester, NY: University of Rochester Press, 1997): 70; Michael Willrich, *Pox: An American History* (New York: Penguin Books, 2011): 71. For references to *The Velveteen Rabbit*, see Margery Williams, *The Velveteen Rabbit or How Toys Become Real* (1922; Rpt. New York: Henry Holt and Company, 1983); John S. Bowman, ed., *The Cambridge Dictionary of American Biography* (Cambridge: Cambridge University Press, 1995), 799. For its relationship to the history of medicine, consult Tomes, *Gospel of Germs*: 61, as well as Naomi Rogers, *Dirt and Disease: Polio Before FDR* (New Brunswick, NJ: Rutgers University Press, 1996): 59.
2. Anne Hardy, "Scarlet Fever," in Kenneth F. Kiple, ed., *Cambridge World History of Human Disease* (New York: Cambridge University Press, 1993): 991. Further consult Eric H. Christianson, "Medicine in New England," in Judith Walzer and Ronald L. Numbers, eds., *Sickness and Health in America: Readings in the History of Medicine and Public Health* (Madison: University of Wisconsin Press, 1997): 54.
3. "Arthur Cosby's Child Dies; Toys from New York Thought to Have Carried Paralysis Germ," *New York Times*, September 4, 1916: 14. Also consult Daniel, *Captain of Death*: 70; Willrich, *Pox*: 71.
4. Roy Porter, *The Greatest Benefit to Mankind: A Medical History of Humanity* (New York: Norton, 1997): 428. Further check Tomes, *Gospel of Germs*: 237.
5. Rogers, *Dirt and Disease*: 16.
6. Tomes, *Gospel of Germs*: 8–9. For prior public health policies, refer to Willrich, *Pox*: 89–91. John Duffy, *The Sanitarians: A History of American Public Health* (Urbana: University of Illinois Press, 1992): 1, 4–7; Peter Haggett, *The Geographical Structure of Epidemics* (New York: Oxford University Press, 2000): 110; Howard Markel, *Quarantine! East*

European Jewish Immigrants and the New York City Epidemics of 1892 (Baltimore: Johns Hopkins University Press, 1997): 3–4; Porter, *Greatest Benefit to Mankind*: 238.

7. J. N. Hays, *Burdens of Disease: Epidemics and Human Response in Western History* (New Brunswick, NJ: Rutgers University Press, 1998): 153. Steven Johnson, *The Ghost Map: The Story of London's Most Terrifying Epidemic—And How It Changed Science, Cities, and the Modern World* (New York: Riverhead Books, 2006).
8. Rogers, *Dirt and Disease*: 16. Also refer to Duffy, *The Sanitarians*: 4; Samuel H. Preston and Michael R. Haines, *Child Mortality in Late Nineteenth-Century America* (Princeton: University of Princeton Press): 7–8; Susan Sontag, *Illness as Metaphor and AIDS and Its Metaphors* (New York: Picador, 1989): 130; Paul Starr, *The Social Transformation of American Medicine* (New York: Basic Books, 1982): 189; Gareth Williams, *Paralyzed with Fear: The Story of Polio* (New York: Palgrave Macmillan, 2013): 54; Terra Ziporyn, *Disease in the Popular American Press: The Case of Diphtheria, Typhoid Fever, and Syphilis, 1870–1920* (Westport, CT: Greenwood Press, 1988): 13.
9. Margaret Byington, *Homestead: The Households of a Mill Town* (1910; Rpt. Pittsburgh: University of Pittsburgh Center for International Studies, 1974): 135 and 24, respectively. Refer also to pages 23, 136–37.
10. Duffy, *Sanitarians*: 201–2.
11. *Report of the Department of Health of the City of Chicago for the Years of 1907, 1908, 1909, 1910*: 99. Preston and Haines, *Child Mortality in Late Nineteenth-Century America*: 208–9; Thomas M. Daniel and Frederick C. Robbins, eds., *Polio* (Rochester, NY: University of Rochester Press, 1997): 8.
12. John R. Paul, *A History of Poliomyelitis* (New Haven: Yale University Press, 1971): 358, 359, 364–65. Italics appear in the original text. Roland H. Berg, *Polio and Its Problems* (Philadelphia: J. B. Lippincott, 1948): 61–63; Jeffrey Kluger, *Splendid Solution: Jonas Salk and the Conquest of Polio* (New York: G. P. Putnum's, 2004): 142. Especially consult Stephen E. Mawdsley, *Selling Science: Polio and the Promise of Gamma Globulin* (New Brunswick: Rutgers University Press, 2016): 5. For a fine summary of morbidity statistics refer to Jason Chu. Lee, "Poliomyelitis in the Lone Star State: A Brief Examination in Rural and Urban Communities" (MA thesis, Texas State University-San Marcos, 2005): 36. Finally, recent research has revealed the "hygiene hypothesis"; that is, growing up less exposed to germs and dust weakens immune systems of young children. Refer to Perri Klass, "Too Clean for Our Children's Good?" *New York Times*, 2017. http://www.nytimes.com/2017/04/17/well/family/too-clean-for-our-childrens-good.html (Accessed April 17, 2017).

13. Porter, *Greatest Benefit to Mankind*: 428 and 433, respectively. Consult pages 434–35, 441–44, as well as Thomas P. Gariepy, "Antisepsis," in J. L. Heilbron, ed., *The Oxford Companion to the History of Modern Science* (New York: Oxford University Press, 2003): 41–42; Evelynn Maxine Hammonds, *Childhood's Deadly Scourge: The Campaign to Control Diphtheria in New York City, 1880–1930* (Baltimore: Johns Hopkins University Press, 1999): 53; J. N. Hays, *The Burdens of Disease: Epidemics and Human Response in Western History* (New Brunswick: Rutgers University Press, 1998): 150, 151, 233–35; Tomes, *Gospel of Germs*: 32–33, 94.
14. Porter, *Greatest Benefit to Mankind*: 39–40, 274, 276, 435. Daniel, *Captain of Death*: 132–33; Gerald L. Geison, "Pasteur's Work on Rabies: Reexamining the Ethical Issues," *Hastings Center Report*, 8 (April 1978): 29–30; Bert Hansen, "America's First Medical Breakthrough: How Popular Excitement About a French Rabies Cure in 1885 Raised New Expectations for Medical Progress," *American Historical Review*, 103 (April 1998): 374–75, 387; Paul A. Offit, *Vaccinated: One Man's Quest to Defeat the World's Deadliest Diseases* (New York: Smithsonian Books, 2007): 35; Ziporyn, *Disease in the Popular American Press*: 13.
15. Hays, *Burdens of Disease*: 151. Also, see pages 149–50, and Duffy, *Sanitarians*: 4; Hammonds, *Childhood's Deadly Scourge*: 53; Bert Hansen, "The Image and Advocacy of Public Health in American Caricature and Cartoons from 1860 to 1900," *American Journal of Public Health*, 87 (November 1997): 1799; Terra Ziporyn, *Disease in the Popular American Press*: 12. Daniel, *Captain of Death*, devotes Chapter 9, "Robert Koch and the Tubercle Bacillus," to this breakthrough.
16. Tomes, *Gospel of Germs*, 45 and 87, respectively. Further see pages 6, 46, 68–69.
17. Hammonds, *Childhood's Deadly Scourge*: 79 and 96, respectively. See pages 71, 73, 80, 97, as well as James Colgrove, *State of Immunity: The Politics of Vaccination in Twentieth-Century America* (Berkeley, CA: University of California Press, 2006): 22; Daniel, *Captain of Death*: 87, 89.
18. *Report of the Department of Health of the City of Chicago for the Years of 1907, 1908, 1909, 1910* (Chicago: Department of Health, 1911): 103; *Biennial Report of the Department of Health of the City of Chicago for 1904–05* (Chicago: Department of Health, 1906): 168, CPL. For examples of state and federal efforts at health education, refer to John C. Burnham, *How Superstition Won and Science Lost: Popularizing Science and Health in the United States* (New Brunswick, NJ: Rutgers University Press, 1987): 58.

19. Hays, *Burdens of Disease*: 232. Refer as well to pages 238–39, 249. Porter, *Greatest Benefit to Mankind*: 459–60.
20. Susan E. Lederer, *Subjected to Science: Human Experimentation in America Before the Second World War* (Baltimore: Johns Hopkins University Press, 1995): 3, 4. Further consult Susan E. Lederer and Michael A. Grodin, "Historical Overview: Pediatric Experimentation," in Michael A. Grodin and Leonard H. Glantz, eds., *Children as Research Subjects: Science, Ethics, and Law* (New York: Oxford University Press, 1994): 5, 8.
21. These paragraphs draw on Williams, *Angel of Death*: 57, 60, 61, 84–85, 90, 93, and 94, respectively. Also consult pages 84, 86, 87, 89, 91, and 127–30. Williams offers the most comprehensive and recent history of smallpox. Further pursue Arthur Allen, *Vaccine: The Controversial Story of Medicine's Greatest Lifesaver* (New York: W. W. Norton Company, 2007): 37; College of Physicians, *History of Vaccines* (Philadelphia: College of Physicians, 2013): 11; Stephen Coss, *The Fever of 1721: The Epidemic that Revolutionized Medicine and American Politics* (New York: Simon and Schuster, 2016): 87, 268; Duffy, *Sanitarians*: 21–22, 27; Elizabeth A. Fenn, *Pox Americana: The Great Smallpox Epidemic of 1775–82* (New York: Hill and Wang: 2001): 32; Sidney A. Halpern, *Lesser Harms: The Morality of Risk in Medical Research* (Chicago: University of Chicago Press, 2004): 20. Emphasis is mine.
22. The preceding paragraphs draw on Coss, *The Fever of 1721*. The quotes are from pages 151, 165, xi, 78, 85, ix, 102, 153, 165, 193, and 195, respectively. See pages 4–6, 28, 42, 44–45, 58–60, 63, 72, 74–75, 81, 84, 90–91, 100–101, 148, 152, 188, 191–92, 194, 196, 269, as well as Allen, *Vaccine*: 25–27, 37; College of Physicians, *History of Vaccines*: 11; Duffy, *Sanitarians*: 27; Fenn, *Pox Americana*: 32–33, 36; Harriet A. Washington, *Medical Apartheid: The Dark History of Medical Experimentation on Black Americans from Colonial Times to the Present* (New York: Doubleday Books, 2006): 49, 71–73; Williams, *Angel of Death*: 96, 101, 110–11, 114–16, 123, 125–26. Margot Minardi, "The Boston Inoculation Controversy of 1721–1722: An Incident in the History of Race," *William and Mary Quarterly*, 61 (January 2004), summarizes the broad impact of this episode. http://www.historycooperative.org/journals/wm/61.1/ (Retrieved April 27, 2004).
23. Duffy, *Sanitarians*: 27.
24. Fenn, *Pox Americana*: 260. Further refer to Coss, *The Fever of 1721*: 273–74.
25. Williams, *Angel of Death*: 196, 197, and 199, respectively. See also pages 175–80, 190, 193, 198.

26. David J. Rothman, *Strangers at the Bedside: A History of How Law and Bioethics Transformed Medical Decision Making* (New York: Basic Books, 1991): 20. Allen, *Vaccine*: 131; Lawrence K. Altman, *Who Goes First? The Story of Self-Experimentation in Medicine* (Berkeley: University of California Press, 1998): 106–7; Halpern, *Lesser Harms*: 12; Williams, *Paralyzed with Fear*: 169–71.
27. Williams, *Angel of Death*: 146, 57, 196–97, 198–99 and 105–6, respectively. Further refer to pages 15, 71–72, 140, 143–44, 151–52, 161–64, 167–68, 172, 175–80, 190, 193, 206, as well as Colgrove, *State of Immunity*: 6–7; College of Physicians, *History of Vaccines*: 11, 13; Coss, *The Fever of 1721*: 86, 275; Daniel, *Captain of Death*: 131–32; Duffy, *Sanitarians*: 28; Fenn, *Pox Americana*: 22, 33; Offit, *Vaccinated*: 32; Porter, *Greatest Benefit to Mankind*: 274; Willrich, *Pox*: 36. According to Lederer and Grodin, "Historical Overview," exposing inoculated children to smallpox—even directly infecting them—represented a common practice through the early 1800s (5). The full title of Jenner's work is *An Inquiry into the Causes and Effects of the Variolae Vaccinate: A Disease Discovered in Some of the Western Counties of England, particularly Gloucestershire, and Known by the Name of the Cow Pox.*
28. Willrich, *Pox*: 36. See as well Colgrove, *State of Immunity*: 6–7; Williams, *Angel of Death*: 206.
29. Willrich, *Pox*: 89–91. Paul A. Offit, *Bad Faith: When Religious Belief Undermines Modern Medicine* (New York: Basic Books, 2015), nicely differentiates vaccination policies: "Mandatory vaccination requires people to receive a vaccine or pay some sort of societal price, such as a fine, or be excluded from school. Compulsory vaccination … requires people to be vaccinated whether they want to or not" (107–8).
30. Willrich, *Pox*: 238, 8, 5, and 8, respectively. Consult pages 10–11, 235–37, as well as "New Plans of the Health Department," *New York Times*, November 20, 1902: 24.
31. Lederer, *Subjected to Science*: 4, 14, and 23, respectively. Additionally, refer to pages 3, 10–13, 15–18, 25.
32. Hammonds, *Childhood's Deadly Scourge*: 13–14, 22, 90–92, 184, 85, 188; Colgrove, *State of Immunity*: 82; History of Vaccines Timeline: www.historyofvaccines.org/content/timelines/diphtheria (Accessed February 13, 2016); Edward A. Mortimer, Jr., "Diphtheria Toxoid," in Stanley A. Plotkin and Edward A. Mortimer, Jr., eds., *Vaccines* (Philadelphia: W.B. Saunders Company, 1994): 49.
33. Colgrove, *State of Immunity*: 82–83. Further refer to Mortimer, "Diphtheria Toxoid": 49.
34. Hammonds, *Childhood's Deadly Scourge*: 181 and 182, respectively. Also refer to pages 178–80. Ruth R. Faden and Tom L. Beauchamp, *A History*

and Theory of Informed Consent (New York: Oxford University Press, 1986), discuss the notion of captive test subjects (15).

35. History of Vaccines Timeline: www.historyofvaccines.org/content/timelines/diphtheria (Accessed February 13, 2016); Mortimer, "Diphtheria Toxid": 50; Offit, *Vaccinated*: 145.
36. Colgrove, *State of Immunity*: 110 and 111, respectively. Further refer to Himan and Orenstein, "A Shot at Protection": 72; Mortimer, "Diphtheria Toxoid": 50; Edward A. Mortimer, Jr., "Pertussis Vaccine," in Plotkin and Mortimer, *Vaccines*: 99, 100.
37. Paul A. Offit, *Deadly Choices: How the Anti-Vaccine Movement Threatens Us All* (New York: Basic Books, 2011): 25.
38. Colgrove, *State of Immunity*: 112. Check, moreover, Offit, *Deadly Choices*: 25.
39. Duffy, *Sanitatarians*: 195. Also consult College of Physicians, *History of Vaccines*: 9; Alan R. Hinman and Walter A. Orenstein, "A Shot at Protection: Immunizations Against Infectious Disease," in John W. Ward and Christian Warren, eds., *Silent Victories: The History and Practice of Public Health in Twentieth-Century America* (New York: Oxford University Press, 2007): 74; Steven G. F. Wassilak, Walter A. Orenstein, and Roland W. Sutter, "Tetanus Toxiod" in Plotkin and Mortimer, *Vaccines*: 68, 70. For Offit, in *Deadly Choices*, DTP represents the official designation, but DPT has become the popular designation (2). Seth Mnookin, *The Panic Virus: The True Story Behind the Vaccine-Autism Controversy* (New York: Simon and Schuster, 2011), takes this a step further, distinguishing between DTP in the UK and DPT in the USA (60). Finally, according to Robert W. Sears, *The Vaccine Book: Making the Right Decision for Your Child* (New York: Little, Brown and Company, 2011), DtaP, a refined version, replaced DTP in the mid-1990s (42).
40. Daniel, *Captain of Death*: 175, 176, 136, 137, and 138, respectively. Consult pages 32, 134–35, 139–41, 171–73, as well as College of Physicians, *History of Vaccines*: 22; Offit, *Deadly Choices*: 56.
41. Gretchen A. Condran, Henry Williams, and Rose A. Cheney, "The Decline in Mortality in Philadelphia from 1870 to 1930: The Role of Municipal Services," in Walzer and Numbers, *Sickness and Health*: 454. Also see pages 452–53, 457–58, 460–63.
42. Colgrove, *State of Immunity*: 48. Further refer to College of Physicians, *History of Vaccines*: 22–23, 49.

CHAPTER 4

Not My Child!

Thirteen-year-old Marcella Gruelle returned from school in 1915 and informed her parents, Johnny and Myrtle, that all students were required to have a smallpox inoculation. They reluctantly granted permission. Several days passed after that simple procedure. When it appeared to be ineffective, the nurse repeated it, this time without her parents' permission. Marcella fell ill several days later and, in spite of being hospitalized, lingered for several weeks with a high fever. The needle used by the nurse apparently had been dirty, causing an infection to settle in her heart. She died on November 8. Her rag doll remained in the studio of her heartbroken father, an illustrator for *Ladies' World*, *McCall's*, and *New York Herald*. That spring before her illness that rag doll, her favorite plaything, had inspired him to secure a patent for the Raggedy Ann doll. He published the first *Raggedy Ann* book three years later, eventually dedicating one entitled *Marcella: A Raggedy Ann Story* in 1929. Johnny Gruelle remained steadfast in his opposition to compulsory immunization. "The doll, with its limp limbs, became a symbol of vaccine-damaged children…."[1]

Early Resistance

Boston became the first American city to require smallpox inoculation in 1827, and Massachusetts became the first state in 1855. However, "[t]he most volatile battleground over vaccination during this period

R. J. Altenbaugh, *Vaccination in America*,
Palgrave Studies in the History of Science and Technology,
https://doi.org/10.1007/978-3-319-96349-5_4

proved to be schools ... since children were a 'captive' population to whom ready access was available." Many parents, angry about the appropriation of their children's lives, flaunted these laws with impunity. Moreover, school building administrators "felt that their primary mission was educating children, not controlling epidemics, and they rarely gave priority to enforcing health laws." Furthermore, their schools received state subsidies based on daily attendance and did not want to jeopardize that revenue stream.[2] Finally, this became a fundamental and bitter contest over who asserted mastery of children's bodies: as one anti-vaccination activist saw it, compulsory inoculations represented "'medical domination of the schools and the children attending them.'"[3]

Legal challenges unfolded at state and federal levels. In 1900 a teacher, employed by the Towanda, Pennsylvania, school district denied admission to David R. Smith because he neither produced a smallpox vaccination certificate nor showed proof of an inoculation scar. That district's school board then charged David's father with breaking the state-mandated attendance law. The state district court, in *Commonwealth v. Smith*, upheld the defendant because he had attempted to send his son to school, reasoning that the immunization requirement barred the child from attending, not the parent. The court did not oppose this health stipulation per se and hinted that the state assembly should legislate the matter to avoid future litigation, a rather ambiguous judicial outcome. A definitive decision grew out of *Jacobson v. Massachusetts*, resting on the "line between police power and individual liberty." Pastor Henning Jacobson's legal odyssey began in March 1902 in the midst of a smallpox epidemic in Cambridge, Massachusetts. He refused that city's board of health order to be re-vaccinated, since he had become extremely ill as a child when he underwent this same procedure in Sweden. "Vaccination laws had been on the books in Massachusetts and other states for decades. But the first legal challenge had not reached a state supreme court until 1890...." The federal supreme court's decision in 1905 became the "authoritative statement of the almost unlimited extent of the police power in the United States." This ruling forced vaccination refusals to rely on local protests and pressure on their state legislatures.[4]

Conflict over this issue was neither an aberration nor did it represent a solely American experience. "Anti-vaccinationism was a worldwide phenomenon in the late-nineteenth and early-twentieth centuries," though deeply imbedded in Western culture. Riots erupted in Montreal in 1885 and Rio de Janeiro in 1904. Although England instituted inoculations for children in 1853, medical historian Roy Porter maintains that,

> anti-vaccination backlash soon followed. Campaigning journals appeared, beginning with Henry Pitman's *Anti-Vaccinator* in 1869, as well as the founding of the National Anti-Compulsory Vaccination League in 1874, leading to civil disobedience which resulted in imprisonment for some. The campaign took a further step forward in 1880 with the London Society for the Abolition of Compulsory Vaccination, which had its moment of glory in 1909 when Parliament rescinded compulsory vaccination.

State coercion exposed the dilemma between "collective protection" and "individual rights."[5]

William Tebb, a leading English opponent, proved instrumental in founding the Anti-Vaccination Society of America in New York City in 1879, introducing organizational coherence to American resistance. Others followed. Pittsburgh industrialist John Pitcairn and Charles M. Higgins founded the Anti-Vaccination League of America in Philadelphia in 1908, combating compulsory vaccination well into the 1930s. Pitcairn's opposition reflected his political view; that is, compulsory vaccination represented medical tyranny. "Higgins was the group's chief spokesman and pamphleteer, writing numerous polemical tracts," among them *The Crime Against the School Child* (1915). The Citizens Medical Reference Bureau, founded in New York City in 1919, maintained a sharp ideological position, opposing "state medicine" upheld by the federal government.[6] It "produced the popular pamphlet *The Facts Against Compulsory Vaccination: It's the School and Not the Child That Is Public*."[7] Harry Bernhardt Anderson, the Bureau's secretary, demanded that city's board of education prohibit "health department doctors from using the city's children as guinea pigs in the experimental use of toxin-antitoxin to produce active immunity against diphtheria." The American Medical Liberty League, headquartered in Chicago, promoted personal freedom in choosing medical treatments (i.e., homeopathy v. allopathy).[8] "Most twentieth-century [homeopaths] opposed state-mandated vaccination, employing medical as well as political arguments against it." Orthodox medicine injured children with these "dangerous" sera, at worst, and, at the very least, "suppressed" children's "natural immunity." These groups published books, pamphlets, and journals, like the *Liberator*, as well as retained their own attorneys and sent delegates to international congresses.[9]

Regardless, the anti-vaccination movement was not monolithic; rather, resistance stemmed from a number of sources, ranging from populist sentiments to religious beliefs, from political ideology to outright fear. In some cases, casual indifference ruled while, in others, compulsory inoculations, particularly with relatively untried sera, generated stiff and organized actions.

Through the late nineteenth century, many families preferred freedom of choice in their medical treatments. Homeopathic treatments, organic ingredients, and chiropractic manipulation led this list. For the most part though, tradition held sway as many people clung to long-held beliefs in the power of folk medicine, with its herbal medicines and homemade salves.[10]

Another option flourished in laissez-faire America. In their desperation to find relief, many clung to miracle remedies and often fell victim to enterprising charlatans who sold seemingly magical elixirs, syrups with a "secret formula," otherwise known as patent medicines. This did not operate as a temporary or marginal enterprise, according to writer Philip J. Hilts; rather, this was a major, unregulated industry marketing more than 15,000 products. Manufacturers used the patent label to ensure that their ingredients remained enigmatic. They numbered from "half a dozen to forty" and included arsenic, cocaine, opium, and sulfuric acid combined with copious amounts of alcohol, up to 40% in one case. Not surprisingly, they failed to heal any ailments or, at worse, caused deaths. They remained popular nonetheless because manufacturers exploited the newest advertising techniques, paying for large newspaper spreads. "The medicinal nostrum makers pioneered promotion and marketing as no other business had to that time." With names such as the "Grand Restorative" and "Wheeler's Nerve Vitalizer," they unabashedly claimed to cure "cancer, scrofula, rheumatism, gout, hepatitis, and syphilis," among others.

Beginning in 1905, a variety of startling exposés spurred attempts to reform this industry. Magazines, like *Ladies Home Journal*, "the most popular … of the era," and newspapers, Hilts adds, ran investigative articles disclosing dangerous chemical content or no remedy whatsoever, as well as retail bribery by these manufacturers. Fueled by Upton Sinclair's food processing revelations in *The Jungle*, public sentiment pushed for oversight. This did not prove easy. After rancorous debates and many sabotage efforts, Congress passed the Food and Drug Act in 1906, with that agency housed in the Department of Agriculture. That law, a "mass of loopholes," proved unenforceable, because its labeling requirement proved toothless. If "no ingredients were listed on the label, there was no offense." With limited fines and punishments, the "secret-ingredient industry, which had fought regulation to the end, found that the new law was really not very burdensome...." Muckrakers continued to point out that the "law still permitted ... the human population of the United States be guinea pigs for all experiments with medicinal drugs." Unimpeded, the "trade in patent medicine ... by the early 1930s was $350 million...."[11]

Many people, especially the poor, generally shunned mainstream health care. First, they feared hospitals as the possible source of contamination. Second, parents resented institutional regulations that sharply restricted access to their children. Third, because they associated hospitals with charity, and the concomitant embarrassment of declaring poverty to qualify for admission, many proud adults chose to preserve their dignity. Fourth, they strongly distrusted public health officials, seeing them intruding into families' lives. This especially applied to vaccinations. During Chicago's 1892 smallpox epidemic, pitched battles ensued with the appearance of inoculation units. "One young surgeon on the vaccination squad had been 'disabled for life' when an agitated tailor shattered his elbow with a bullet." In 1898, an outraged mob in Wilmington, North Carolina, rioted over forced inoculations.[12]

Immigrants represented another source of resistance. Germans during Brooklyn's 1893–1894 smallpox epidemic opposed vaccination, reflecting their attitude toward compulsion in their former homeland. Italians also refused but their decision stemmed from fear. Brooklyn's health department ignored these views. In April 1894, it sent fifty-six vaccinators to schools and administered 27,000 inoculations, provided two dozen free immunization clinics, arranged for doctors to visit "more than 200 factories and other places of business ... to vaccinate employees," and conducted "house-to-house sweeps in areas adjacent to cases that were discovered." By August, the Brooklyn health department had given 225,000 vaccinations, "close to three quarters were done house-to-house." These strong-armed tactics spawned the Brooklyn Anti-Vaccination League led by homeopathic doctors. This organization filed suit against unlawful quarantines and won. Others pursued lawsuits claiming "either assault or wrongful death as a result of vaccination."

All of the cases were either overturned or thrown out when they reached the appellate courts, clearing Brooklyn's health department of any wrongdoing.[13]

Many religious beliefs also confronted allopathy. This represented a contest for authority between religion rooted in static notions and science based on change, between the sanctity of the human body, protecting the soul, and the physical reality of research, treating it as a clinical instrument to expand knowledge. The Christian Science movement relied on a holistic approach, combining spiritual and physical well-being to facilitate self-healing. Fundamentalist Christians believed that inoculations corrupted the human body, a creation of God. Moreover, a distinct "anti-medicine" attitude developed among Seventh-Day Adventists. Finally, Mormons passed "laws restricting the 'deadly poisons' of orthodox remedies and, in particular, championed the right to resist compulsory smallpox vaccination."[14] In either case, they deemed "compulsory vaccination" as a "violation of religious freedom."[15]

Americans further responded in quiet, unaccountable ways. Personal, individual actions consisted of "concealing sick family members at home, forging vaccination certificates, or simply dodging their legal duty to be vaccinated." The decentralized nature of schooling assured uneven enforcement. In late nineteenth-century Kentucky at least one-third of children lacked immunization. Private, urban physicians commonly provided false smallpox inoculation certificates for a fee.[16]

General uncertainty about vaccines and sera themselves raised serious doubts. Side effects could cover a broad range of symptoms. Most treated individuals suffered from mild headaches or low-grade fevers. Death could also occur. In 1895, seventeen-year-old Bertha Valentine died in New York City within minutes after receiving an injection of the new diphtheria antitoxin. A medical investigation ensued and exonerated it. The antitoxin did not kill patients per se, but their health represented the best mark of success or failure; that is, if they were already physically weakened by the disease, the inoculation would surely kill them. Moreover, based on anecdotal evidence at best or hearsay at worst, many believed that smallpox vaccinations caused infantile paralysis.[17]

In the end, most opponents were not even devout anti-vaccinationists. Many were simply parents who pragmatically considered risks and made decisions accordingly. They weighed the immediate threat of illness and death of their children from potentially bad vaccine batches against the possibility of contracting a disease sometime in the future; the former appeared more concrete than the latter. Certainly, smallpox immunization had existed in the USA since 1800, but "vaccine manufacturing did not emerge as a commercial industry until the 1870s, with the shift from 'humanized' to 'bovine' virus." By the turn of the century, the entire pharmaceutical industry was in transition as it began to forge "close ties with government health departments and universities." Nevertheless, production remained largely unregulated. "Just seven states had laws providing for some supervision of the vaccine manufactured or used in the state." This resulted in a highly uneven concoction, from inert to contaminated. Toward the end of 1901, "[p]opular distrust of vaccine surged ... as newspapers across the country reported that batches of tetanus-contaminated diphtheria antitoxin and smallpox vaccine had caused the deaths of thirteen children in St. Louis, four in Cleveland, nine in Camden, and isolated fatalities in Philadelphia, Atlantic City, Bristol (Pennsylvania), and other communities," causing parents to grow alarmed more so about vaccines than diseases.[18]

Support for vaccinations in the USA thus experienced a long and dramatic ebb and flow. Based on the success of Edward Jenner's smallpox treatments in England, this disease was virtually eradicated in the USA during the first half of the nineteenth century. With a false sense of security, people began to ignore immunization; this led to an outbreak in 1870. Apathy only represented part of the picture, however. Opposition blossomed through many organizations, embracing human rights, citing the Bible, exaggerating potential risks, and claiming profiteering. This movement scored some serious successes. Many states enacted measures to outlaw compulsory smallpox inoculations for school-aged children. This included Minnesota in 1903, Utah in 1907, and Massachusetts in 1908. Between 1918 and 1921, Arizona, California, Maine, North Dakota, Washington, and Wisconsin followed suit. "These were victories for Christian Scientists and others who opposed the fact the state could, in any capacity, scientifically regulate or assume authority over the bodies of their children."[19] As late as 1926, *The Quest*, an anti-vaccination pamphlet, printed and distributed out of Brooklyn, shrilly warned: "Americans! Wake up before it is too late! Get your eyes open to the enormous crime of compulsory vaccination."[20] Although its organizations had begun to wane, decades-long resistance had left its mark. Ultimately the "United States would remain, in the words of one the nation's preeminent public health experts, 'the least vaccinated of any civilized country.'"[21]

The health establishment struck back. Medical journals published numerous editorials about and news snippets of anti-vaccination campaigns between 1906 and 1922, marking a relentless struggle during that period. Several applauded legislative and judicial victories in Chicago and Pennsylvania over Christian Scientists and parents as well as other anti-vaccination groups who opposed smallpox immunization for school enrollment. One praised the efforts of the Council on Public Education of the American Medical Association (AMA) for promoting the virtues of wise health and medical practices. One piece questioned the psychological stability of an anti-vaccination judge in New York. Others saw resistance rooted in small-town provincialism, likened complaining parents to criminals, and called anti-vaccination organizers outright liars. In a Darwinist twist, a 1909 editorial in *American Medicine* assumed a superior I-told-you-so attitude when several anti-vaccination "propagandists" died of smallpox: "The vaccinated are fittest for survival where the infection of smallpox abounds." Finally, *The Survey* published a short, sympathetic item about a Chicago judge who upheld the school superintendent's policy of barring unvaccinated children from attending school. That judge based his decision on the *Jacobson* ruling and added: "Public officials have a right to be guided by what science has demonstrated to be as near the truth as truth can be ascertained and science has come to the conclusion universally that vaccination is a preventive of smallpox…." Medical science not only marshaled the power of the American legal system but invoked the imprimatur of "truth" itself. No further debate seemed necessary.[22]

Public health authorities felt no compunction about executing draconian measures to carry the vaccination campaign directly to children's families. Hiram M. Read, Seattle's health commissioner, in 1919, ordered quarantine signs posted on the homes of unvaccinated students. An intransigent mother defiantly tore one from her home; the police promptly arrested her. The Public School Protective League of Seattle, a newly formed anti-vaccination organization, filed a complaint with the school board on her behalf as well as subsidized her legal expenses. The misdiagnosis of one student and the death of another recently immunized only intensified parent anger. Nevertheless, with the ever-increasing number of smallpox cases, Read persisted in inoculating volunteers and maintained the quarantine policy.[23] Likewise, the city of Chicago prohibited public and private school principals from admitting any child who did not provide a "certificate signed by the Commissioner of Health or any physician duly licensed by the State Board of Health."

Violators would be fined between $10 and $200 for each offense.[24] Municipal health departments also became extremely intrusive, evidenced by two 1922 publications by the Citizens Medical Reference Bureau, titled *Protest Against Sending Nurses Into Homes of School Children to Urge Medical Treatment* and *Against Using Public Schools to Promote the Schick Test, and Toxin-Antitoxin*. Both objected to public health nurses visiting families to "educate and persuade" parents to give their children smallpox and diphtheria inoculations as well as examine their adenoids and tonsils.[25]

Anti-vaccination sentiment eventually spawned federal intervention to establish public confidence in the safety of vaccines and antitoxins. In 1902, Milton J. Rosenau, following tests in the Hygienics Laboratory revealing contamination, officially reported "that defective vaccine was a national problem that required a national solution." In July of that year, President Theodore Roosevelt signed the Biologics Control Act that "established a system of licensing and inspection for all biologics sold in interstate commerce or imported from abroad," as well as designated the newly named US Public Health and Marine-Hospital Service (later the US Public Health Service) to oversee licensing and inspection. "Four years later … Congress would enact another, much better remembered statute modeled closely after it, the Pure Food and Drug Act." The Hygienics Laboratory, which tested manufacturers' samples, expanded and became the National Institute (later Institutes) of Health in 1930. "The new inspection regime saved compulsory vaccination at its moment of greatest crisis in the United States."[26]

Children's Rights

A less acknowledged, but related and serious, concern emerged amidst this furor: Child welfare itself symbolized a comprehensive concept during the Progressive Era. Henry Bergh, the privileged son of a wealthy New York City shipbuilder, seemed an unlikely yet important part of this development. He attended Columbia University but abandoned his studies to travel through Europe. While there, President Abraham Lincoln appointed Bergh as Acting Consul to St. Petersburg in 1862. He resigned two years later due to poor health and returned to New York. By 1866, he had solicited sufficient funds and ample support to secure a state charter for the American Society for the Prevention of Cruelty to Animals (ASPCA).

This reflected a cultural transition that unfolded around the time of the Civil War. “Urban Americans had begun to perceive some animals … as family members and child substitutes….” Instead of cats and dogs used as “mousers and hunters,” they became “pets.” These emotional attachments caused “experimenting on them … to be seen as ‘inhumane.’” This gave birth to a “strident anti-vivisection movement” embodied in the ASPCA.[27]

From an office occupying two attic rooms in New York City, Bergh filed the first charges for cruelty to animals. Subsequent campaigns addressed abuses of livestock and workhorses, sought slaughterhouse reforms, as well as opposed cock and dog fights, successfully lobbying that state’s legislature for laws. Rising to a national figure, he too shepherded in other groups, namely from Boston and Philadelphia. That Society’s members elected him as its first president, serving until his death in 1888. By 1924, its permanent headquarters occupied a four-story building located on Madison Avenue.

Eldrige T. Gerry, one of Bergh’s staunchest supporters, became the ASPCA’s legal counsel. Together they founded the Society for the Prevention of Cruelty to Children (SPCC) in 1874, becoming a part of Progressive Era’s child-saving crusade. By 1924, its headquarters resided in a six-story building on Fifth Avenue.[28]

Ultimately separate campaigns to protect animals and children merged institutionally. The use of animals in laboratory settings had to be curbed. Anti-vivisectionists “cautioned repeatedly that the failure to enact legislative safeguards against immoral animal experimentation would lead directly to human experimentation.”[29] The SPCC consisted of autonomous, local branches in most major cities, but the New York City organization proved to be particularly active. Its highly moralistic, missionary zeal involved swooping into poor and immigrant communities to rescue—as they perceived it—abused, exploited, or neglected children, operating shelters, and working with the juvenile court system. These seemingly divergent interests found a common home in the American Humane Association (AHA) founded in 1874.

What had begun as general opposition to animal cruelty extended to scientific experimentation in the 1880s. Delegates at its 1895 meeting in Minneapolis agreed to the following resolution: “All experimentation upon living animals we consider unnecessary, unjustifiable, and morally wrong.” Two years later, the AHA broadened its campaign to include humans, revealing an "overlap between the anti-vaccination and anti-vivisection [experimentation] causes." In 1888, Gerry served as president of the AHA. Meanwhile, Bergh opposed vaccinations, seeing them as crass commercial ploys to bilk the public for profits. He represented, for researcher William J. Shultz writing in a 1924 retrospective, "the first American anti-vivisection protagonist of the anti-vaccination movement." By 1922, 539 active societies affiliated with the parent AHA: 175 focused solely on animals; 57 on children; and 307 included both. "The term 'humane,'" Schultz chronicled, "... includes both animal and child protection."[30] The stated mission of *The Quest* left no doubt: it "opposed to vaccination and cruel vivisection and in defense of our children and our animals...."[31]

Numerous incidences spurred this movement. In 1899, the AHA published a pamphlet, "Human Vivisection: A Statement and an Inquiry." It defined human vivisection as "the practice of subjecting human beings, men, women and children, who are patients in hospitals or asylums, to experiments involving pain, mutilation, disease or death, for no object connected with their individual benefit, but entirely for scientific purpose." This represented a moral problem that defied religion (i.e., Christianity), "Justice and Humanity (sic)." It also reproduced Senate Document No. 78 which described several international and domestic experiments, demonstrating this as common knowledge. Institutionalized orphans, one Viennese doctor declared, "were cheaper than animals" for use in medical experiments. An American in an 1898 issue of the *New England Journal of Medicine* objectified test subjects as mere "material." Attendees at the 1899 AHA meeting noted that "not a single scientific society in our country has ever made the faintest protest against the atrocious subjection of infants to mutilation, to inoculation with loathsome and sometimes fatal disease, or to any other form of human vivisection."[32] For the AHA, the only safeguard involved the complete cessation of all medical investigations.

A pamphlet, "The Reality of Human Vivisection: A Review," penned by James M. Brown, AHA president in 1901, stridently rebutted testimony by William W. Keen, former AMA president, before a 1900 Senate Committee investigating vivisection. Brown reiterated the AHA's position, condemning all human experimentation on moral and ethical grounds, especially uncontrolled and unmonitored research. He described the "... inoculation of innocent children with foul disease, the grafting of cancers into the healthy breasts of unconscious women ...

[and] the inoculation of hospital patients with yellow fever...." Brown too soundly condemned "experiments upon sick and dying children in an 'Infants Hospital' ... in the last throes of death that were sometimes used as 'material.'" He claimed that many of these experiments had "no pretense of 'curing.'"[33]

Nothing drew the ire of anti-vivisectionists more than the 1911 luetin experiment. Hideyo Noguchi, at the Rockefeller Institute for Medical Research, developed luetin, an inactive extract derived from syphilis, to create a diagnostic tool to detect "hidden or latent cases." Following tests on laboratory rabbits, he administered it to 400 human subjects at the Good Samaritan Hospital, King's Park State Hospital, Randall's Island Asylum, and Rockefeller Hospital. He found that 254 had syphilis while the remaining 146 served as "various controls." The medical profession applauded his study. The New York Anti-Vivisection Society and the Vivisection Investigation League publicly protested, stressing "Noguchi's ethical violation of the personal rights of the health of children and ailing adults he had used in the luetin test and the potentially harmful results of a diagnostic experiment with the serum of a 'loathsome' disease." Critics saw consent as pivotal. Rockefeller Institute representatives countered that luetin was safe, justified by the fact that Noguchi and other physicians had injected that substance into themselves before using it on other subjects. Moreover, since the children Noguchi tested were wards of the New York City Board of Charities, institutional permission provided legal cover.[34]

Anti-vivisectionist efforts proved relentless. In 1910, *Cosmopolitan* magazine published an exposé by New York anti-vivisectionist Diana Belais. She described tests conducted on orphans at the St. Vincent's Home in Philadelphia, where doctors injected tuberculosis directly into their eyes. Some suffered temporary blindness. Anti-vivisectionists also opposed the inoculation of military recruits during World War I, pushing for voluntary participation. New York social worker Konrad Bercovici, in a 1921 article published in the *Nation*, revealed how two doctors induced scurvy in children residing at the Hebrew Orphan Asylum. Then the doctors reversed this condition by changing their diets, only repeating that cycle after they had recovered. This study lasted over two years and involved 150 children. Using such institutional settings, researchers could control all of the variables involved with human experimentation. Finally, anti-vivisectionists lobbied the New York state legislature, which considered proposals in 1914 and 1923 aimed at curbing animal and child experimentation. Fierce resistance from the AMA and life insurance companies throttled these initiatives. The anti-vivisection movement "reached its crescendo from the 1890s through 1917, and then died out in the 1920s."[35] Left unfettered, researchers continued to experiment on children. Early vaccine development thus became inextricably linked to the unregulated use of children as human subjects.

Building Public Trust

On April 26, 1954, six-year-old Randy Kerr of McLean, Virginia, received the first injection of Jonas E. Salk's experimental polio vaccine. "V-Day," as it was officially designated, thus began. Two hundred thousand adult volunteers, 60,000 physicians, nurses, public health officials, and 64,000 educators oversaw the entire enterprise. Between April and June, 1.8 million elementary students lined up at their school buildings to receive their jabs—the largest medical research project using children as test subjects.[36] What had changed to mobilize so many resources and generate so much conformity? Simply put, society had become medicalized.

As the USA grew more urbanized, with ever denser populations and greater possibilities of many and different epidemics, as bacteriology emerged, with the promise of combating diseases, a medical conversion unfolded, enlightening people about the benefits of hygiene and the miracles of medicine. An epistemological shift in popular culture occurred, portraying scientists as heroes to adults and children. Spanning a thirty-year period, beginning in the 1920s, an inter-generational phenomenon, this informal, uncoordinated campaign gradually legitimated health professionals and their institutions, and built unqualified trust in the wisdom of science, helping to set the stage for this famous experiment. Taking a broader, cultural view and the roles of various institutions therein offers a dynamic and contextually rich explanation of the vaccine trials of the 1950s, unveiling complex and fluid relationships between medical research, public health policies and actions, as well as the fate of American children.

By the early twentieth century, the medical community had begun to recognize the complexities, risks, and potentially tragic outcomes of immunization campaigns. And it needed to overcome deeply imbedded suspicions as well as allay fears. Compulsion represented an early and crudely executed approach; coercion alone would not ultimately win the day. Medical practices had to be legitimated and faith in new-found cures had to be cultivated. This paradigm shift involved an extensive campaign to induce the public to accept modern health notions and practices. Two avenues of conversion existed: informal and formal persuasion and education. The former supplanted the need for police power while the latter acted to internalize consensus.[37]

Notes and Sources

1. Arthur Allen, *Vaccine: The Controversial Story of Medicine's Greatest Lifesaver* (New York: W. W. Norton Company, 2007): 99. Further see Patricia Hall, *Johnny Gruelle: Creator of Raggedy Ann and Andy* (Gretna, LA: Pelican Publishing Company, 1998): 18, 87, 89, 91–92, 103, 153; Paul A. Offit, *Deadly Choices: How the Anti-Vaccine Movement Threatens Us All* (New York: Basic Books, 2011): 135–36; Josepha Sherman, *Johnny Gruelle and the Story of Raggedy Ann and Andy* (Hockessin, DE: Mitchell Lane Publishers, 2006).
2. James Colgrove, *State of Immunity: The Politics of Vaccination in Twentieth-Century: America* (Berkeley: University of California Press, 2006): 72–73. Moreover, consult pages 7, 9; Allen, *Vaccine*: 60; Stephen T. Woolworth, "Conflict, Collaboration, and Concession: A Study of the Rise and Fall of Medical Authority in the Seattle Public Schools, 1892–1922" (PhD dissertation, University of Washington, 2002): 93.
3. James Colgrove, "Science in a Democracy: The Contested Status of Vaccination in the Progressive Era and the 1920s," *Isis*, 96 (2005): 172. John Duffy, "School Vaccination: The Precursor to School Medical Inspection," *Journal of the History of Medicine*, 33 (July 1978): 352–53.
4. Michael Willrich, *Pox: An American History* (New York: Penguin Books, 2011): 299, 308, and 332, respectively. Refer also to pages 285, 325; Colgrove, "Science in a Democracy": 170; Colgrove, *State of Immunity*: 33–34, 38–39, 42; *Jacobson v. Massachusetts*, 197 U.S. 11; 25 S. Ct. 358; 49 L. Ed. 2d643 (1905); *Commonwealth v. Smith*, 9 Pa. D.R. 625 (1900).
5. Roy Porter, *Greatest Benefit to Mankind: A Medical History of Humanity* (New York: W. W. Norton Company, 1997): 420. Further consult Colgrove, *State of Immunity*: 2; E. P. Hennock, "Vaccination Policy Against Smallpox, 1835–1914: A Comparison of England with Prussia and Imperial Germany," *Social History of Medicine*, 11 (April 1988): 51, 54, 58, 60; Willrich, *Pox*: 37, 39, 40.
6. Willrich, *Pox*: 253 and 54, respectively. Consult page 52, 254–55, as well as Anti-Vaccination League of Philadelphia, Pennsylvania., ca. 1900–16

[ref.# 23]; Anti-Vaccination Society of America, New York, etc., 1885–1898 [ref.# 24]; Keen, William Williams, 1837–1932 [ref.# 488], COP; Colgrove, "Science in a Democracy": 173–74.

7. Offit, *Deadly Choices*: 134.
8. Colgrove, "Science in a Democracy": 176–77. Also see Colgrove, *State of Immunity*: 2, 9, 44, 52–54, 56, 103; James Colgrove, "Foot Soldiers Against Infectious Diseases: Nurses, Families, and Immunization in the Twentieth Century," *Pediatric Nursing*, 33 (September–October 2007): 450; J. N. Hays, *Burdens of Disease: Epidemics and Human Response in Western History* (New Brunswick: Rutgers University Press, 1998): 223, 224–25; Porter, *Greatest Benefit to Mankind*: 389–390; Willrich, *Pox*: 40, 253–55.
9. Anne Taylor Kirschmann, "Making Friends for 'Pure' Homeopathy: Hahnemanniansand the Twentieth-Century Preservation and Transformation of Homeopathy," in Robert D. Johnston, ed., *The Politics of Healing: Histories of Alternative Medicine in Twentieth-Century North America* (New York: Routledge, 2004): 33. Also see James M. Brown, "The Reality of Human Vivisection: A Review" (Boston, 1901). Anti-Vivisection Papers, Rockefeller University Archives, Record Group 600–1. Box 16, Folder #1. RAC; Willrich, *Pox*: 255.
10. Colgrove, *State of Immunity*: 52, 58–59; Hays, *Burdens of Disease*: 223, 225–229, 230–31; Anne Taylor Kirschmann, *A Vital Force: Women in American Homeopathy* (New Brunswick, NJ: Rutgers University Press, 2004); Porter, *Greatest Benefit to Mankind*: 393, 395, 396 and "What is Disease" in Roy Porter, ed., *The Cambridge Illustrated History of Medicine* (Cambridge: Cambridge University Press, 2001): 113 16; John Pickstone, "Medicine, Society, and the State" in *Cambridge Illustrated History of Medicine*: 316; Porter, *Greatest Benefit to Mankind*, 281–82, 284–85; Regina Morantz-Sanchez, *Sympathy and Science: Women Physicians in American Medicine* (Chapel Hill: North Carolina University Press, 2000): 17; Edward Shorter, "Primary Care," in *Cambridge Illustrated History of Medicine*: 126, 153; Willrich, *Pox*: 256–57, 260; Woolworth, "Conflict, Collaboration, and Concession": 93; Russell Viner and Janet Goldman, "Children's Experiences of Illness," in Roger Cooter and John Pickstone, eds., *Medicine in the Twentieth Century* (Australia: Harwood Academic Publishers, 2000): 579–80.
11. These two paragraphs include quotes and information from Philip J. Hilts, *Protecting America's Health: The FDA, Business, and One-Hundred Years of Regulation* (New York: Alfred A. Knopf, 2003): 24, 48, 25, 47–48, 75, 61, 59, 61, and 75, respectively. Also see pages 23, 26–27, 49–53, 54, 56–57, 60. For a humorous overview of patent medicine, consult Lydia Kang and Nate Pedersen, *Quackery: A Brief History of the Worst*

Ways to Cure Everything (New York: Workman Publishing Company, 2017).

12. Willrich, *Pox*: 83. Further check pages 86–88, as well as Evelynn Maxine Hammonds, *Childhood's Deadly Scourge: The Campaign to Control Diphtheria in New York City, 1880–1930* (Baltimore: Johns Hopkins University Press, 1999): 43; Woolworth, "Conflict, Collaboration, and Concession": 93. For a treatment of general immigrant resistance to medical treatments during this period, see Naomi Rogers, *Dirt and Disease: Polio Before FDR* (New Brunswick, NJ: Rutgers University Press, 1996).
13. Colgrove, *State of Immunity*: 22, 29, and 30, respectively. Likewise, consult pages 21, 25–26, 28, 32.
14. Porter, *Greatest Benefit to Mankind*: 276 and 395, respectively. See also pages 281, 395, 396, and "What Is Disease": 115, as well as Colgrove, *State of Immunity*: 56–57; Colgrove, "Science in a Democracy": 177–78; Hays, *Burdens of Disease*: 224–27.
15. Woolworth, "Conflict, Collaboration, and Concession": 92–93.
16. Willrich, *Pox*: 13. Refer as well to pages 52, 67.
17. Colgrove, "Science in a Democracy": 167; Hammonds, *Childhood's Deadly Scourge*: 122, 125, 133–134.
18. Willrich, *Pox*: 180, 187, 167–68, respectively.
19. Woolworth, "Conflict, Collaboration, and Concession": 236–37. Consult pages 234–35, 238–42, as well as Colgrove, *State of Immunity*: 64; Martin Kaufman, "The American Anti-Vaccinationists and Their Arguments," *Bulletin of the History of Medicine*, 41 (September–October, 1967): 463–64, 465–66, 467–68, 470, 473, 474–75, 478.
20. *The Quest: Against Vaccinations and Cruel Vivisection*, 1 (December 1926). Excerpted from the front matter of this anti-vaccination journal. No pagination. Anti-Vivisection Papers, Rockefeller University Archives, Record Group 600-1. Box 13, Folder #4. RAC.
21. Willrich, *Pox*: 338. Consult page 261, as well as Kaufman, "The American Anti-Vaccinationists and Their Arguments": 478.
22. The first quote in this paragraph is from "The Slow Decadence of the Anti-Vaccination Propaganda," *American Medicine*, 15 (June 1909): 291, and the second can be found in "Vaccination and Parents' Rights," *The Survey*, 47 (February 1922): 718. Editorials and news notes appeared anonymously in "Conference of Anti-Vaccinationists," *American Medicine*, 12 (October 1906), 449; "Pennsylvania Wins Vaccination Fight," *American Medicine*, 12 (October 1906): 449–50; "Defeat of the Anti-Vaccination Bill," *Boston Medical and Surgical Journal*, 170 (May 1914): 847; "Compulsory Vaccination," *American Medicine*, 15 (June 1909): 219–20; "The Anti-Vaccinationists," *American Medicine*,

16 (November 1910): 558; "An Anti-Vaccination Judge," *American Medicine*, 20 (October 1914): 613–14. Further consult Colgrove, "Science in a Democracy": 182–83.

23. Woolworth, "Conflict, Collaboration, and Concession": 238–42.
24. *Biennial Report of the Department of Health of the City of Chicago for 1904–05* (Chicago: Department of Health, 1906): 171–72, CPL.
25. James Colgrove, "Foot Soldiers Against Infectious Disease": 450.
26. Willrich, *Pox*: 195, 206–7, 208–9, respectively. See as well pages 170, 197, 206–7; Colgrove, "Science in a Democracy": 169.
27. Robert Baker, *Before Bioethics: A History of American Medical Ethics from the Colonial Period to the Bioethics Revolution* (New York: Oxford University Press, 2013): 236 and 237, respectively. Refer to page 265, as well as Sidney H. Coleman, *Humane Society Leaders in America: With a Sketch of the Early History of the Humane Movement in England* (Albany, NY: American Humane Association, 1924): 34–36, 38.
28. Coleman, *Humane Society Leaders in America*: 40–42, 44–45, 48–50, 53, 75, 77, 79, 254–55.
29. Susan E. Lederer, "Hideyo Noguchi's Luetin Experiment and the Antivivsectionists," *Isis*, 76 (March 1985): 32.
30. William J. Schultz, *The Humane Movement in the United States, 1910–1922* (1924: Rpt. New York: AMS Press, 1968): 14 and 144–45, respectively. See pages 11, 41, 141–43, 155, 162–72, 196, as well as Baker, *Before Bioethics*: 237; Colgrove, *State of Immunity*: 59, 103; Susan E. Lederer, *Subjected to Science: Human Experimentation in America Before the Second World War* (Baltimore: Johns Hopkins University Press, 1995): 29, 50, 52–72; Susan E. Lederer and Michael A. Grodin, "Historical Overview: Pediatric Experimentation," in Michael A. Grodin and Leonard H. Glantz, eds., *Children as Research Subjects: Science, Ethics, and Law* (New York: Oxford University Press, 1999): 11–14. The AHA's 1895 resolution is from American Humane Association, "AHA Report" (1895: Rpt. Memphis, TN: General Books, 2012): 4.
31. *The Quest*. No pagination. RAC. Further check Susan E. Lederer, "Children as Guinea Pigs: Historical Perspective," *Accountability in Research*, 10 (2003): 5.
32. American Humane Association, "Human Vivisection: A Statement and an Inquiry" (Chicago: American Humane Association, 1899): 1, 2, 4, 8, 11, respectively. Generally consult pages 5–10, as well as Schultz, *The Humane Movement*: 153–55, 158, 161. Baker, *Before Bioethics*, points out that vivisection serves as a "Generic Name for Any Form of Research Conducted on a Living Animal, or Nonhuman or Human" (264).
33. James M. Brown, "The Reality of Human Vivisection: A Review" (Boston, 1901), 1, 7, 17, 23, folder 1, Box 16, RG 600-1

Anti-Vivisection Papers, Rockefeller University Archives, RAC. Italics are in the original text.

34. Lederer, "Hideyo Noguchi's Luetin Experiment and the Antivivsectionists." All of the quotes in this paragraph can be found on pages 35–36. Refer also to pages 34–35, 38, 41, 43–44.
35. Baker, *Before Bioethics*: 263. Also consult pages 264–73, as well as Lederer, *Subjected to Science*: 74; Lederer, "Children as Guinea Pigs": 5–7; Schultz, *The Humane Movement in the United States*: 153, 158, 161.
36. Examples of traditional interpretations include Liza Dawson, "The Salk Polio Vaccine Trial of 1954: Risks, Randomization and Public Involvement," *Clinical Trials*, 1 (2004): 128; Aaron E. Klein, *Trial by Fury: The Polio Vaccine Controversy* (New York: Charles Scribner's Sons, 1972); David M. Oshinsky, *Polio: An American Story* (New York: Oxford University Press, 2005); John R. Paul, *A History of Poliomyelitis* (New Haven: Yale University Press, 1971); Nina Gilden Seavey, Jane S. Smith, and Paul Wagner, *A Paralyzing Fear: The Triumph Over Polio in America* (New York: TV Books, 1998); Daniel J. Wilson, *Living with Polio: The Epidemic and Its Survivors* (Chicago: University of Chicago Press, 2005).
37. Willrich, *Pox*: 338. Colgrove, *State of Immunity*, devotes Chapter 1, "Between Persuasion and Compulsion," to these approaches.

PART II

Friendly Persuasion

CHAPTER 5

Invisible Bugs Are Bad for You

Science reshaped public health policies, but "social attitudes" operated as a powerful counter force.[1] To change these outlooks, vaccine advocates targeted two institutions: the home and the school. They "saw a great need for popular health education, especially for women and children." Such "[m]ass health education" received "high priority in the interwar decades. Leaders in both public health and organized medicine keenly appreciated the value of health propaganda … and sought to bring popular health education under closer medical supervision." This informal process remained decentralized and laissez-faire but nonetheless purposeful and effective.[2] Although no grand scheme existed, nor did a single organization initiate it, some played key roles. Generally speaking, it represented a natural by-product of significant medical breakthroughs, coupled with technological refinements and the expansion of mass media. This information became accessible to all segments of society, readily affordable, if not free.

Popular Culture

Mass media during the first half of the twentieth century operated as a comprehensive educational experience for the general public, especially raising the visibility and concomitantly the status of "health soldiers," their work and the scientific world in general. "[M]edical history images and stories came to be widely disseminated in popular books and

R. J. Altenbaugh, *Vaccination in America*,
Palgrave Studies in the History of Science and Technology,
https://doi.org/10.1007/978-3-319-96349-5_5

magazines, commemorations, Hollywood films, children's literature, radio dramas, schoolbooks, corporate advertising, and the then-brand-new genre of comic books."[3]

Paul de Krui became a prolific author. As a young microbiologist at the Rockefeller Institute for Medical Research, he began his writing career in 1922 by publishing a series of articles for *Century Magazine* and *Harper's*, "spoofing" in his words, "the high priests of medical science." Simon Flexner, that Institute's director, saw it as a slight and promptly fired him. De Kruif's writing took a more serious and popular turn with the publication of *Microbe Hunters* four years later, providing nonfiction accounts of a dozen pioneer scientists. Writing these stories for the general public, de Kruif brought them to life and made their historical discoveries concrete. He also collaborated with Sinclair Lewis on his Pulitzer Prize-winning novel, *Arrowsmith*.[4]

Lewis's 1925 character study introduced a "new kind of hero." Medical researchers, brilliant, devoted, relentless, and selfless, toiled in their laboratories, tediously unlocking the secrets of diseases. Through his protagonist, Martin Arrowsmith, Lewis leads readers on an odyssey through medical school, private practice, public health, and ultimately the science laboratory. Medical students during the late nineteenth and early twentieth centuries prepared themselves for parochial lives, becoming nothing more than practitioners bent on securing lucrative careers. Lewis portrayed self-absorbed instructors who lectured as much about the color and style of office furniture young physicians should purchase to impress their patients as they did on human anatomy. Rare were the intellectuals in Arrowsmith's academic world. Max Gottlieb, his German-trained mentor, represented the lone exception. He exhausted himself as he labored nightly in his laboratory, alienated his colleagues who he viewed as trite, narrow-minded, and pompous, and sacrificed financial security in order to pursue pure research. He spawned Arrowsmith who found little satisfaction working first as a country doctor, then a public health official, and in a profitable city practice. He ultimately found fulfillment, arriving full circle to work with Gottlieb who had since moved on to a new private, research laboratory—analogous to the Rockefeller Institute. Arrowsmith subsequently discovered a miracle drug in his laboratory and bravely used it to save lives on a plague-ridden Carribean island. The story ended as he sacrificed marriage and wealth to pursue more life-saving discoveries.[5]

Lewis's message was clear. Medical researchers served as the enlightened protectors of humankind, elevating the health profession. They valiantly fought against diseases using science to discover their causes as well as their cures. They stood in sharp contrast to country doctors who participated in small-town gossip, employed folksy bedside manners, applied tried-and-true experience to help patients, and eschewed emerging scientific knowledge, suitable only for *eggheads*. Scientists strived for something higher than big-city doctors who merely catered to an affluent clientele in order to accumulate wealth and achieve social status. Laboratory investigators were indeed a cut above the rest. By the 1920s and 1930s, they had emerged as the scientific nobility when their discoveries began to receive public acclaim and respect. The impact of these writers cannot be underestimated. Sinclair Lewis's *Arrowsmith* (1925) inspired many readers to become doctors while Paul de Kruif's *Microbe Hunters* led them to pursue research.

De Kruif gained fame as an author of thirteen books between 1922 and 1962. Successful fund-raising activities by Franklin D. Roosevelt's National Committee for the Celebration of the President's Birthday resulted in the formation of the Infantile Paralysis Research Commission in 1935, with $250,000 to award as research grants. Jeremiah Milbank served as chair and de Kruif as secretary, who recruited pioneer virologist Thomas M. Rivers at the Rockefeller Hospital as a committee member. According to Rivers, de Kruif became the "most important member of the [President's Birthday Ball] advisory committee…." Through his recommendations, it distributed grants to scientists at Johns Hopkins and the University of Pennsylvania, and the Harvard Infantile Paralysis Commission and Yale Poliomyelitis Commission.[6]

De Kruif blended this growing, public faith in laboratory pursuits with poliomyelitis investigations. In his article, "What Science Is Doing," appearing in the 1938 edition of the *President's Birthday Magazine*, de Kruif described the power and credibility of research in the battle against this dreaded disease by assuring readers that "infantile paralysis is not a mystery." Public patience with and faith in science would be rewarded: "Seemingly theoretical" pursuits will generate "practical" solutions. De Kruif even defended failed attempts since they too expanded knowledge about this virus. In a clinical tone, he briefly described the unsuccessful Park-Brodie "formalinized vaccine" experiments, subsidized by the Birthday Ball Commission. He likewise alluded to the tragic Kolmer trials.

Both failures contributed new pieces to the larger puzzle, he concluded: "In short, it pays to find out what is not so as well as what is so."[7]

In addition to fiction and nonfiction, wide press coverage greeted the development of life-saving equipment, such as the artificial respirator in 1929. Philip Drinker, its inventor, completed his undergraduate degree at Princeton University and graduate work in chemical engineering at Lehigh University before accepting a position at Harvard University's School of Public Health. The Rockefeller Institute funded this project with money donated by the Consolidated Gas and Electric Company of New York, "whose chief concern was with asphyxiation due to [mining] accidents; its rescue squads were regularly called upon to administer artificial respiration to victims of carbon monoxide poisoning, drowning, and drug overdose."[8] He conceived of using an electric pump, alternating positive and negative pressure, to force air into patients' noses and mouths to expand their lungs and then expel it. In his initial experiment, Drinker paralyzed a cat with curare and placed it in a small, experimental respirator. He observed it breathing. With this initial breakthrough, he commissioned a local tinsmith to fabricate an adult-sized tank. Drinker attached two vacuum cleaners and used a garage mechanic's creeper to slide subjects into it. Philip, his brother, and a physiologist each used it successfully. Meanwhile, a pediatrician at Boston Children's Hospital consulted with Drinker about developing a special, artificial environment for premature babies. When he visited the children's ward to familiarize himself with the problem, Drinker witnessed, for the first time, its young patients strangling to death from polio-induced paralysis. Moved by this agonizing scene, he further refined his prototype and moved it to that facility. The first clinical trial occurred in 1928 when a paralyzed eight-year-old girl passed out because of a lack of oxygen. They called Drinker to start the machine which resuscitated her; he "stood there and cried" out of joy. Although she died shortly thereafter from pneumonia complications, he quickly built several more using discarded casket boxes. The press declared it the "new robot for artificial breathing." It promptly became known as the Drinker Respirator and the Drinker Lung. One newspaper reporter labeled it the iron lung, a name that stuck. Drinker achieved worldwide fame as his sister, Catherine Drinker Bowen, recalls: "The *Peiping and Tienstin Times* described it, as did the Paris *L'Illustration*. At Chicago's World Fair, the iron lung drew crowds to the Hall of Science. In his daily comic strip, Dick Tracy rushed a child to the machine." Magazine and newspaper articles hailed Drinker as "Champion of the handicapped, the half-shy professor, a rangy engineer..." and headlines exclaimed, "Thousands owe him their lives."[9] Philip Drinker became a medical celebrity.

Magazines, between the 1910 and 1955, also played a crucial role portraying "scientists as national heroes." Science historian Marcel LaFollette describes its impact: "Until the rise of television in the late 1950s, mass magazines such as *The Saturday Evening Post* and *Cosmopolitan* were information sources about the world of science." During that period, "the combined monthly circulation of these periodicals ranged from 3.2 million to almost 12 million." Topics included biomedicine, engineering, researchers' biographies, and major breakthroughs, featuring some of the most prominent scientists. In their presentations, "writers for popular magazines constantly celebrated a myth of scientific differentness," stressing their unique "genius." They possessed "extraordinary" intellectual abilities, curiosity, modesty, and persistence. Images often depicted them wearing spectacles, causing them to appear "wondrous wise."[10]

Life, a popular, weekly magazine that presented high-quality photographs of major events and individuals from 1936 to 1972, served as the "most potent force in shaping Americans' visual images of themselves and the world at large." Its editors created popular icons, from movie stars to new technology. "It supported biomedical research by presenting it in interesting ways, by making clear the importance of research funding, and by cheerfully celebrating the use of animal and human experiments to generate life-saving science and technology." The September 1942 issue devoted an article and famous photograph of Elizabeth Kenny tenderly holding a child in a diaper, paralyzed by infantile paralysis, as she performed physical therapy. As historian Bert Hansen summarizes,

> *Life* taught the American public about science and medicine with an unprecedented richness of detail packaged in an effective format. Part of that format was memorable photography. Another ... was well-conceived and elegantly designed charts and drawings. Another ingredient was entertainment; humor, cleverness, surprise, and sheer visual beauty leavened the learning in *Life*'s approach to medicine ... [and] accounted for its achievements as an informal science teacher for Americans.[11]

Magazines, during the first half of the twentieth century, ensured "a rosy halo surrounded science, an image of good intentions and good will, an image purportedly worthy of unquestioning public trust."[12]

A new medium exploited this theme. When "one-third of the nation entered a movie theater every week," Hollywood producers featured doctor films during the 1930s and 1940s.[13] The American Medical Association lobbied studios to project the field in a positive light. Therefore, both "medical and media professionals participated in dissolving the boundaries between entertainment and education.... The popularity of the 1931 screen adaption of American novelist Sinclair Lewis's novel *Arrowsmith* ... signaled the beginning of a golden age for American medicine on screen."[14] Subsequent productions included *The Story of Louis Pasteur* (1936), *The White Angel* (1936), about Florence Nightingale, and *The Great Moment* (1944), highlighting William T. G. Morton's discovery of ether as an anesthetic. "Although the film *Sister Kenny* (1946) featured a contemporary pioneer of polio therapy rather than a historical figure, it closely resembled the historical films...."[15]

Radio brought medical heroes into listeners' very living rooms through weekly programs like *Devils, Drugs, and Doctors* and *The Human Adventure*. Radio audiences eavesdropped on recreated momentous, medical episodes like Edward Jenner conversing "with the two children he was vaccinating" against smallpox. Other radio programs introduced a cavalcade of scientific notables, like Edith Cavell, Paul Ehrlich, and Walter Reed.[16] Informational programs existed as well. In 1921, the "United States Public Health Service initiated a biweekly radio health talk, and ... the New York State health office broadcast weekly talks on 'Keeping Well.'"[17]

Corporate advertising and sponsorship represented an additional influence. Pharmaceutical companies, like E. R. Squib and Sons and Parke, Davis and Company, utilized magazine ads, featuring medical pioneers. The Wyeth Pharmaceutical Company commissioned a "series of six paintings," "Pioneers of American Medicine" in 1939, while Bayer Aspirin ran eighteen newspaper ads from 1942 through 1944 marking "important medical discoveries" Coca-Cola produced posters titled "Famous Doctors." The widespread marketing of health-care products, like aspirin, laxatives, and vitamins as well as wellness advice, such as articles in popular magazines or health books, fed a consumer desire to purchase good health through over-the-counter medications and medical information.[18] The Metropolitan Life Insurance Company supported

New York state's 1926 diphtheria immunization campaign by running full-page ads in popular magazines while its field agents distributed pamphlets. Metropolitan further published "booklets about historical figures under the rubric 'Health Heroes.'" It produced other informative pamphlets about healthy practices, especially promoting vaccinations. "Quite a number of the company's publications targeted children, with coloring books, nursery rhymes, games, and puzzles." Finally, it produced filmstrips and movies for classroom use.[19] The motive that drove these life insurance companies stemmed from good public relations—that is, to be seen as a "caring, responsible corporate citizen." Most certainly though they had self-serving reasons, good health reduced payouts to policyholders' relatives.[20]

Although the ubiquitous nature of this cultural mosaic molded public views about medical science, infantile paralysis cut across them more than any other disease. As medical historian Naomi Rogers characterizes it, "A didactic and prescriptive literature, this polio material sought to create consumers of science and medicine." The Illinois Department of Health issued a free 1941 pamphlet, titled "Things You Want to Know About Infantile Paralysis." Primarily aimed at families, it provided basic background, care instructions, and upheld President Roosevelt as an inspiration for children with physical disabilities. It heralded science as *the* solution with its optimistic tone. In 1946, the Metropolitan Life Insurance Company distributed *Common Childhood Diseases*, a thirty-five-page "reference book, for ... use in the home." It promoted the immunization of children against smallpox and typhoid as well as hygienic practices for all family members. The fourteen diseases it covered included poliomyelitis. Sister Elizabeth Kenny and her successful physical therapy with paralyzed children received special attention. Another example of a commercial endeavor came from the Lysol Company in 1950 when it distributed the *Handbook on Infantile Paralysis*. Piggybacking on the notion of scientific motherhood, it urged them to keep their houses "hospital clean." Of course, Lysol's disinfecting agents held the imprimatur of science, or so it claimed. It also provided information about Kenny's therapy and the gamma globulin prophylactic. Alton Blakeslee, an Associated Press science reporter, in 1949 wrote a pamphlet, *Polio Can Be Conquered*. He portrayed a highly promising picture of recovery through Kenny's therapy and lauded the National Foundation for Infantile Paralysis for its March of Dimes fund-raising efforts, financial assistance to families of ill children, and research agenda, which appeared promising. A year later Turnley Walker's autobiography, *Rise Up and Walk*, described his polio illness, hospitalization, and paralysis experiences. He maintained a personal, inspirational tone as well as hailed the National Foundation for Infantile Paralysis in its broad agenda confronting this disease. Finally, in 1955, Dorothy and Philip Sterling wrote *Polio Pioneers: The Story of the Fight Against Polio*. This "children's science story" lionized scientists who held the secrets to defeating this dreadful virus.[21] These "popular images and values helped raise funds for the March of Dimes and its research projects ... [as well as] shaped reactions to the polio vaccine trials in the early 1950s."[22]

This health education process created an aura of trust. As polio researcher John R. Paul recalled, public health authorities were stymied during outbreaks. "Having issued weak guidelines to the public, warning that children should stay away from crowds, and quarantining households, there was little else they could do but watch as the number of cases mounted." In reality, while medical scientists remained powerless to fight this disease, the public perceived them as their champions. The "arrival of consultants from some distant place gave the impression that at least *something* was being done. As a rule, not much else than the bolstering of public morale was expected from the visiting investigators."[23]

Scientists as Role Models

Comic books, introduced in the late 1930s, created an insatiable demand. Often stereotyped as depicting superheroes, medical historian Bert Hansen points out, their scope transcended that genre. Series titles included *Calling All Girls*, *It Really Happened*, *Real Heroes*, and *Trail Blazers*, among others, emphasizing the lives of brave aviators, great American Presidents, fearless explorers, and talented athletes. They also portrayed the tedious work of medical researchers embarking on adventures fraught with personal dangers but ultimately saving lives as they transported young readers into laboratories and the tropics. They consumed this entertaining and cheap reading matter, enthralled by witnessing people being cured of fatal diseases. Science, ever progressing and awe inspiring, represented the focus. This formative reading audience saw medical heroes portrayed as diligent, selfless, tireless, visionaries, and virtuous. Comics were intended to be "educational, yet they were generally not stodgy and were often rather fresh, even raw, like the action stories and adventures with which they competed for space and attention on the newsstands." William H. Park of New York City's diphtheria campaign became a popular hero. The story of the discovery and use of penicillin, "the miracle drug," appeared in 1944 on the heels of its sensational public release. And of course they documented the entire history of infantile paralysis.[24]

For Hansen, comics portrayed "stories about [this] children's disease as upbeat," dwelling on "success stories." The February 1944 issue of *True Comics* centered on a "hardy young soldier, Philip Hawco." Set against the background of World War II, his "army buddies" exclaim, "We can't believe you've ever had infantile [paralysis], Phil!" He recounts his personal trial, the "loneliness he had experienced as a cripple [sic] at home until he was sent to a special hospital for free treatments, [where] he recovered completely." He links this to the larger picture, beginning with the 1916 epidemic, recounting the founding of Warm Springs (Georgia) Institute for Rehabilitation and the National Foundation for Infantile Paralysis, describing the effectiveness of Kenny's therapy, and extolling March of Dimes' celebrity fund-raising events. "A remarkably rich and coherent account of three decades of polio's history … in only eighteen panels on four pages."

This cheap, thin, and colorful reading matter became universal, with monthly circulation of *True Comics* ranging from 325,000 to 560,000 during the early 1940s, entertaining and informing a generation of impressionable readers. As Hansen concludes, this popular literature contributed to "medicine's mid-twentieth-century 'golden age.'"[25]

World War II

American military needs fostered the development of new therapies, further boosting public confidence. War exigencies "fused together a coalition of government, university, and industry scientists" who ushered in remarkable advances in medicine, generating a "revolution in the drugs which doctors prescribed for patients."[26] Military leaders especially feared a repetition of the infamous influenza epidemic of World War I, which felled 46,992 members of the armed services, almost matching the 50,385 combat casualties. An aggressive, and ultimately victorious, military effort could not be sustained with a high casualty rate due to diseases. Massive government subsidies funded research into vaccines and other weapons against contagions. "In January 1941, the surgeon general announced the creation of the Board for the Investigation and Control of Influenza and Other Epidemic Diseases, which became known as the Armed Forces Epidemiological Board." Rockefeller Institute scientists Alphonse Duchey, Thomas Francis, Jr., and Frank Horsfall had already been "working on a vaccine for twenty years." World War II accelerated that effort. Keeping troops on battlefields instead of in hospitals became conflated with research on other antigens, including those for Japanese encephalitis B virus, meningitis, and tetanus. They also developed dichlorodiphenyltrichloroethane (DDT) powder. When sprayed on humans, it killed disease-carrying lice. This proved to be the ultimate weapon against typhus as well as mosquito-borne malaria for troops fighting in the Pacific's tropical climates.

Penicillin became the wonder drug to attack bacterial infections, ensuring that tens of thousands of soldiers and sailors survived their wounds. While Oxford University scientists developed the prototype in 1940, American government researchers refined and stabilized it. The Office of Scientific Research and Development not only oversaw the Manhattan Project but directed the penicillin program, bringing it to production status by Merck and Company for testing purposes. The first human trials began in the spring of 1942. It proved an unqualified success, and by the end of the war, many companies manufactured this miracle drug on a mass scale.[27]

Mothers' Work

Twentieth-century women, like their nineteenth-century forebears, held sole responsibility for the family's health. As engineers, like Frederick Winslow Taylor, imposed efficiency production methods in the workplace, it inserted the clock into the nursery. Such rigid rules would produce the "mechanical baby"—that is, according to one child authority, "an efficient little machine" that would make "less trouble" for its mother. This systematic process regulated bathing, feeding, toilet training, and walking within a rigid timetable. A steady stream of expert-advice books, from Henry Chapin, *The Theory and Practice of Infant Feeding, with Notes on Development* (1902) to S. Josephine Baker, *Healthy Babies: A Volume Devoted to the Health of the Expectant Mother and the Care and Welfare of the Child* (1920), promoted this new middle-class mantra.[28]

Modern middle-class mothers would no longer operate through intuition or simply mimic the practices of their mothers. Childrearing now was couched in scientific parlance. As medical historian Nancy Tomes notes, "... personal health practices fell into the realm of housecleaning, childcare, and food preparation, domains traditionally designated women's work...." Pamphlets and popular women's magazines not only sanctioned the benefits of medical science but universalized much of this by touting the virtues of hygiene and sanitation, as well as provided practical ideas for accomplishing this important domestic function. One such manual suggested how women could, if necessary, convert their abodes into "'home hospitals' for the care of contagious illness...." They had to burn all of the patient's clothes and belongings to prevent the spread of diseases, since germs inhabited them for months and possibly years. Articles in women's magazines admonished them for not using disinfectants in excessive amounts. "Domestic hygiene authors" even blamed women if family members became ill because they possessed the power to prevent it. Physicians and child experts developed charts for mothers to monitor the growth and weight gains of their children to measure progress within a normative world. In sum, guided by educational, psychological, and medical experts, proponents of scientific motherhood maintained the goal of protecting babies and properly raising children; they endeavored to do this by introducing mothers to the gospel of cleanliness.[29]

Various private organizations facilitated this through adult education activities. The Child Study Association, established in 1896, sponsored parent-study groups, disseminated information through lectures and publications, and arranged national conferences. The Visiting Nurse Society of Philadelphia, incorporated in 1887, involved affluent women who, as volunteers, tended to the poor. These health missionaries approached their duties as a moral, uplifting process that, they believed, "brought comfort, cleanliness, and personal care into the homes of the sick poor and helped them to follow the doctor's advice," as well as taught them "'cleanliness and hygiene.' By 1895, that Society's staff had grown to eleven nurses, the annual number of nursing visits to 13,748, and the year's expenditures to $8280."[30] Hospitals, insurance companies, and settlement houses also sent visiting nurses to "poor urban households." They not only provided health care for the ill but taught habits of cleanliness. Finally, during the early 1920s, the American

Medical Association "created a separate Bureau of Health and Public Instruction, and in 1923 began to publish its own lay magazine, *Hygeia*, the forerunner of *Today's Health*."[31] The American Public Health Association, in 1923, "established a health education and publicity section for members," targeting mothers.[32]

Municipal health officials peppered poor, working-class, and immigrant communities with hygiene messages. Designating flies as disease vectors, they embarked on a nationwide eradication campaign, offering bounties for them. A twelve-year-old boy in Worcester, Massachusetts, won $100 "when he delivered ninety-five quarts of flies." The *New York Times* reported how "other cities and towns offered prizes for the best essays written by school children as to the dangers of flies and how to get rid of them."[33] These agencies especially targeted women by tailoring their hygiene messages in a dictum format. New York City's Bureau of Child Hygiene, founded in 1908, distributed a pamphlet, *Ten Commandments for Keeping Baby Well.* Between 1907 and 1910, it also used "Schoolgrams" to emphasize student well-being with little lessons like: "'Skidoo' from the boy or girl with 'a little sore throat'" or "Keep that pencil out of your mouth–it may have scarlet fever, diphtheria or typhoid fever germs on it."[34] Aimed at reducing infant mortality rates, Chicago's Department of Health issued "Healthgrams," glib little phrases that dripped with jingoism:

> Educate the mother and save the child.
> In saving the child you are saving the state....
> The greatest menace to the nation: A childless home....
> Tender-aged children in factories mean a crippled citizenship....
> As a national industry, raising strong, healthy human beings should be as profitable as raising fine breeds of livestock.[35]

Beginning in 1911, it sponsored parent-training classes for girls at that city's settlement houses. These "Little Mothers' Clubs" catered to girls in grades six through eight, preparing them to tend to their younger siblings and become mothers of healthy children, covering topics like home sanitation, baby care, first aid, and nursing a child through an illness. By 1916, Chicago's health department sponsored 22 such courses with 341 sessions that accommodated 9775 girls, and also worked with local Parent Teacher Associations to implement its health education program.

Visiting public nurses further trained mothers in infant care, conducting 106,150 "home calls" in 1939 alone.[36]

Federal involvement in mothers' education grew from three distinct pieces of legislation. The Children's Bureau, the "brainchild of Lillian Wald, a nurse and founder of the Henry Street Settlement in New York City, and child-labor activist Florence Kelly," was created in 1912 as part of the U.S. Department of Labor.[37] It collected health statistics, advocated for child welfare within the federal government, and coordinated state and local service agencies serving children and mothers. The Bureau initially focused on infant mortality, sponsoring "National Baby Weeks and 'better babies' contests … to highlight infants' health needs."[38] But it also pushed parent education, accomplishing this through its publications, namely *Prenatal Care* (1913) and *Infant Care* (1914), which became informal conduits for "scientific norms of child rearing as they were being formulated by physicians, social workers, psychologists, and educators."[39]

Nothing exemplified these activities more than the annual Indiana State Fair, hosting one of the "most vibrant health agencies in the nation," the Better Babies Contest. One of that state's "most anticipated happenings, … physicians and psychologists affiliated with the State Board of Health's Division of Infant and Child Hygiene … weighed, measured, and tested" contestants entered by their proud mothers. Judges evaluated "infants like livestock … with scorecards that tallied the level of physical health, anthropometric traits, and mental developments." This competition represented the culmination of health lessons disseminated through "radio talks, mothers' classes, hygiene films, consultation clinics, and statistical inquiries and reports." The learning process continued at that Fair as mothers "watched nurses demonstrate proper infant feeding techniques, collected free pamphlets, such as the *Indiana Mothers' Baby Book*, or perused displays about nutrition and the virtues of sterilized and sparkling bathrooms and kitchens."[40]

Washington assumed a more active role with the implementation of the Sheppard–Towner Maternity and Infancy Protection Act of 1921, supporting local community health education efforts through $7 million in subsidies. Wisconsin, with its grant, created the Child Health Special, "a trailer staffed by nurses and physicians from the state's Bureau of Maternal and Child Health." It traveled "from village to town to crossroads, offering health clinics in rural areas."

The Social Security Act of 1935 provided a plethora of services including, but not limited to, subsidies to needy, single-parent families, and a variety of child services. It not only revived but expanded Wisconsin's rural health education efforts. Public health nurses visited the homes of "pregnant women, new mothers, and children" as well as conducted lessons in "preventive medicine" at the "local Homemakers Council, Girl Scout troop, or high school class."[41]

The number of mass-circulation magazines aimed exclusively at women and devoted to parenting exploded in the 1920s. *American Motherhood,* Babyhood, *Good Housekeeping, Ladies' Home Journal, Modern Priscilla, Parents' Magazine,* and *Women's Home Companion* fit into the emerging culture of scientific motherhood, serving as forums to promote nutritional needs, such as vitamins, and reinforcing the "precepts of household bacteriology" through advertisements and advice articles written by experts.[42] The commercialization of pharmaceutical products, historian Rima D. Apple adds, began as early as 1870 with the introduction of infant formula and its accouterments: bottles, rubber nipples, and sterilizers. Promotion tapped a combination of "fear and faith." Ads and articles in the popular press attributed high infant mortality rates to poor nutrition. Physicians and experts worried that a mother's poor health or weak constitution could, in fact, deprive babies of adequate nutrition, literally starving them. Breast milk also varied among individuals as well as over time in the same mother. But formula would eliminate these risks.

Infant food producers' national marketing campaigns, Apple continues, employed a three-pronged strategy. First, they purchased advertising space in medical journals. Second, they sent pamphlets, citing information that drew on scientific studies, to doctors' offices. Third, these companies gave physicians free samples to try on their children or give to patients. As a result, the infant formula business grew from "nothing to large-scale industrial firms," with national stature and international distribution. Nestlé's Milk Food, a Swiss company, produced and distributed infant formula to North and South America, Europe, and Australia. Major U.S. producers included Doliber and Goodale Company in Boston and Horlick's Food for Infants and Invalids in Racine, Wisconsin.[43]

Pharmaceutical corporations, which manufactured and sold vitamin supplements to food processors, promoted the need to the general public to include these in diets. "[N]utrition-related advertising," explains

Apple, became "extensive in the 1920s and 1930s...." Marketing food products, like chocolate drinks and bread, focused on vitamin content. This strategy also relied on the dual ploys of fear and benefits. The former, a "scare" tactic, invoked a tone of immediacy by warning mothers about the dire harm they caused their children's health by failing to use a vitamin-enriched food product. The latter claimed, often unsubstantially, how children's health improved after they consumed vitamin-enriched food products.

Advertising during the interwar era intensified its use of so-called scientific and medical authority. Any mother who ignored this tried-and-true knowledge was guilty of nothing less than outright neglect. The marketing of hygienic products, operating as another informal education mechanism, exercised a profound impact on domestic science, claiming abilities to kill germs as well as prevent them from growing in the first place. This list proved endless: cleaning disinfectants, first-aid antiseptics, hand soaps, home water filters, mouthwashes, refrigerators, tin cans, tin foil, vacuum cleaners, and waxed paper. White porcelain bathroom sinks and toilets and ceramic wall tile provided impervious surfaces, repelling bacteria.[44]

Ultimately, when a child contracted an illness or disease, it reflected poor parenting skills: They failed to employ hygienic practices or secure vaccinations. This approach relied on guilt, at best, and dereliction, at worst. "The rise of health education was strongly influenced by a broader trend that was transforming American civic life during this period: the growth of advertising, marketing, and public relations."[45]

All Together Now

Taken collectively, the gospel of health conferred an omnipresent mantle on science. "Scientists and journalists alike," Marcel LaFollette stresses, "used the term *science* interchangeably to refer to the research process, the body of knowledge, ***and*** the professional community of scientists; moreover, they often anthropomorphized the concept, making science into a living, growing thing that could 'do' things, could 'act,' could even 'assert.'" This represented an interactive, dynamic process: "The face of science ... was not only the face that scientists wanted presented but also one that Americans wanted to see as their own." These researchers paid close "attention to their public images," belying their "shy and modest" stereotypes. They controlled the "communications content, structure, and tone, and urged colleagues to be similarly cautious"—optimism and progress undergirded all of it. Thus, while the informal education process remained largely uncoordinated, it did not represent the product of happenstance; most journalists served as "agents of transmission," as mere conduits of information from medical and scientific experts to the general public.[46]

The campaign against diphtheria revealed the critical role of public schools. With ready access to young children, it appeared the perfect institution to exploit for testing and immunization. The glimmer of a long-term strategy had begun to take shape. In many ways, it operated as a prototype for the mass polio experiment and immunization efforts in the 1950s. Something deeper had occurred, though. Diphtheria, more often than not, had proven fatal for children. The Behring Serum, injected into a sick child, "made the difference between life and death." This left a profound "image" with the public: "the physician as healer, capable of intervening in the course of a life-threatening illness." Coupled with "[Louis] Pasteur's treatment for rabies ... [Robert] Koch's discoveries of the tuberculosis and cholera organisms," and [Joseph] Lister's introduction of antiseptics, "the laboratory was coming to be seen as the source of medicine's new explanatory powers...."[47]

The war effort further elevated medicine and researchers. "By 1945," Arthur Allen writes, "many of the deadly infectious diseases that haunted the tenements and farms of turn-of-the-century America–diphtheria, scarlet fever, whooping cough, and smallpox—were no longer commonplace tragedies of childhood. Although vaccination was only partially responsible for this bright turn of events, it enjoyed the same status as other weapons in the arsenal of scientific medicine—antibiotics, isolation, sanitation, and better disease treatment in general." Moreover, the next generation of medical researchers grew out of this context. The "Armed Forces Epidemiological Board included [Albert] Sabin, [Jonas E.] Salk, and Thomas Francis, Jr., working on influenza, Joseph Stokes, Jr., on measles, John Enders on mumps, [Thomas] Rivers and Joseph Smadel on typhus, and [Max] Theiler and John Fox on yellow fever." Among these as well were "Joseph Melnick, David Bodian, Dorothy Horstmann, and John Paul [who all] deepened the foundations of polio research." Finally, the American military accomplished through routine induction procedures what health departments and the public schools had failed to do through the compulsory immunization. Eleven million troops had been successfully vaccinated during World War II and, after returning home, wanted their "children to have the same protection."[48] By this point in time, the health infrastructure of the public schools would easily accommodate this desire.

Notes and Sources

1. John Duffy, *The Sanitarians: A History of American Public Health* (Urbana: University of Illinois Press, 1992): 4.
2. Nancy Tomes, *The Gospel of Germs: Men, Women, and the Microbe in American Life* (Cambridge: Harvard University Press, 1998): 242. Refer also to Bernard Bailyn, *Education in the Forming of American Society* (Chapel Hill: University of North Carolina Press, 1960), and Lawrence A. Cremin, *The Wonderful World of Ellwood Patterson Cubberley: An Essay on the Historiography of American Education* (New York: Teachers College Press, 1965), *Traditions of American Education* (New York: Basic Books, 1977), and *American Education: The National Experience, 1783–1876* (New York: Harper and Row, 1980).
3. Susan E. Lederer, *Subjected to Science: Human Experimentation in America Before the Second World War* (Baltimore: Johns Hopkins University Press, 1995): quoted on page 130. The second quote is in Bert Hansen, "Medical History for the Masses: How American Comic Books Celebrated Heroes of Medicine in the 1940s," *Bulletin of the History of Medicine* 78 (2004): 149. Further consult Marcel C. LaFollette, *Making Science Our Own: Public Images of Science, 1910–1955* (Chicago: University of Chicago Press, 1990): 19, 41, 49, 51–52, 67–76, 106–8.
4. Paul de Kruif, *The Sweeping Wind: A Memoir* (New York: Harcourt, Brace and World, Inc., 1962): 38. Also see pages 25, 30–31, 45, 49–51, 58–59, 69–71, 78–77, 80–99, and 103, as well as *Microbe Hunters* (New York: Harcourt, Brace, & Jovanovich, 1926), and Charles E. Rosenberg, "Martin Arrowsmith as Hero," *American Quarterly* (Fall 1963): 448–49.
5. Rosenberg, "Martin Arrowsmith as Hero": 447. Sinclair Lewis, *Arrowsmith* (New York: Harcourt, Brace & World, 1961). LaFollette, *Making Science Our Own*, places the character Arrowsmith firmly in the category of a scientific hero: "To Lewis, the scientific hero possessed integrity and dignity as well as the restlessness of the pioneering spirit" (108).
6. Thomas R. Rivers, *Reflections on a Life in Medicine and Science* (Cambridge: Massachusetts Institute of Technology, 1967): 180–81. See the note to page 181. Also consult Arthur Allen, *Vaccine: The Controversial Story of Medicine's Greatest Lifesaver* (New York: W. W. Norton Company, 2007): 172; De Kruif, *Sweeping Wind*: 177–78; Hansen, "Medical History for the Masses": 152; Jeffrey Kluger,

Splendid Solution: Jonas Salk and the Conquest of Polio (New York: G. P. Putnum's, 2004): 52; Susan E. Lederer, "Medical Ethics and the Media: Oaths, Codes, and Popular Culture," in Robert B. Baker, et al., eds., *The American Ethics Revolution: How the AMA's Code of Ethics Has Transformed Physicians' Relationships to Patients, Professionals, and Society* (Baltimore: Johns Hopkins University Press, 1999): 92; and John R. Paul, *History of Poliomyelitis* (New Haven: Yale University Press, 1971): 306.

7. Paul de Kruif, "What Science Is Doing," *President's Birthday Magazine*, 1, #3 (1938): 20, 21, 58, 60, folder 113, box 14, Medical Interests series, RG 2, Office of the Messrs. Rockefeller, Rockefeller Family Archives, RAC.
8. David J. Rothman, "The Iron Lung and Democratic Medicine," in David J. Rothman, ed., *Beginnings Count: The Technological Imperative in American Health Care* (New York: Oxford University Press, 1997): 42–43. See page 44, as well as Anon., "Landmark Perspective: The Iron Lung," *Journal of the American Medical Association* 255 (March 1986): 1176; Lynne M. Dunphy, "'The Steel Cocoon': Tales of the Nurses and Patients of the Iron Lung, 1929–55." *Nursing History Review* 9 (2001): 11; Thomas M. Daniel and Frederick C. Robbins, eds., *Polio* (Rochester, NY: University of Rochester Press, 1997): 116; Paul, *History of Poliomyelitis*: 328, 329–30, 333; Regina Woods, *Tales from Inside the Iron Lung (and How I Got Out of It)* (Philadelphia: University of Pennsylvania Press, 1994): 125.
9. Catherine Drinker Bowen, *Family Portrait* (Boston: Little, Brown and Company, 1970): 242. Bowen's use of a newspaper quotes also appear on page 242. Refer as well to pages 225, 236, 237–38, 240–43, and "Landmark Perspective": 1477; James H. Maxwell, "The Iron Lung: Halfway Technology or Necessary Step?" *Milbank Quarterly* 64 (1986): 6–7; Howard Merkel, "The Genesis of the Iron Lung: Philip Drinker, Charles F. McKhann, James L. Wilson, and Early Attempts at Administering Artificial Respiration to Patients with Poliomyelitis," *Archive of Pediatric and Adolescent Medicine* 148 (November 1994): 1174–80.
10. LaFollette, *Making Science Our Own*: vii, 3, 24, 67, and 75, respectively. See also pages 25, 51, 70–71, 76. LaFollette chose as her sample eleven American magazines, "nationally circulated and moderately priced" and aimed at "general audiences" (23–24). She splits them into two groups: general content and public affairs, covering distinct reading audiences (see pages 27–37). She includes *The American Magazine*, *The American Mercury*, *The Atlantic Monthly*, *The Century Illustrated Monthly Magazine*, and *Colliers*, among others (24).

11. Bert Hansen, *Picturing Medical Progress from Pasteur to Polio: A History of Mass Media Images and Popular Attitudes in America* (New Brunswick, NJ: Rutgers University Press, 2009): 208–9 and 252, respectively. Further consult pages 207, 225, 234–35.
12. LaFollette, *Making Science Our Own*: 6.
13. Hansen, *Picturing Medical Progress*: 137. Further check Leslie J. Reagan, Nancy Tomes, and Paula A. Treichler, *Medicine's Moving Pictures: Medicine, Health, and Bodies in American Film and Television* (Rochester, NY: University of Rochester Press, 2007): 2.
14. Susan E. Lederer and Naomi Rogers, "Media," in Roger Cooter and John Pickstone, eds., *Medicine in the Twentieth Century* (Amsterdam: Harwood Academic Publishers, 2000): 487–88 and 491–92, respectively. Lederer and Rogers (pages 493–95) specifically point out how the medical establishment monitored the film industry. First, it ensured that it presented physicians in a favorable light. Second, it opposed any hint of socialized medicine. Third, "[i]n addition to graphic depictions of surgical and medical practice, censors challenged the portrayal of such volatile issues as abortion and euthanasia." Fourth, it made sure that technical aspects were accurately represented. See also Reagan, Tomes, Treichler, *Medicine's Moving Pictures*: 1.
15. Hansen, "Medical History for the Masses": 153. Also consult Hansen, *Picturing Medical Progress*: 138–39, 141, 147; Lederer and Rogers, "Media": 487. Reagan, Tomes, Treichler, *Medicine's Moving Pictures*: 7–8.
16. Hansen, "Medical History for the Masses": 153. Refer likewise to page 154; Hansen, *Picturing Medical Progress*: 155, 158; LaFollette, *Making Science Our Own*: 19; and Reagan, Tomes, Treichler, *Medicine's Moving Pictures*: 1.
17. Terra Ziporyn, *Disease in the Popular American Press: The Case of Diphtheria, Typhoid Fever, and Syphilis, 1870–1920* (Westport, CT: Greenwood Press, 1988): 23.
18. Hansen, *Picturing Medical Progress*: 165–66. Consult pages 160, 162, 167–68, as well as Rima D. Apple, "The More Things Change: A Historical Perspective on the Debate Over Vitamin Advertising in the United States," in John D. Ward and Christian Warren, eds., *Silent Victories: The History and Practice of Public Health in Twentieth-Century America* (New York: Oxford University Press, 2007): 193–206; and Nancy Tomes, "Merchants of Health: Medicine and Consumer Culture in the United States, 1900–1940," *Journal of American History* 88 (September 2001): 533, 535.
19. Hansen, *Picturing Medical Progress*: 162 and 164, respectively. Further refer to James Colgrove, *State of Immunity The Politics of Vaccination in*

Twentieth-Century America (Berkeley, CA: University of California Press, 2006): 86, 88, 111.

20. Colgrove, *State of Immunity*: 89. Further consult Hansen, *Picturing Medical Progress*: 162. For the concept and process of the popularization of science in American society, see Tomes's insightful historiographical discussion in *Gospel of Germs*: 13–20. The educational process, both informal and formal, resulted in a "fundamental" shift in medical attitudes and "personal behaviors" (14).
21. Naomi Rogers, "Polio Can Be Conquered: Science and Health Propaganda in the United States from Polio Polly to Jonas Salk," in *Silent Victories*: 82, 87, 89, 90, 91, and 96, respectively. Refer also to pages 88, 91–93, 94.
22. Hansen, "Medical History for the Masses": 190.
23. The earliest and, many consider, most respected work on the subject of polio is Paul, *History of Poliomyelitis*: 208. The emphasis is the author's.
24. Hansen, "Medical History for the Masses": 159, and 176, respectively. See also pages 154–55, 158–60, 167–68, 176–78. LaFollette, *Making Science Our Own*, expands the term "hero" on pages 106–7, referring to these role models as "creative, inspired, dedicated, and self-reliant...." (107).
25. The quotes in the prior paragraph are from Hansen, *Picturing Medical Progress*: 194–95, while this paragraph cites Hansen, "Medical History for the Masses": 162 and 183, respectively. Refer to page 160, as well as Bert Hansen, "True-Adventure Comic Books and American Popular Culture in the 1940s: An Annotated Research Bibliography of the Medical Heroes," *International Journal of Comic Art* 6 (Spring 2004): 117–47.
26. Philip J. Hilts, *Protecting America's Health: The FDA, Business, and One-Hundred Years of Regulation* (New York: Alfred A. Knopf, 2003): 95.
27. Allen, *Vaccine*: 137–38. Further consult pages 139–42, and Hilts, *Protecting America's Health*: 102–4.
28. Daniel Beekman, *The Mechanical Baby: A Popular History of the Theory and Practice of Child Raising* (Westport, CT: Lawrence Hill & Co., 1977): 109–10. See pages 111–12, 117–19, 121, 126. In chronological order, Henry Chapin, *The Theory and Practice of Infant Feeding, with Notes on Development* (New York: W. Wood and Company, 1902); Anna G. Noyes, *How I Kept My Baby Well* (Baltimore: Warwick & York, Inc., 1913); Mary L. Read, *The Mothercraft Manual* (Boston: Little, Brown, and Company, 1916); Samuel Visanska, *Better Babies: A Guide to the Practical Care of the Mother and Young Child* (Atlanta, GA: Foote & Davis Company, 1917); S. Josephine Baker, *Healthy Babies: A Volume Devoted to the Health of the Expectant Mother and the Care and Welfare of*

the Child (Minneapolis, MN: Federal Publishing Company, 1920); and Jay Mechling, "Advice to Historians on Advice to Mothers," *Journal of Social History* 9 (Fall 1975), cautions about relying solely on child rearing manuals, making four points: First, these tracts represented ideal advice rather than actual parental behavior (47). Second, literacy played a significant role since affluent and privileged adults could afford and read them but it cannot be assumed that poorer ones could either purchase or read and comprehend them (47). Third, parenting represents a primary skill, i.e., internalized from one's own parents, but expert "advice" operates as "secondary socialization" and is therefore less effective (49). Fourth, and finally, Mechling concludes with a sobering question: Did child rearing manuals truly reflect dominant cultural values toward child rearing? (51).

29. Tomes, "Merchants of Health": 16, 53, and 65, respectively. See pages 9, 60–62, as well as Rima D. Apple, "'They Need It Now': Science, Advertising and Vitamins, 1925–1940," *Journal of Popular Culture* 22 (1988): 71; Beekman, *The Mechanical Baby*: 114, 128–31; Evelynn Maxine, Hammonds, *Childhood's Deadly Scourge: The Campaign to Control Diphtheria in New York City, 1880–1930* (Baltimore: Johns Hopkins University Press, 1999): 200; Aaron E. Klein, *Trial by Fury: The Polio Vaccine Controversy* (New York: Charles Scribner's Sons, 1972): 5–6; Molly Ladd-Taylor, *Raising a Baby the Government Way: Mothers' Letters to the Children's Bureau, 1915–1932* (New Brunswick, NJ: Rutgers University Press, 1986): 36; Susan Levine, *School Lunch Politics: The Surprising History of America's Favorite Welfare Program* (Princeton: Princeton University Press, 2008): 12–13; and Marc Shell, *Polio and Its Aftermath: The Paralysis of Culture* (Cambridge: Harvard University Press, 2005): 75.
30. Barbara Bates, *Bargaining for Life: A Social History of Tuberculosis, 1876–1938* (Philadelphia: University of Pennsylvania Press, 1992): 231–33. Refer to page 238, as well as Guy M. Whipple. Editor. Preschool and Parental Education. *National Society for the Study of Education Yearbook* 14, no. 28 (Bloomington, IL: Public School Publishing Company, 1929): 38–39, 107.
31. Tomes, *Gospel of Germs*: 187 and 242, respectively. Also see pages 183, 186, 188.
32. Colgrove, *State of Immunity*: 93.
33. "Greatest Anti-Fly Crusade Ever Known Is Beginning," *New York Times*, May 19, 1912, Magazine Section, p. 11.
34. *Report of the Department of Health of the City of Chicago for the Years of 1907, 1908, 1909, 1910*: 45. These "Schoolgrams" followed the same format as "Healthgrams" (77). *Report of the Department of Health of the City of Chicago for the Years of 1911 to 1918 Inclusive*: 558, CPL.

35. "Healthgrams," *Report of the Department of Health of the City of Chicago for the Years of 1907, 1908, 1909, 1910*: 77, CPL.
36. *Report of the Department of Health of the City of Chicago for the Years of 1911 to 1918 Inclusive*: 559, CPL. Refer to Appendix A, "Syllabus of Course Given to Little Mothers' Clubs," for details about this course of study, pages 592–95; *Annual Report of the Board of Health of the City of Chicago for 1939* (Chicago: 1940): 27; *Annual Report of the Board of Health of the City of Chicago for 1950*: 5, 13; *Annual Report of the Board of Health of the City of Chicago for 1951*: 11, CPL. Also refer to Tomes, *Gospel of Germs*: 140.
37. Ladd-Taylor, *Raising a Baby the Government Way*: 6. Also refer to Barbara Beatty, Emily D. Cahan, and Julia Grant, eds.,. *When Science Encounters the Child: Education, Parenting, and Child Welfare in Twentieth-Century America* (New York: Teachers College Press, 2006): 6–7, 10; and Carole A. Estabrooks, "Lavinia Lloyd Dock: The Henry Street Years," *Nursing History Review* 3 (1995): 144.
38. Viviana A. Zelizer, *Pricing the Priceless Child: The Changing Social Value of Children* (New York: Basic Books, 1985): 29. See also Rima D. Apple, "Educating Mothers: The Wisconsin Bureau of Maternal and Child Health," *Women's History Review* 12 (2003): 561–62; Barbara Finkelstein, "Uncle Sam and the Children: A History of Government Involvement in Child Rearing," in N. Ray Hiner and Joseph M. Hawes, eds., *Growing Up in America: Children in Historical Perspective* (Urbana: University of Illinois Press, 1985): 261.
39. Finkelstein, "Uncle Sam and the Children": 261. Refer also to Apple, "Educating Mothers": 561.
40. Alexandra Minna Stern, "Better Babies Contests at the Indiana State Fair: Child Health, Scientific Motherhood, and Eugenics in the Midwest, 1920–35," in Stern and Markel, eds., *Formative Years*: 122, 121, 137, and 121, respectively.
41. Apple, "Educating Mothers": 562 and 564, respectively. Additional background about the topics in this and the preceding three paragraphs can be gleaned from Finkelstein, "Uncle Sam and the Children": 262–63; Ladd-Taylor, *Raising a Baby the Government Way*: 24; Margarete Sandelowski, "Making the Best of Things: Technology in American Nursing, 1870–1940," *Nursing History Review* 5 (1997): 12; Stern and Markel, eds., *Formative Years*: 9; Zelizer, *Pricing the Priceless Child*: 29.
42. Tomes, "Merchants of Health: Medicine and Consumer Culture": 141. Also see page 140.
43. Rima D. Apple, "'Advertised by Our Loving Friends': The INFANT Formula Industry and the Creation of New Pharmaceutical Markets, 1870–1910," *Journal of the History of Medicine and Allied Sciences* 41

(January 1986): 23 and 8, respectively. Refer to pages 3–4, 5–7, 9–12, 14, 17, as well as Apple, "They Need It Now": 66, 70; Colgrove, *State of Immunity*: 91.

44. Apple, "They Need It Now": 69 and 72, respectively. Further consult pages 66, 70–73, 76–79, and Tomes, *Gospel of Germs*: 162–71.
45. Colgrove, *State of Immunity*: 93 and 94, respectively. Further refer to pages 97–98.
46. LaFollette, *Making Science Their Own*: 5–6, emphasis is in original. Other quotes are from pages viii, 1, 52, and 23, respectively. Also see pages 3–4.
47. Charles E. Rosenberg, *The Care of Strangers: The Rise of America's Hospital System* (Baltimore: Johns Hopkins University Press, 1987): 159 and 160, respectively.
48. Allen, *Vaccine*: 120, 174, and 118, respectively.

CHAPTER 6

Schoolhouse Medicine

Haven Emerson, director of New York City's health department, addressed 2000 school board officials, building principals, supervisors of hygiene, and teachers at Washington Irving High School on September 14, 1916. In the waning weeks of that city's devastating polio epidemic, he urged the promotion of health "among teachers and students," supporting annual physical examinations of children and their instructors, and teaching personal hygiene. For him, the purpose of schooling was clear: "The graduating diploma should be a certificate of physical as well as mental preparedness to continue to learn and to serve. We look forward," he continued, "to the time when graduation will be a privilege withheld until the pupil can present a clean record as to physical fitness and an understanding of the rudiments of personal hygiene sufficient to protect him against the strains, trials and exposures of self support and the burdens of establishing a family and household."[1]

As public health goals shifted from treating illnesses to preventing them, "planned interventions" represented a more socially efficient approach and began to dominate policy. The "medicalization of society," in addition to informal efforts, included formal activities to ensure the internalization of these values, following a path of regulation and imposition, and eventually persuasion. Two separate, but overlapping, influences shaped this process. First, municipal and state health departments invoked the imprimatur and power of medical science to protect schoolchildren and their communities. Second, public education, with

R. J. Altenbaugh, *Vaccination in America*,
Palgrave Studies in the History of Science and Technology,
https://doi.org/10.1007/978-3-319-96349-5_6

ever-expanding enrollments and increasing attendance, became a compulsory and central institution in children's lives. Progressive school reforms, emerging as part of broader business, municipal, political, and social crusades between 1890 and 1920, turned health concerns into educational practices.[2]

As children crowded into classrooms in unprecedented numbers, more of them seemed to become ill. Concerns arose among public health observers about the impact of the school plant itself as the cause, what historian Richard A. Meckel terms "school diseases." Fomite theories held that building contents, such as books and dust, served as disease vectors. The lack of modern toilets also produced a miasma that spread contagion. To protect schoolchildren, urban health departments imposed quarantines, mandated hygienic practices, and ordered sanitation measures. That agenda began to shift during the 1880s. Informed by germ theory, health officials now worried about students, surrounded by infected classmates, carrying these contagions back to their communities and sparking citywide epidemics. These modified policies stressed "surveillance, exclusion, and compulsory immunization."[3]

Health Problems, Both Big and Small

Who ultimately held jurisdiction over the well-being of schoolchildren? Stephen Woolworth's historical analysis of Seattle's 1895 scarlet fever outbreak sheds light on the social conflicts surrounding this question. That city's board of health prohibited the re-admission of previously ill students for the "remainder of the academic year." It further ordered the district to shutter two buildings with the highest infection rates. When school officials refused, the health department, without warning early one morning, sent one of its uniformed employees along with a city police officer to post quarantine notices on each of the two schools. In both cases, several hundred surprised and angry principals, teachers, and schoolchildren rushed the buildings. Additional police arrived, took administrators into custody, and posted guards. Seattle's health department had trumped the school board. When the board of health demanded that the buildings be closed and disinfected, school board members complied. Officials further ordered all schoolbooks burned to eliminate the possible source of additional contamination; the school board again acquiesced. Finally, the health department mandated that all school buildings be connected to Seattle's sewer system; education authorities once more agreed. "The city's schools were securely positioned as sites of public health regulation." Once health officials had legitimatized their ability to dictate how school buildings were used, they then extended their reach "to include the health of individual students." The medical world had begun to encroach on school authority. As general public health policy grew more encompassing, the public school likewise shifted "from a site of public health regulation (i.e., sanitary inspections) to one of public health enforcement (i.e., medical examinations and vaccinations) and acculturation (hygienic principles)." A similar pattern would unfold in other American cities.[4]

Medical Inspections

"Schools are excellent spreading places for infection," Chicago's Department of Health declared in its 1910 annual report. Its authors reasoned that the "number of cases of infectious diseases always decreases when the schools close for the summer.... The number of children who get together on the streets constitutes a smaller unit than the school unit. The contact within the street unit is not as close as it is in the school unit; therefore the school is a means of spreading contagion." This could have serious repercussions, according to the health department. Hosting infected children not only posed a hazard for other students but threatened the larger community. However, that report provided the solution: "... the schoolroom is the only place available for the inspection of school children."[5]

Medical examinations began in the 1890s when the "Boston Board of Health appointed fifty physicians ... to visit all schools and examine any students ill or complaining of illness." They checked about 5000 children and found "58 cases of diphtheria, 19 of scarlet fever, 42 of measles, 17 of whooping cough, 55 of mumps, and 7 cases of congenital syphilis."[6] New York City's health board established a similar policy in 1897. Chicago's Department of Health in 1910 enumerated three reasons for them. They exposed illnesses or other health-related problems, determined if those maladies were contagious, and ordered school doctors or nurses to initiate measures to protect other children. The entire process began with classroom instructors who separated and isolated pupils they suspected of being ill, a highly subjective selection process by non-medical personnel. In addition to noting children's ages, heights, and weights, medical personnel checked their eyes, nose, teeth, throat, and general "physical development." They sent copies of their findings to the health department, classroom teachers, and parents. When instructors or medical professionals sent sick children home, that city's department of health stipulated that they had to be "visited by the school doctor or the school nurse, as the family, thinking the illness of small moment, will probably neglect the necessary quarantine provisions." Medical inspections grew at a rapid pace. While the Division of School Medical Inspection of Chicago's health department conducted none in 1907, it performed 45,365 in 1908; this figure soared to 123,897 the following year.[7] In 1910, the number of city school districts that adopted medical inspections climbed from 44 to 312; by 1915, that figure had jumped to 500. In addition to contagious diseases, examiners found noncontagious ailments, like hearing and speech problems, infected and swollen glands, malnourishment, rickets, as well as sight impairments. This marked a profound expansion of the function and reach of public health: "By bringing physicians into contact with children in school, inspections had the effect of introducing a new body of knowledge from which to view children and their bodies in school settings."[8]

Medical inspectors also checked children for evidence of smallpox inoculations. Boston, as early as 1827, required it for all schoolchildren. Massachusetts became the first state to mandate it for school-aged children in 1855 and four years later created the first state board of health to enforce inoculation. Connecticut, New Jersey, New York, and Rhode Island soon followed. The New York City Board of Health "reported with pride in 1875 that its 'vaccinating corps' had 'performed 17,505 vaccinations in the schools.'" Chicago's board of health inaugurated an inspector corps in 1868. Inoculation as an admission prerequisite became the primary strategy by public health departments to control epidemics—that is, schools became a key instrument to shield the public from smallpox epidemics.[9]

New York City's health department took this one step further, envisioning the public school system, with its captive audience, as the ideal vehicle to promote diphtheria immunization. Moreover, as science historian Evelyn Maxine Hammonds stresses, "[a]uthority figures, such as principals and teachers, served as intermediaries between the health department and parents." With financial support from the American Red Cross and permission from New York City's board of public education, William H. Park and Abraham Zingher, in 1921, resumed their pre-war "mass Schick testing and administration of toxin-antitoxin to students."

This took place during the worst year for diphtheria in the twentieth century, with 206,000 cases and 15,520 deaths nationwide. However, unlike before, when they used hospitalized and institutionalized children, Park and Zingher had to secure "affirmative consent" from parents and guardians before they could initiate any procedures. First, they obtained each building principal's approval; second, organized informational meetings with teachers; and third, sent consent forms home with students. Teachers compiled lists of parent signatures. School personnel also arranged a classroom in each building for physicians and nurses, and physically shepherded students through this routine. Zingher estimated that each team tested between 500 and 600 children an hour. While educators approached this operation without question, and even enthusiastically in some cases, parents proved to be less cooperative. Zingher estimated that less than 25% gave permission. Immigrants proved to be especially resentful of public health experts meddling with their cultural traditions and undermining their parental authority. In the end, Park and Zingher used a total of 180,000 schoolchildren: half served as a control group, receiving neither the test nor toxin–antitoxin (TAT), and half underwent the Schick test and were given TAT when needed. The outcome proved hopeful: The control group contracted diphtheria at four times the rate for the tested and immunized group.[10]

Hermann M. Biggs, former director of New York City's bacteriology laboratory and Park's mentor, became state commissioner of health in 1914. "After a decade of trials in New York City schools, [Biggs] launched an ambitious statewide campaign to eliminate diphtheria by 1930." With $15,000 from the Metropolitan Life Insurance Company, that health department employed the "newest techniques of mass persuasion–newspaper advertisements, billboards, motion pictures, staged publicity events, colorful placards using emotional appeals to parental duty and sentiment–[to] motivate the public." Public school educators required students to draw posters and write essays, with awards for the best submissions, and gave them "gold stars and badges after receiving their injections." New York City's health department sent mobile centers to poor communities, inaugurated "healthmobiles," and opened temporary clinics at public beaches and in local hospitals. In 1933, the entire city celebrated when the one-millionth child received a diphtheria vaccination. City officials commemorated this milestone with a three-block parade by "several thousand schoolchildren" from an "Upper East Side clinic" to a "health pageant held in Central Park."

There music performers and ethnic dancers entertained them while "speeches by politicians and civic leaders drew attention to the importance of protecting children against one of the most common infectious disease."[11] Both morbidity and mortality rates plummeted. "In New York, the death rate had peaked at 785 per 100,000 in 1894; by 1920, it had dropped to under 100. By 1940, with 60 percent of pre-school children immunized, diphtheria deaths had become a thing of the past."[12] Equally important, the "popular perception that diphtheria immunization was safe and effective would greatly influence new vaccines against other illnesses."[13]

Public perceptions toward vaccinations, in general, reflected confidence. The U.S. Public Health Service, upon completing a survey of 8758 families in 130 localities in eighteen states and all regions, including rural villages and large cities, between 1928 and 1931, reported generally higher rates of "artificial immunization" for children, amounting to 43% for diphtheria by age nine and 66% for smallpox by age sixteen. Treatments for tetanus continued to be relegated to injuries, even then only 7% received attention, though much higher for children than adults.[14] These figures increased within a decade. Leona Baumgartner, director of New York City's Bureau of Child Hygiene, reported that a nationwide poll of adults in 1941 indicated 80% of mothers surveyed wanted their children to receive at least one immunization shot. They only expressed uncertainty about when their children should be inoculated, with that survey reporting ages ranging from less than a year to six. Baumgartner optimistically concluded that the "public for the most part seems ready for immunization procedures." However, waves of summer polio epidemics caused parents to hesitate about submitting their children to diphtheria and smallpox inoculations; they believed that TAT compromised their immune systems, making them more susceptible to infantile paralysis.[15]

The Whole Child

Progressive school reformers, who acted as child savers in every sense of the phrase, included health in their educational agenda. "The new 'whole' child constructed out of law, medicine, and psychology had a strong hand in shaping, and in turn was shaped by, the modern politics of maternalist reform."[16] Indeed, the Progressive Era is considered the "age of the child." Reformers envisioned urban public school buildings operating as community centers, offering evening English-language classes for immigrants, year-round playgrounds for children, neighborhood entertainment and events in auditoriums, and local athletic contests on playing fields. These physical plants provided gymnasia and swimming pools where children could gain access to informal outlets for exercise. School-sponsored lectures for adults covered house cleaning, hygiene, nutrition, and tuberculosis prevention.

Progressive education consisted of four crosscurrents: administrative progressives, who focused on the centralization of the decision-making process for efficiency reasons; scientific progressives, who introduced standardized tests and psychological services; social reconstructionists, who saw the public school system in general and teachers in particular as spearheading broad, social reforms; and child developmentalists, who projected students as the center of the educational process. These various tenets found common ground with a general concern for child welfare. The National Education Association (NEA), the leading professional educational organization, institutionalized these tenets in 1918, consolidating them into the *Cardinal Principles of Education*, establishing new and far-reaching public school curriculum guidelines. These included academic, aesthetic, and vocational subjects as well as health and physical education. By the 1920s, this extended to the health of preschool and high school students, and suburban, small-town, and rural schools.[17]

Two textbooks, used in teacher preparation programs during the first half of the twentieth century, illuminate the scope of that agenda: Helen Leslie Coops, *Health Education in Elementary Schools: Activities, Materials, and Methods*, and Charles C. Wilson, *Health Education, A Guide for Teacher and Text for Teacher Education*. Wilson's book was published through four editions (1924, 1930, 1941, 1948) by the NEA and endorsed by the American Medical Association (AMA). Both authors treated health education as an integral part of broader social progress, conforming to Progressive thinking—that is, part of broader "community health" programs that encompassed the family physician and dentist, hospitals, the Parent Teachers Association, and public health agencies. Personal and public health appeared integrated and inseparable. The school's curriculum operated as a continuum, beginning at the elementary level. Comprehensive topics touched all aspects of students' lives: anatomy, community health services, diseases, emotional well-being, exercise, first aid, hearing, hygiene, nutrition, physical fitness, public health, rest, sanitation, sex education, sleep, teeth, social relations, and vision. Coops, in particular, prescribed biographies of famous medical researchers, such as Edward Jenner, Robert Koch, and Louis Pasteur, while Wilson suggested a unit on communicable diseases, focusing on infantile paralysis, be taught at the beginning of each school year, following summer polio epidemics. At the secondary level, home economics, physical education, and science courses incorporated characteristics of health education. Finally, both authors included vaccinations as part of their health gospels.[18]

Organizations adroitly dovetailed with the school's program. In 1911, the AMA supplied a "health bibliography to health educators in the public schools...."[19] The Progressive Education Association produced study guides for popular movies like *Arrowsmith* (1931) and *The Story of Louis Pasteur* (1936). By 1946, with record numbers of polio cases, the National Foundation for Infantile Paralysis/March of Dimes supplemented the health curriculum with a thirty-two-page booklet, *Teacher's Guide in the Use of a High School Unit on Poliomyelitis*, intended for tenth-grade biology instructors. It suggested that teachers introduce the topic within the broader context of viruses. The ensuing discussion clarified terms. Laboratory work followed, with the students referring to two-by-two inch prints of poliovirus, in addition to charts displaying the nervous system. It concluded with a true/false list to stimulate discussion, more laboratory activities and accompanying student workbook pages, and an examination. The National Foundation also published a sixteen-page reference book, *Poliomyelitis: A Source Book for High School Students*. Intended for biology classes, it maintained a comprehensive scope by providing a brief history of the disease, presenting a statistical profile of it, clarifying various types, outlining the causes of the virus, revealing the source of infections, describing treatments, hospitalization, and rehabilitation—promoting its Warm Springs rehabilitation facility in Georgia—and supplying background about March Dimes campaigns.[20] In sum, health transcended a simple course or an area of study; it shaped the entire school environment, often incorporated into other academic subjects. Wilson saw health education occurring "in every room of the school plant and at all times of the day...."

Moreover, all physical activities within the school building, formal instruction, informal recess, and intra- and extramural sports programs contributed to the school's overall health emphasis. Although they occurred at different times and in different places, like playgrounds, gymnasiums, playing fields, and swimming pools, they did not stand alone. First, physical education offset the sedentary and stressful regimen of academic work through relaxing breaks, contributing to emotional and intellectual well-being. Second, it developed strength and endurance, assuring fit bodies. Third, it taught teamwork which, it was assumed, spilled over into positive social interactions. Fourth, this "total health program," as Wilson termed it, "requires a wide variety of personnel–doctors, nurses, dentists, teachers, supervisors, health co-ordinators, guidance counselors, psychologists…."[21] The presence of these health professionals became routine, with their facilities integrated into school architecture.

One-Stop Service

Treatments signaled another stage in the ascent of medical authority. The Seattle public education system serves as a case in point. In 1914, school officials appointed Ira Brown, a former U.S. Army physician, as medical director and began to operate its own infirmary, housed in the district's administration building, rendering physical examinations, diagnostic services, vision and dental checkups, as well as minor surgical procedures. "With the opening of the medical department and clinic," Stephen Woolworth asserts, "Seattle became the first school district in the United States to sponsor and provide comprehensive medical treatment for children"—that is, an in-house medical infrastructure, consisting of "a school medical department complete with a lead medical inspector, full-service clinic, and nursing staff." Further, Brown's broad health education program introduced "physical exercises" into the school curriculum and medical examinations for students participating in school athletic programs. Finally, he put into place the "Little Mothers' League" for sixth- and eighth-grade girls, a class devoted to infant care taught by school nurses.[22]

Since tuberculosis most often struck during childhood, schools nationwide emphasized prevention by training children in "healthful habits."[23] In 1917, the National Tuberculosis Association through its Modern Health Crusade shaped that curriculum. "The messages contained in Crusader curricula, whether manifested in pamphlets, lectures, films, or plays, emphasized citizenship, middle-class hygiene, and morality. By 1919, three million American children were involved" in this campaign, turning them into "health monitors and educators for the families and communities."[24]

Children deemed susceptible to consumption, usually malnourished or sick, received treatment at open-air schools. Advocates believed that a "plentiful supply of rest, hearty food, and cold fresh air could not only prevent those infected from developing the disease but also help those afflicted effect a spontaneous healing of tuberculosis lesions, thus conquering the disease or at least rendering it inert." Providence, Rhode Island, opened the first such institution in the USA in 1908: "like their [Progressive] counterparts elsewhere in the nation, Providence education and health officials began advocating the expansion of the fresh-air program not only as a continuing strategy to combat tuberculosis in children but also as a means of facilitating the education of children whose physical frailty was impeding their academic progress." The school district constructed a classroom with glass walls that, except for driving rain and blowing snow, remained open while children sat bundled in warm clothing and blankets. They studied the same curriculum as their healthy counterparts, but one that was individualized and presented at a lighter pace. Exercise and rest supplemented academic studies. Teachers fed them hot soup twice a day to restore their health, and a physician examined them weekly. Other cities quickly adopted this approach: Boston in 1908, Chicago, Rochester, and Hartford in 1909, and New York City in 1910. The open-air school movement peaked in the 1920s with 150 urban school districts utilizing them.[25]

School nurses represented another form of medical science imbedded in the schools. Philanthropies and public health policies introduced nurses into American schools, tending to and instructing children. The Henry Street Settlement, founded in 1903 by Lillian D. Wald, and located on New York City's Lower East Side, a densely populated tenement area, dispatched visiting nurses to provide direct care in homes for a variety of diseases, ran milk dispensaries, administered first aid, and rendered postpartum care. In 1913 alone, they conducted 188,214 home visits. Chicago's board of health saw public nurses as the "principal medium of contact with the American family…." Meanwhile, municipal public health departments sent nurses into schools to inspect pupils for "evidence of contagion," assist doctors while they conducted physical examinations, give diphtheria and smallpox vaccinations, help with dental procedures, and visit children's homes. Using the public schools as their staging area, they spearheaded the delivery of mass medical services, but on a largely irregular basis with minimal follow-up.[26]

To ensure consistent care and fully integrate them into the school environment, administrators began to hire nurses and incorporate medical facilities into their buildings. Public health nurses practiced what became known as "social medicine," fulfilling two key functions: therapeutic, offering care and monitoring recovery; education, sharing information about diet and hygiene.[27] School nurses thus served "first in the line of defense against child neglect in the city and a visible link between the home and the urban school." New York City's board of education led the country in the early 1900s, hiring twenty-seven nurses; by 1910, eighty different municipal school districts had followed suit while the American Red Cross introduced them into rural districts in 1912.[28] In Seattle, they examined students and maintained office hours. Chicago reported an increase in in-school treatments by nurses, involving cases of ringworm, tonsillitis, and tuberculosis, from 2635 in 1908 to 24,547 in 1910. They also conducted year-round family visits, jumping from 17,089 in 1908 to 69,646 in 1910, to instruct parents about their children's physical and nutritional care. "[S]chool nurses were therefore instrumental in not only the development of school medical inspections, but assembling an even broader interventionist and regulatory child health and welfare program in the schools," spearheading what would become a nationwide health campaign. "The nurses were the conduit through which the knowledge of applied medical science flowed from the technical expertise of physicians through the schools then out into the homes of the city."[29]

The Doctor Is In

Paul G. White, speaking before the New York Odontological Society on January 18, 1910, dismissed parental objections to student inspections: In health, as in crime, the public good superceded individual rights. His rhetoric, at times bombastic and full of mixed metaphors, likened such measures to a call to war. And prevention provided the best protection. This could be achieved "only when a generation of American citizens has been systematically instructed in the principles of hygiene and sanitation."[30]

Dental services followed the same pattern as other medical initiatives—that is, it began as an external effort and then gradually became assimilated into public school culture. New York City's Bureau of Child

Hygiene, beginning in 1913, maintained clinics working in conjunction with the schools while Chicago's health department ran student dispensaries with dentists on staff. By 1920, free or low-cost oral care appeared common. Elizabeth Beatty, a Bridgeport, Connecticut, dentist described the increased demand on such services by parents who could not afford to have their children treated in private practice. Public schools there housed clinics that focused on the "prevention of dental decay" through a host of measures. First, its "dental corps" taught students about the importance of oral health. Second, it gave instruction about care. Third, it operated on first and second graders to fill small cavities, with most parents signing permission slips. Fourth, it undertook parent education to demonstrate how they can oversee the health of their own children's teeth. In Tonawanda, New York, the dispensary consisted of four beds (where doctors could also conduct non-orthodontal procedures, such as removing students' adenoids and tonsils). The Nassau County school system in New York relied on "mobile dental units," each outfitted with a reclining chair and X-ray machine, instruments and supplies, and of course a dentist, traveling to different school buildings. In January 1920, this service took care of 322 children and conducted a total of 1750 "corrections," covering extractions and fillings. Participants received a button they pinned to their clothes proudly proclaiming "Clean Teeth–Nassau County." St. Louis school superintendent, John W. Withers, speaking before the 1919 Missouri State Dental Association described public school dental clinics in that city. Based on successful experiences in Boston, New York City, and Rochester, New York, philanthropies funded the free, but segregated, dental clinics in St. Louis and some rural areas of Missouri. From kindergarten through eighth grade, they stressed prevention through education sponsoring toothbrush drills. Withers further advanced, in clear Progressive terms, how public schools had to transcend simple academic instruction. One of these new functions involved it as a "protective" institution, which oversaw multiple activities promoting healthy children.[31]

Pittsburgh illustrates how medical services steadily expanded. Between 1911 and 1920 that district changed the name of the Department of Physical Training to the Department of Hygiene, marking the adoption of a broad array of responsibilities, combining communicable disease control, health and physical education instruction, health services, medical examinations, dental and eye clinics, and open-air schools. In 1934, it introduced homebound teachers as well as "sight conservation classes." Over the next six years, it inaugurated "hard-of-hearing classes," hired a "nurse-audiometer operator," began to instruct "cerebral palsied children," and introduced tuberculosis tests. By 1940, thirty-nine school medical inspectors conducted 70,000 annual examinations, the "central medical diagnostic clinic" had a staff of six, and the district ran a total of thirteen dental clinics and a "refraction clinic" with "two ophthalmologists" on staff. All of this existed in addition to the Department of Special Education which specifically treated "slow learners" as well as students with physical disabilities. Instruction for the latter existed, the district's superintendent explained in 1954–1955, "through homebound teachers at nine centers in elementary and high schools, hospitals, and institutions."[32] Later, as the 1950s waned, Pittsburgh's superintendent summarized the profound change in the school system's medical role: "... the emphasis has shifted from disease control to preventive medicine and educational health guidance. School health programs have moved closer to the problem of dealing with the 'whole child.'"[33]

Mother Knows Best

Progressives sought a closer connection between school and home as the social mission of public education expanded. They perceived the family as an "educational agency" but one that did not consistently provide a healthy environment or convey acceptable role models. As a "teaching unit," it was disorganized and the educational process "haphazard," as evidenced by a 1929 analysis by the National Society for the Study of Education (NSSE). Children younger than six comprised more than 33% of total deaths in the USA. Moreover, between 1921 and 1925, from 40 to 60% of school-aged children had some "physical defect." Dental problems and rickets dominated, with malnourishment ranging from 20 to 80%. This study attributed all of these problems to the "failure of parents" who neglected hygienic environments and nutritious diets. "Educating parents" would eliminate these health deficiencies.

Parent education and health instruction remained virtually indistinguishable in the eyes of reformers. Adult family members shaped their children's impressions through their own hygienic habits. In order to deliver a positive message, each family's outlook needed to be *improved*. This required finesse, as the NSSE proclaimed: "The effective way to reform the family is not found in a direct attack upon its faults but in a relationship between it and other social institutions which will automatically reveal to the family its faulty practices."[34]

The National Congress of Mothers (NCM), founded in 1897, provided an ideal vehicle, formalizing parent–school relationships, promoting parent education, and advocating "scientific motherhood and child welfare reform." Concerns for children's health, according to historian William W. Cutler, grew from this, becoming a "national campaign." It formed a committee in 1903 to study the needs of delinquent, dependent, and disabled children, "lobbied for legislation to eliminate tainted milk, urged every board of health to establish a child hygiene department, and in 1909 created one of its own that focused on infant mortality." Health education and physical fitness extended this agenda. The NCM's scope continued to evolve, as did its name becoming the National Congress of Mothers and Parent Teacher Associations in 1908, advocating "child study, domestic science, and health education." The third name change in 1924 to the National Congress of Parents and Teachers (PTA) cemented linkages between local, state, and national organizations. Five years later, it established the Department of Parent Education which sponsored "training programs" to "transform parents"—that is, to develop their "child rearing" skills.[35] And the "new medicine held a central place in … early-twentieth-century parent education for both its theoretical findings and its practical applications. No annual convention of the PTA was complete without several presentations by doctors and nurses on the urgency of incorporating recent medical discoveries into everyday child care."[36]

Parent education broadened the public school's mission, pushing it toward becoming a "social service station." In 1925, the PTA inaugurated the "'summer round-up' of schoolchildren." This effort to "identify and correct such common problems as swollen adenoids and poor vision soon became a popular program at the grassroots level." The AMA, federal Bureau of Education, NEA, and U.S. Children's Bureau endorsed this program. By 1931, it conducted physical examinations of 75,000 children in forty-four states.[37]

This "scientized/technologized" approach assuredly applied to poliomyelitis. Preparedness became prevention through sanitation and hygiene—at least, that is what was believed. During some of the worst epidemics in the late 1940s and early 1950s, literary scholar Jacqueline Foertsch posits, it empowered women within an otherwise highly domesticated culture by positioning "mothers as chief scientist and the mastermind behind 'germ warfare' in every home…."[38]

Center Stage

A medical presence, in one form or another, became universal in the public schools. Two broadly focused movements converged to promote health care and disease prevention, strongly influencing American attitudes toward medical science in a rapidly growing urban and diverse society. Taken together, they facilitated the 1954 polio experiments on schoolchildren. This phenomenon proved to be broader and more ubiquitous than the reflexive actions of anxious parents or the product of Madison Avenue tactics by the National Foundation for Infantile Paralysis. The roots of the medicalization of schooling thus reside as much in medical history as in educational history. The infrastructure, involving visiting physicians and school nurses, grew out of the fields of medical science and public health. The subject matter, encompassing health classes and physical education, emerged from the Progressive education movement. The culmination of these forces resulted in the most significant medical and educational event in American history.

Notes and Sources

1. Emerson is quoted in "New School Health Work. Dr. Emerson Tells Teachers What Is Expected of Them," *New York Times*, September 14, 1916: 10. Further, see Heather Munro Prescott, *A Doctor of Their Own: The History of Adolescent Medicine* (Cambridge: Harvard University Press, 1998): 9, 44–46.
2. Roy Porter, *The Greatest Benefit to Mankind: A Medical History of Humanity* (New York: Norton, 1997): 633 and 635, respectively. In mapping this process, Stephen T. Woolworth, "Conflict, Collaboration, and Concession: A Study of the Rise and Fall of Medical Authority in the Seattle Public Schools, 1892–1920s" (PhD dissertation, University of Washington, 2002), uses a sequence of sanitation, medical inspections, and clinics, while Richard A. Meckel, *Classrooms and Clinics: Urban Schools and the Protection and Promotion of Child Health, 1870–1930* (New Brunswick: Rutgers University Press, 2013), relies on four overlapping periods.
3. Meckel, *Classrooms and Clinics*: 5. Also see pages 6, 8, 10–14.
4. Woolworth, "Conflict, Collaboration, and Concession": 64, 25, and 67, respectively. Further check pages 25, 34–35, 40, 48–51, 53–54, 60–62. Also consult Michael Kort, "The Delivery of Primary Health Care in American Public Schools, 1890–1980," *Journal of Social*

Health 54 (December 1984): 453; and Meckel, *Classrooms and Clinics*: 17–19. Finally, Bernard Harris, *The Health of the Schoolchild: A History of the School Medical Service in England and Wales* (Buckingham: Open University Press, 1995), provides a splendid, comprehensive history of the British school medical service, which, in some ways, followed the same pattern in the States.

5. *Report of the Department of Health of the City of Chicago for the Years of 1907, 1908, 1909, 1910*: 44. CPL. See also James Colgrove, *State of Immunity: The Politics of Vaccination in Twentieth-Century America* (Berkeley, CA: University of California Press, 2006): 10; and Meckel, *Classrooms and Clinics*: 38–39, 57–59, 61–62.
6. John Duffy, *The Sanitarians: A History of American Public Health* (Urbana: University of Illinois Press, 1992): 182–83.
7. *Report of the Department of Health of the City of Chicago for the Years of 1907, 1908, 1909, 1910* (Chicago: Department of Health, 1911): 44. See page 167, as well as *Report of the Department of Health of the City of Chicago for the Years of 1911 to 1918 Inclusive*: 592. Refer also to pages 495–507. CPL. Further consult Duffy, *Sanitarians*: 183; William J. Reese, *Power and the Promise of School Reform: Grass-Roots Movements During the Progressive Era* (Boston: Routledge & Kegan Paul, 1986): 226; and Woolworth, "Conflict, Collaboration, and Concession": 69.
8. Woolworth, "Conflict, Collaboration, and Concession": 116. Also consult pages 65–67, 97, and Meckel, *Classrooms and Clinics*: 67.
9. John Duffy, "School Vaccination: The Precursor to School Medical Inspection," *Journal of the History of Medicine* 33 (July 1978): 344–46, 350–51. Also see pages 348–49, as well as Prescott, *A Doctor of Their Own*: 40–41; Samuel H. Preston and Michael R. Haines, *Fatal Years: Child Mortality in Late Nineteenth-Century America* (Princeton: University of Princeton Press, 1991): 13–14; Woolworth, "Conflict, Collaboration, and Concession": 25, 66–69, 97–100, 116, 119–21, 126, 132, 150–53, 255–58, 260, 265, 269–74.
10. Evelyn Maxine Hammonds, *Children's Deadly Scourge: The Campaign to Control Diphtheria in New York City, 1880–1930* (Baltimore: Johns Hopkins University Press, 1999): 198–99, 13, and 83, respectively. Also see pages 84, 103, 182–83, 188, 197, and History of Vaccine Timeline, www.historyofvaccines.org/content/timelines/diphtheria (Accessed February 13, 2016); and Meckel, *Classrooms and Clinics*: 186–87.
11. Colgrove, *State of Immunity*: 81–82, 90, and 81, respectively. Refer to pages 83–84, 90–91, as well as Hammonds, *Childhood's Deadly Scourge*: 6, 7, 8, 14, 178, 166–67, 182–84, 188, 204–5, 207–8, 211–13. Hammonds provides an excellent treatment of the public relations campaign by New York's health department on pages 98–116. Also

see Bert Hansen, "New Images of a New Medicine: Visual Evidence for the Widespread Popularity of Therapeutic Discoveries in America After 1885," *Bulletin of the History of Medicine* 73 (1999): 665–70; Meckel, *Classrooms and Clinics*: 163–64; and Paul Starr, *The Social Transformation of American Medicine* (New York: Basic Books, 1982): 185–86.

12. Porter, *Greatest Benefit to Mankind*: 439. For a solid analysis of the New York City diphtheria campaign, refer to Hammonds, *Childhood's Deadly Scourge*, especially page 216 for additional statistics on the reduction in morbidity and mortality rates. Further check History of Vaccine Timeline; Wade W. Oliver, *The Man Who Lived for Tomorrow: A Biography of William Hallock Park* (New York: E. P. Dutton & Co., 1941): 111; Terra Ziporyn, *Disease in the Popular American Press: The Case of Diphtheria, Typhoid Fever, and Syphilis, 1870–1920* (Westport, CT: Greenwood Press, 1988): 35, 41–42. Especially see pages 42–46 for a description of that news coverage.
13. Colgrove, *State of Immunity*: 109. Also refer to pages 87, 104–5.
14. Selwyn D. Collins, "Frequency of Immunizing Procedures of Various Kinds in 9000 Families Observed for 12 Months, 1928–1931," *American Journal of Public Health and the Nation's Health* 25 (1935): 1221. Also check pages 1223–24.
15. The first quote is from Collins, "Frequency of Immunizing Procedures": 1221. Also check pages 1223–24. For the second quote, see Leona Baumgartner, "Attitude of the Nation Toward Immunization Procedures: A Study Based on a Public Opinion Poll Made in 1941," *American Journal of Public Health* 33 (March 1943): 257. Further consult pages 257, 258, 259.
16. Richard S. Lowry, "Lewis Hine's Family Romance," in Caroline F. Levander and Carol J. Singley, eds., *The American Child: A Cultural Studies Reader* (New Brunswick, NJ: Rutgers University Press, 2003): 194. Further consult Mary Cable, *The Little Darlings: A History of Child Rearing in America* (New York: Charles Scribner's Sons, 1975): 165–66, 168, 175–77.
17. For a general overview of these changes, see Chapters 4–8 in Richard J. Altenbaugh, *The American People and Their Education: A Social History* (Upper Saddle, NJ: Merrill/Prentice-Hall, 2003). Further refer to Herbert M. Kliebard, "*Cardinal Principles of Secondary Education*," in Richard J. Altenbaugh, *Historical Dictionary of American Education* (Westport, CT: Greenwood Press, 1999): 67–68; Craig Kridel, "Progressive Education Association," in Altenbaugh, *Historical Dictionary*: 303–4; Meckel, *Classrooms and Clinics*: 164–66, 190–91; and Charles C. Wilson, *Health Education: A Guide for Teachers and*

a Text for Teacher Education. 4th ed. (Washington, DC: National Education Association, 1948): 2–3.

18. Wilson, *Health Education*: 78. 77, 128, 137, and 237–39, respectively. Additional information may be gleaned from pages ix, 5–6, 7, 63, 64–65, 77, 79, 116–18, 119–28, 137, 191, 214–15, 217, 219, 246–48, as well as Helen Leslie Coops, *Health Education in Elementary Schools: Activities, Materials, Methods* (New York: A. S. Barnes & Company, 1950): 36, 43–49, 53, 156–227. I thank Michael Cleary, a health and physical education colleague at Slippery Rock University, for pointing out and then loaning me the Coops and Wilson books.
19. Ziporyn, *Disease in the Popular American Press*: 17. Also check John C. Burnham, *How Superstition Won and Science Lost: Popularizing Science and Health in the United States* (New Brunswick, NJ: Rutgers University Press, 1987): 59–60.
20. National Foundation for Infantile Paralysis, *Teacher's Guide in the Use of a High School Unit on Poliomyelitis* (New York: NFIP, 1946) and *Poliomyelitis: A Source Book for High School Students* (New York: NFIP, 1946), NYAMA.
21. Wilson, *Health Education*: 161–62.
22. Woolworth, "Conflict, Collaboration, and Concession": 121 and 25, respectively. Further consult pages 119–21, 126, 132, 150–53.
23. Barbara Bates, *Bargaining for Life: A Social History of Tuberculosis, 1876–1938* (Philadelphia: University of Pennsylvania Press, 1992): 274–75. Additionally, refer to page 280. I am not arguing from a social efficiency perspective but attempting to illustrate the growth and institutionalization of medical hegemony in the public schools. The simultaneous increasing influence of medical science and the expanding role of public schooling provided the environment for this development.
24. Cynthia A. Connolly, *Saving Sickly Children: The Tuberculosis Preventorium in American Life, 1909–1970* (New Brunswick, NJ: Rutgers University Press, 2008): 80.
25. Richard A. Meckel, "Combating Tuberculosis in Schoolchildren: Providence's Open-Air Schools," *Rhode Island History* 53 (August 1995): 96, 95, and 98, respectively. See pages 91–94, as well as Bates, *Bargaining for Life*: 274–78; Colgrove, *State of Immunity*: 50–51; Connelly, *Saving Sickly Children*: 2, 41; and Meckel, *Clinics and Classrooms*: 108–11.
26. *Annual Report of the Board of Health of the City of Chicago for 1933* (Chicago: 1934): 23 and 26, respectively. See also pages 24, 26, 27. CPL. Further check Carole A. Estabrooks, "Lavinia Lloyd Dock: The Henry Street Years," *Nursing History Review* 3 (1995): 144, 146;

Meckel, *Classrooms and Clinics*: 62–63; and Woolworth, "Conflict, Collaboration, and Concession": 67, 79.

27. Charles E. Rosenberg, *Care of Strangers: The Rise of America's Hospital System* (Baltimore: Johns Hopkins University Press, 1987): 316. For additional information consult page 314.
28. William W. Cutler, III, *Parents and Schools: The 150-Year Struggle for Control in American Education* (Chicago: University of Chicago Press, 2000): 144. Also refer to Estabrooks, "Lavinia Lloyd Dock": 146; Woolworth, "Conflict, Collaboration, and Concession": 79–80.
29. Woolworth, "Conflict, Collaboration, and Concession": 86 and 136, respectively. Further see pages 81, 117, 134–35. Cutler, *Parents and Schools*, makes a similar assertion. Prevention became the main function of school nurses. They taught children about personal hygiene and their parents health education. "In the eyes of the nursing profession, she remained the most effective link between the home and the school, interpreting the child's family life to the teacher, while explaining the school to the home" (146–47). Finally, consult *Report of the Department of Health of the City of Chicago for the Years of 1907, 1908, 1909, 1910*: 168.
30. Paul G. White, "The Practical Application of Oral Hygiene to the School Curriculum," *Dental Cosmos* 52 (1910): 864–65.
31. Elizabeth Beatty, "Operative Work in the School Dental Clinic as Preventive Dentistry," *Dental Cosmos* 59 (December 1917): 206–7; "School Clinic at North Tonawanda," *Dental Cosmos* 60 (September 1918): 840; W. J. Multer, "Dental Education in the Schools of Nassau County, New York," *Dental Cosmos* 62 (May 1920): 578, 575, 576, respectively; John W. Withers, "The Dental Clinic and the Public Schools," *Dental Cosmos* 61 (October 1919): 1015, 1016, 1017, 1022; William Z. Hill, "The Care of Children's Teeth," *Dental Cosmos* 62 (May 1920): 1379–84. Further consult "Work of the Department of Health's Dental Clinics for School Children," *Dental Cosmos* 61 (October 1919): 1032; and *Report of the Department of Health of the City of Chicago for the Years of 1911 to 1918 Inclusive*: 558.
32. *Annual Report of the Superintendent of the Pittsburgh Public Schools, 1954–55* (Pittsburgh: Board of Public Education, 1955): 18–19. See also page 14. CLP.
33. *Annual Report of the Superintendent of the Pittsburgh Public Schools, 1957–58* (Pittsburgh: Board of Public Education, 1958): 45. CLP. Also refer to Kort, "Delivery of Primary Health Care in American Public Schools": 454; Clarence A. Perry, "The Community-Used School," in the *National Society for the Study of Education Yearbook*, vol. 10 (Bloomington, IL: Public School Publishing Company, 1911): 64–65; and Porter, *Greatest Benefit to Mankind*: 641.

34. Guy M. Whipple, ed., "Preschool and Parental Education," *National Society for the Study of Education Yearbook* 14, no. 28 (Bloomington, IL: Public School Publishing Company, 1929): 55, 71, 58, and 85, respectively. Additional information may be gleaned from pages 56–58, 60, 62–63, 79–83, 109–10, 112–14, 278–79, 320, 329–30.
35. Cutler, *Parents and Schools*: 104, 106, 33, and 48 respectively. I have drawn additional information from pages 25, 30, 33, 43–45, 48.
36. Steven L. Schlossman, "Before Home Start: Notes Toward a History of Parent Education in America, 1897–1919," *Harvard Educational Review* 46 (August 1976): 446.
37. Cutler, *Parents and Schools*: 129 and 50, respectively. Refer to pages 106–7, 143, 148, as well as Colgrove, *State of Immunity*: 93–94.
38. Jacqueline Foertsch, *Bracing Accounts: The Literature and Culture of Polio in Postwar America* (Teaneck, NJ: Farleigh Dickinson University Press, 2008): 45. Further, see pages 43, 46, 50.

CHAPTER 7

Capstone Events

Infantile paralysis in the immediate post-war years wreaked havoc on young children, their families, and American society. On August 6, 1945, the *New York Times* ran an article with the headline, "Trenton Enforces 'Polio' Quarantine." That city's commissioners adopted this harsh measure because of mounting infections. It banned "children under eighteen years old from churches, Sunday schools, theatres, stores, playgrounds, parks, and gatherings of any kind." Restaurant owners turned away families with children and the zoo "was all but deserted." Trenton's children had become epidemic refugees.[1] At the same time, across the continent in rural Minnesota, seventeen-month old Ron Zemke contracted polio. He would remain hospitalized until 1948, spanning acute and convalescent care. By 1951, polio represented the "chief cause of orthopedic impairments among young people" and accounted for 20% of all physical disabilities, sparking a "serious social problem."[2] Nothing stopped this virus. Between 1943 and 1951, morbidity rates increased by 45%. "By 1952, polio had reached truly epidemic proportions … and more children died of it than any other infectious disease." The following year witnessed a record 57,628 cases of infantile paralysis. Local health officials from the Yukon to Montgomery, Alabama, issued quarantines, banned public gatherings, closed public swimming pools, and shuttered movie theaters. Parents burned their infected children's toys and disposed of their comic book collections.[3] Medical science had achieved significant

R. J. Altenbaugh, *Vaccination in America*,
Palgrave Studies in the History of Science and Technology,
https://doi.org/10.1007/978-3-319-96349-5_7

breakthroughs in its ongoing "'conquest' of epidemic infectious diseases," gaining public trust and respect, but poliomyelitis proved a nemesis for physicians.[4]

Nevertheless, desperate parents seeking protection for their children readily turned to science for answers. Their faith remained unshaken. A profound transformation in public attitudes toward medical care had occurred. Positive information and images promulgated through informal channels, like advertisements, comic books, magazines, the military, movies, and novels, plus the systematic inculcation of good health and hygiene habits in the public schools ensured that children and adults embraced medical knowledge and practitioners. As a result, the civilian population, of all ages, convinced about the safety and efficacy of vaccines accepted them, and trusting parents willingly surrendered their children as test subjects for polio experiments. Two events leading to this point characterized this new attitude.

Trouble in the Big Apple

On April 5, 1947, the *New York Times* declared: "Smallpox in City." A month earlier, Eugene La Bar, an American businessman, and his wife had arrived on a bus originating in Mexico City, where they had resided for twenty years. Although they had purchased tickets for Maine, their final destination, Eugene did not feel well and decided to remain in New York City for a few days, checking into a midtown hotel. Feeling worse and developing a rash, Eugene entered Bellevue Hospital, but dermatologists did not readily recognize his symptoms. When his condition worsened, and suspecting he had a potentially infectious illness, they transferred him to Willard Parker Hospital for Contagious Diseases. Eugene died on March 10. Doctors initially attributed his death to "bronchitis with hemorrhages." However, subsequent tests, confirmed by the US Army Medical School Laboratory in Washington, DC, revealed he had died of smallpox. It was too late: two patients at that hospital had already contracted it.

The *Times* announcement, by city officials, triggered the total mobilization of public institutions and personnel. The New York City health department "provided free vaccinations at each of its twenty-one neighborhood health centers and sixty child health stations. The thirteen municipal hospitals provided the service in their outpatient clinics." Eighty-four police stations also served as inoculation sites. Each day thousands began to queue at dawn, with rain often drenching them as they waited for hours.

Tensions mounted on April 16, when the *Times* proclaimed "Vaccinations Stop." In spite of the US Navy and Army donating 780,000 units and the city's health department producing 400,000 units, vaccine supplies suddenly dried up because manufacturers had failed to fulfill their commitments. Mayor William O'Dwyer, advised by Thomas M. Rivers, "a Board of Health member and head of the Rockefeller Institute for Medical Research, … attended a meeting during which O'Dwyer virtually locked the pharmaceutical representatives into City Hall until they agreed to comply with his demands." Mass immunization resumed, and on April 17 alone that city's health workers jabbed 500,000 people. Beginning on April 21 at 9:00 a.m., all of that city's public and parochial schools gave inoculations, but only those students who had consent forms received them.

This entire episode ended anticlimactically a month later when a tiny article, buried deep inside the *Times*, officially declared the end of free vaccinations. Over 6 million residents had been inoculated. As medical historian Judith Walzer Leavitt notes, "More people had been vaccinated in a shorter period of time than ever before in history." A flood of media messages and lingering wartime cooperation accounted for some of the public's overwhelming positive response—no opposition whatsoever had been vocalized. This reflected the culmination of a great deal of work, as Leavitt concludes: "The task was enormous. That it proceeded systematically and methodically is a testament to how far public health in general and smallpox control in particular had progressed in New York by the middle of the twentieth century."[5] Decades of medical proselytism, not legal pressure, confrontation, or ridicule, had won the day.

A Scientific Miracle?

An apparent breakthrough preventative for infantile polio affirmed this new faith in medical science. Gamma globulin (GG)—blood plasma rich in antibodies—promised the first post-war defense against this disease's summer outbreaks. Although only offering temporary protection, its development involved tens of thousands of children as human subjects.

According to Stephen E. Mawdsley's historical analysis, it took twenty years for this treatment to reach fruition. Pediatrician Joseph Stokes, Jr., at Children's Hospital in Philadelphia, dabbled with a prototype in 1932. During World War II, as director of the Measles and Mumps Commission for the Armed Forces Epidemiological Board, "he undertook a series of studies with GG to explore its protective effect, if any, against influenza, measles, mumps, and hepatitis."[6] He found it both safe and effective as a prophylactic. This led him to consider its application to infantile paralysis. Stokes and T. F. McNair, another pediatrician at that hospital, quietly experimented on children in two Pennsylvania summer camps in 1941. The result, Stokes vaguely states, "was not unfavorable."[7] An additional experiment occurred in 1944 when Harvard researchers used a scarce amount of gamma globulin on a few children. Wartime shortages of blood supplies and medical personnel delayed any further initiatives.

A low-profile trial took place on August 26, 1945, at St. Vincent's Orphanage, a Roman Catholic institution, near Freeport, Illinois. In the waning days of a polio outbreak there, "state public health officers and scientific consultants" implemented an "efficient, assembly-line approach … as a team of two nurses prepared syringes, while two physicians conducted the injections." This "private experiment," Mawdsley continues, involved 427 children, with ages ranging from "nursery to high school," that orphanage's total population. "As wards of the orphanage, the child subjects would be compelled to participate." Thomas Francis, Jr., Professor of Epidemiology, University of Michigan School of Public Health, designed an injection-control protocol, with half receiving saline and half GG serum. The results of the study were "inconclusive" because orphanage administrators had refused to grant researchers permission to draw blood, both before and after injections.[8]

Scattered private attempts continued to unfold. During Houston's severe polio outbreak in 1948, "pediatricians, demoralized by their limited means to stem the epidemic, openly administered GG to family members and close contacts in the hope that it would protect against infection." Officials at National Foundation for Infantile Paralysis (NFIP), Mawdsley points out, did not want to be left out of the GG race. "Attending this epidemic on behalf of the NFIP was an epidemiological consultant named William McDowell Hammon, who witnessed the extensive use of gamma globulin and was intrigued by its potential."

Hammon had attended Allegheny College in Western Pennsylvania, completed his medical studies at Harvard University in 1936, and with National Foundation funding, worked on encephalitis and infantile paralysis at the University of California-Berkeley in 1940. During the war, he became a close acquaintance of Thomas Francis. In 1950, Hammon moved to the University of Pittsburgh where he served as head of the Department of Epidemiology and Microbiology in the Graduate School of Public Health. "Hammon's more senior appointment and parallel interest in polio prompted tensions with Jonas E. Salk, [also employed by that university and funded by the NFIP] which were exacerbated by their differing theories about prevention."[9]

Meanwhile, during the late 1940s, Stokes had lobbied Harry M. Weaver, the National Foundation's research director, about the use of gamma globulin to control infantile paralysis, but Foundation leadership, according to Mawdsley, remained divided in its support. A frustrated Stokes "privately encouraged Hammon to step forward and submit a research proposal to the NFIP." Since he already worked as a Foundation consultant, Hammon held credibility with Basil O'Connor, that organization's director, and his staff. Moreover, supporting GG as a stopgap would justify the Foundation's mission and account for public donations; at the same time, it would stimulate March of Dimes campaigns.[10]

Hammon met with the National Foundation's Committee on Immunization on May 17, 1951, at New York City's Commodore Hotel. Weaver had formed it to operate as a "peer review forum and a means to steer polio research …" David Bodian, John Enders, Thomas Francis, John R. Paul, Albert B. Sabin, Jonas Salk, and other prominent virologists served as members. During this inaugural meeting, committee members debated the merits of a field trial, focusing on three points. First, Hammon's design skipped any laboratory research, jumping to a clinical trial; that is, injecting children without prior, controlled results made that committee's members uneasy. Second, the use of a "placebo-controlled clinical study" appeared to be an additional bone of contention. Third, the amount of each GG dose remained unresolved. After hours of deliberation, that committee rejected Hammon's proposal. "Despite criticism, Hammon and NFIP officials were far from defeated." After two months of further debate and heavy pressure by O'Connor, committee members unanimously agreed to support a pilot study involving 5000 children. However, they remained at an impasse over the protocol. Salk suggested that Hammon alone be allowed to determine the experiment's design. As Mawdsley explains it, Hammon faced a daunting task: "Such an endeavor would not be simple, since a mass placebo-controlled study had not been undertaken on civilians before."[11]

The National Foundation embarked on an elaborate set of preparations, much of it unprecedented. Hammon, O'Connor, and an attorney developed a one-page consent form. It consisted of a brief descriptive paragraph of the study followed by blank lines for the child's name, a parent's signature and address, and witness's signature. Local newspapers would print the consent form so parents could complete and sign it ahead of time, ensuring "streamlined clinic enrollment." It corresponded to the permission form Salk and the Foundation would employ during subsequent local and national vaccine trials. The Foundation's public relations office planned a methodical marketing campaign. Mawdsley describes how it had to maintain a delicate balance: "At the local level, media coverage had two agendas: the study had to be attractive enough to entice parents, but tempered enough to avoid implying that GG was effective. At the national level, the coverage had to address widespread interest, yet discourage parents from seeking private GG injections from family physicians." In so doing, it released carefully worded press statements, prepared "radio broadcast scripts," and staged "controlled photographic content." Hammon and Stokes, as key members of the research team, selected the site. They preferred to execute the trial in a small town because it afforded them more influence. For example, small-town doctors appeared more cooperative with their smaller, family practices, making it easier to build trust among participants. Further, this setting offered a modicum of operational privacy, away from "medical detractors or prying journalists." In sum, "by favoring a small-town context, trial administrators perhaps intended to re-create the 'wall of silence' prevalent in institutionalized medical experimentation." They chose Provo, Utah, and surrounding Utah County, the scene of an ongoing polio epidemic. Hammon began by securing the support of state and county medical societies, state health department, and Provo's mayor and school superintendent.[12]

Injections began promptly at 9:00 a.m. on September 4, 1951. As Mawdsley points out, "Hammon and NFIP officials' decision to establish injection clinics within schools and civic buildings smoothed the transition from the laboratory to the neighborhood. As civic buildings were conveniently situated at the center of community life, the clinics benefitted from the visibility of prominent locations." One hiccup occurred when, in the beginning, some parents enrolled their children at more than one site in the hopes of compensating for the placebo. Hammon took immediate measures to discourage this. The impact on the statistical analysis remains unknown. The medical staff rewarded children with lollipops to minimize crying as they sucked on them passing by long lines of other anxious, waiting children. The operation proved highly successful. The protocol designated 5000 as the human subject target, but the turnout exceeded expectations. "For researchers and the NFIP, the pilot study was … a sociological victory. It proved that civilians would willingly sponsor and serve as subjects for medical experimentation." In particular, this represented the crowning glory, a testament that "consent forms, injected placebo controls, potential health risks, evidence of participant pain, long queues, and the absence of protective assurances would not deter" parents from submitting their children for medical experimentation.[13] Desperation accounted for some of this response, but trust in science contributed. The medical outcome appeared equally positive: "Of the 2,871 children who received the true gamma globulin, one contracted paralytic polio. Of the 2,860 who received the counterfeit shot, four got sick."[14] These results were "extremely encouraging," for Stokes, but the number tested proved statistically inconclusive.[15]

Word of the Utah experiment quickly reached the general public. The September 1951 issue of *Life* magazine devoted a story to it, supplemented with dramatic images of children and their parents waiting in lines to receive injections. *Life*'s editors portrayed the sheer scale of this effort with a photograph of empty GG vials piled waist high in a pyramid shape. When Hammon met with the NFIP's immunization committee, once again at the Commodore Hotel, on December 4, he reported the Provo pilot study an unqualified success. "The presentation justified continuation of the experiment by focusing on a need for more data and the promise of improvement [in the protocol]," refining marketing and public relations, as well as perfecting clinic processing.

In lieu of a small town setting, Houston and surrounding Harris County offered a convenient opportunity to complete the pilot study. Its size, facilities, and more importantly its ongoing epidemic seemed, Mawdsley adds, to lend itself as a suitable test site. Injections began on July 2, 1952, at eight locations. Because of segregation, only one African American school was included. This number of sites soon became inadequate to meet the overwhelming demand. "Confusion mounted" and many "parents were turned away." Frustrated white parents, willing and able to pay for it, deluged their pediatricians for GG jabs, avoiding the experiment altogether. Hammon publicly discouraged such "bootlegging." Others submitted their children as test subjects but also asked their pediatricians for an injection in case they had received the placebo. As whites trickled away to private sources, Hammon "extended the duration of the study by two days and added two black-only injection clinics" in hopes of meeting his target figure of 35,000. He fell short and quickly relocated the trial to another ongoing epidemic site. Sioux City, Iowa, and surrounding Woodbury County appeared "an ideal choice for a proving ground," fulfilling all of the criteria of a "manageable community." Pursuing a cohort of 15,000, he "amended the protocol to enroll those from age one to eleven years, instead of two to eight in Utah County and one to six in Harris County." Injection sites opened on July 21, and clinic staffers and volunteers jabbed 15,595 children in fewer than six days. "Most importantly, statistical significance was finally achieved with a total enrollment of 54,772."[16]

Hammon and the National Foundation declared the experiment a success. "Of the … children included in the final assessment, 90 had succumbed to paralytic polio; within this group, 26 were recipients of gamma globulin while 64 received inert gelatin [i.e., the placebo]." Hammon though had "cleansed" and "trimmed" the data to purportedly account for the "effect of uncontrollable factors." The Foundation, according to Mawdsley, launched a publicity blitz in the immediate aftermath heralding its role through a documentary, *Operation Marbles and Lollipops* (1952), tempering any skepticism. "NFIP producers went to great lengths to portray families in the film as informed actors and not exploited subjects in the service of medical science. Decades of antivivisectionists' criticism of celebrity medical films, such as … *Yellow Jack* (1938), inspired 1950s producers to incorporate themes of free will as concomitant with human experimentation." This production seemed to anticipate the massive Salk polio trials as it projected smiling parents and children contributing to the advancement of medical science and ensuring better health for all. "Producers thus fashioned a palatable account of the GG field trials where contented families and efficient medical professionals enjoyed a favorable encounter."[17]

Thomas Rivers appeared less sanguine. As a virologist who headed the National Foundation's Virus Research Committee, a Rockefeller Hospital administrator, and had been involved with the Birthday Ball Commission, he condemned the Houston experiment. Private physicians destroyed the validity of the entire enterprise by surreptitiously injecting many of their patients: "… they shouldn't have done this. They thought they were doing good, but they weren't." At that time, however, he remained mum—at least publicly—about this anomaly.[18]

The Foundation subsidized a GG program for "thousands of school children," covering the costs of needles, nurses, physicians, and syringes.[19] It also contracted with Cutter Laboratories, in Berkeley, California, to manufacture gamma globulin. Finally, discussions among Harry Weaver, Joseph Stokes, and Jonas Salk occurred between November and December of 1952 over a permission form. Weaver preferred the NFIP's version while Stokes favored the "Release and Receipt" document used and legally approved at Children's Hospital of Philadelphia. Salk, conducting early trials of his experimental vaccine, deferred to Weaver. In the end, Weaver used the Foundation's "Form of Consent," focusing on liability (i.e., to "protect a physician from a lawsuit").[20] This matter would generate additional consultation as Salk's own field trials proceeded.[21]

Anxious parents saw the gamma globulin prophylactic for polio as a scientific miracle. A *Newsweek* article, "Polio: 1953," appeared on July 20, praising this stopgap measure amidst that summer's raging epidemic. Anyone who had been in contact with an infected individual received it, which the National Foundation deemed an "emergency basis." Other parents resorted to purchasing the serum, even at exorbitant prices, in order to spare their children the horrors of this virus.[22] Two months later, an unsympathetic *Newsweek* article, "Panic Triumphant," criticized protests in New York City. A group of parents, who resided in Queens, panicked with news that a neighbor infected with polio had been rushed to the hospital. Forty of them descended on Queens General Hospital to request injections for them and their children. Unsuccessful and frustrated there, they marched to that borough's department of health and demanded serum. Commissioner John F. Mahoney attempted to calm them but remained steadfast. Fifteen of them, now even angrier, refused to disperse; they remained at his office all night, later joined by twenty additional demonstrators. Mahoney finally caved in and supplied them with gamma globulin. Meanwhile, a group of Brooklyn parents also picketed their health department.[23]

Other problems arose. In spite of GG treatment, field reports indicated that some children still contracted the virus. "It appeared that the supposed four-week window of polio protection, commencing the first week after injection, was not guaranteed." On the world stage, scientific experts raised doubts about GG's ability to control polio.[24]

Finally, Oveta Culp Hobby, Secretary of the US Department of Health, Education, and Welfare (later split into the departments of Education and Health and Human Services), commissioned a formal study of GG's efficacy. She appointed eighteen members to the National Advisory Committee for the Evaluation of Gamma Globulin in the Prophylaxis of Poliomyelitis, including most notably Thomas Francis, William Hammon, and John Paul. During the summer of 1953, 235,000 children were inoculated in twenty-three communities, pinpointed because of serious polio outbreaks, scattered throughout Alabama, Florida, Illinois, Kentucky, Michigan, Minnesota, Missouri, New York, North Carolina, Tennessee, Virginia, and Wisconsin. "Most of this gamma globulin was made available to the Nation [sic] by the National Foundation for Infantile Paralysis and the American Red Cross."[25] US Public Health Service staffers coordinated data collection through the Centers for Disease Control (CDC).

After assessing the results at its headquarters in Atlanta, the CDC's final study, published in 1954, found that the "methods of analysis of carefully compiled and extensive data on the use of gamma globulin in these epidemic areas and populations, where it might have been expected to be effective, did not yield statistically measurable results." Hammon refused to endorse the study's conclusion and embarked on a speaking tour praising GG's effectiveness. The National Foundation also launched a marketing campaign to avoid embarrassment and a potential hit on March of Dimes fund-raising activities, mainly relying on testimonials from firsthand witnesses, especially from county health officers who had participated in the pilot studies. Finally, Basil O'Connor simply ignored the CDC's report and pledged to supply gamma globulin for mass injections in 1954.[26] The American public's perception of GG likewise defied the facts since it continued to clamor for this perceived panacea. Moreover, the imminent, nationwide Salk vaccine trial diverted the public's attention from GG's disappointing results. Trust in science in general and the National Foundation in particular outweighed such questions. As historian James Colgrove concludes, "favorable public attitudes towards the concept of immunization, fueled in large measure by advances in scientific medicine during and after World War II, set the stage for the most high profile medical saga of the twentieth century: the development of a polio vaccine."[27]

The gamma globulin episode encapsulated how, in a few decades, parents had shifted from a general skepticism of science, protesting against smallpox and diphtheria vaccinations, to submitting their children as test subjects and demanding injections of its latest discovery. Education had helped forge a formidable consensus. Further, the entire gamma globulin enterprise very much appeared as a dress rehearsal for the forthcoming Salk vaccine trials. Designs for the GG and vaccine injection at public school buildings were quite similar. More importantly, as Mawdsley concludes, the GG "project was a marketing success that boosted the preeminence of the National Foundation … and launched an era of large placebo-controlled trials."[28]

A Successful Crusade?

Both New York City's 1947 mass smallpox inoculations and the nationwide 1952 GG injections reflect significant public faith in medical science. Key questions remain, however: How could millions of Americans be so easily mobilized for the 1954 national polio vaccine experiment? How could it be done so efficiently? Historiography points to fear of this virus and the NFIP's public relations strategy as explanations. That is certainly borne out by the facts. But as organized as that campaign was the National Foundation alone could not marshal millions of parents on a nationwide basis, especially against the backdrop of the anti-fluoridation sentiment during the 1950s.

In 1949 and 1950, the US Public Health Service and American Dental Association, respectively, endorsed fluoridated drinking water, based on "preliminary reports from field trials in Grand Rapids, Michigan…. From 1953 to 1955, awareness of fluoridation spread, leading to adoptions in the South, East, and in areas of the Midwest…." Resistance manifested itself locally since the authority to add fluoride to the water system was decided and implemented at that level. Although a seemingly decentralized movement, a strong national network existed. From 1955 to 1988, a periodical, *The National Fluoridation News*, was published. Organizations included the American Foundation for Homeopathy, Massachusetts Citizens Rights Association, National Committee Against Fluoridation, and National Health Guardian. A highly vocal group of conservatives also railed against this violation of "individual liberties" and warned how it would lead to "totalitarianism." Since the "government, rather than a doctor, was prescribing medication … [f]luoridation was a form of socialized medicine…." Finally, a very small segment believed that it operated as a part of a "Communist plot to destroy America," echoing the sharp rhetoric of the notorious McCarthy era. The Soviet Union, they claimed, wielded fluoride as a weapon. Red saboteurs used the "fluoridation machinery at water treatment plants to deliver a lethal dose … to the unsuspecting public.… Others believed fluoride would poison its victims slowly, causing mental weakness, cancer, or sterility, and when the United States no longer had enough healthy young men to defend itself, the Soviet Union would invade." They equated this to a mass destruction scenario, analogous to the potential devastation wrought by Soviet atomic weapons. In sum, the fluoridation of drinking water was downright anti-American.[29]

Many of these assertions should have resonated with the American public. As it did some thirty years earlier, the Citizens Medical Reference Bureau, a staunch anti-vaccination organization, battled fluoridation, opposing government intervention into the private, medical affairs of individuals. "[M]any anti-fluoridationists also objected to vaccinations, in particular the polio vaccine."[30]

In spite of this shrill campaign, most parents submitted their children as human subjects for the polio vaccine trial. Granted, some may have seen opposition to fluoridation as a the antics of a fringe group and simply dismissed its shrill rhetoric. Other parents may have decided that the long-term threat of tooth decay proved more abstract and far less threatening to children's lives than a regular, summer peril like poliomyelitis. Fear, of course, fed a sense of urgency. Medical science though offered them protection.

This confidence, whether shaped through informal or formal education, certainly existed, but it is difficult to discern where one started and the other ended. Regardless, it appeared, ubiquitous.[31] As historians Russell Viner and Janet Golden point out:

> … fundamental changes occurred in the twentieth century that in turn transformed all children's experiences with medicine. These changes were in part quantitative: the number of contacts between children and health workers increased dramatically in the early decades of the century. They were also qualitative: medicine changed the way children experienced their lives–at home, at school, in their bodies, in their minds, in sickness, and in health.

As that century unfolded, children fell under a wide net of public health supervision involving "baby milk stations, settlement houses, health visitors, visiting nurses ... traveling vaccination clinics, school dentistry, social work...."[32]

It came as no surprise that *place* served as a crucial variable regarding the national, polio vaccine trial. Juveniles dominated elementary school buildings, providing an ample supply of human subjects. Through the first half of the twentieth century, public schools also grew to become institutions that oversaw the welfare of children, this included health needs. Therefore, bringing this experiment to school buildings, which children and parents saw as locations of safety and trust, made sense. Receiving injections there certainly connoted less of a threat or concern than in a laboratory, an unfamiliar and scary setting for young subjects. The stage had been set as Jonas Salk and the National Foundation entered with great fanfare.

Notes and Sources

1. "Trenton Enforces 'Polio' Quarantine," *New York Times*, August 6, 1945: 17. For epidemic refugees, consult Richard J. Altenbaugh, *The Last Children's Plague: Poliomyelitis, Disability, and Twentieth-Century American Culture* (New York: Palgrave Macmillan, 2015): 16–20.
2. George G. Deaver, "A Study of the Adjustment of 500 Persons Over Sixteen Years of Age with Disabilities Resulting from Poliomyelitis." Reprinted from *New York Medicine*, 7 (April 1951): no pagination. Further consult Sydney A. Halpern, *Lesser Harms: The Morality of Risk in Medical Research* (Chicago: University of Chicago Press, 2004): 140; Charles L. Mee, *A Nearly Normal Life* (Boston: Little, Brown & Co., 1999): 5–6; Thomas M. Rivers, *Reflections on a Life in Medicine and Science* (Cambridge: Massachusetts Institute of Technology, 1967): 498. Finally, for Zemke's story, see Diane Zemke, *Polio: A Special Ride?* (Minnetonka, MN: Diagnostic Center of Learning Patterns, Inc., 1997): 8, 10, 18.
3. Elizabeth W. Etheridge, *Sentinel for Health: A History of the Centers for Disease Control* (Berkeley: University of California Press, 1992): 67.
4. Ruth R. Faden and Tom L. Beauchamp, *A History and Theory of Informed Consent* (New York: Oxford University Press, 1986): 151–52.
5. For the entire 1947 New York City smallpox episode, consult Judith Walzer Leavitt, "'Be Safe. Be Sure.' New York City's Experience with Epidemic Smallpox," in Judith Walzer Leavitt and Ronald L. Numbers, eds., *Sickness and Health in America: Readings in the History of Medicine and Public Health* (Madison: University of Wisconsin Press, 1997): 407, 408 and 409, respectively. Refer as well to James Colgrove, *State*

of Immunity: The Politics of Vaccination in Twentieth-Century America (Berkeley: University of California Press, 2006): 74–78, 79–80; Michael Willrich, *Pox: An American History* (New York: Penguin Books, 2011): 33. Especially see the following *New York Times* articles: "Smallpox in City, Inoculation Urged," April 5, 1947: 21; "Many Vaccinated Against Smallpox," April 6, 1947: 27; "Smallpox Victim Hotel Guest Here," April 10, 1947: 27; 'Vaccinations Stop," April 16, 1947: 27; "Half Million Here Vaccinated," April 17, 1947: 23; "War on Smallpox Waged in Schools," April 21, 1947: 29; "Smallpox Danger Over," May 3, 1947: 15.

6. Stephen E. Mawdsley, "Fighting Polio: Selling the Gamma Globulin Field Trials, 1950–1953" (PhD dissertation, University of Cambridge, 2011): 38–39. Further consult pages 21, 37. Mawdsley has written the most complete, thorough, and compelling study of the gamma globulin episode, updating it with *Selling Science: Polio and the Promise of Gamma Globulin* (New Brunswick: Rutgers University Press, 2016). I cite both. Refer as well to Arthur Allen, *Vaccine: The Controversial Story of Medicine's Greatest Lifesaver* (New York: W. W. Norton Company, 2007): 19; Aaron Klein, *Trial by Fury: The Polio Vaccine Controversy* (New York: Charles Scribner's Sons, 1972): 60; Rivers, *Reflections on a Life in Medicine and Science*: 469, 475.
7. "Infantile Paralysis: Confidential Information Presented to Board of Managers on 4/24/52": 2–3. Joseph Stokes, Jr., Papers, Folder 6: National Foundation for Infantile Paralysis (Hereafter NFIP), APS. Also see Mawdsley, *Selling Science*: 16.
8. Mawdsley, *Selling Science*: 14, 21, and 22, respectively. Further consult Mawdsley, "Fighting Polio": 31, 35, 47, 49–52.
9. Mawdsley, "Fighting Polio": 53 and 71, respectively. Also peruse pages 55–59, 70, and Mawdsley, *Selling Science*: 26, 33, 101.
10. Mawdsley, *Selling Science*: 33–34. Moreover, check pages 35–36, and Mawdsley, "Fighting Polio": 72–74.
11. Mawdsley, "Fighting Polio": 101, 105, 106, and 122, respectively. See also pages 102, 103–4, 110, 112, and Mawdsley, *Selling Science*: 44–46.
12. Mawdsley, *Selling Science*: 82, 65, 66, and 56, respectively. Further check pages 54–55, 64, 77–80. Mawdsley includes a copy of the actual consent form on page 65. For the Watson form, refer to Box No. 1, File Folder No. 6, Salk Vaccine, RG 42, WIA.
13. Mawdsley, "Fighting Polio": 185 and 196–97, respectively. See also pages 184, 193–94, as well as Mawdsley, *Selling Science*: 86, 89. For primary documentation, consult MOD, Medical Programs Records; Series 5: Committees; Vaccine Advisory Committee, 1951–1960, and *Gamma Globulin Advisory Committee, 1944–1955.* Further check Allen, *Vaccine*: 177; Thomas M. Daniel and Frederick C. Robbins, eds.,

Polio (Rochester, NY: University of Rochester Press, 1997): 13–14; Charlotte DeCroes Jacobs. *Jonas Salk: A Life* (New York: Oxford University Press, 2015): 96–97; Jeffrey Kluger, *Splendid Solution: Jonas Salk and the Conquest of Polio* (New York: G. P. Putnum's, 2004): 143, 146; Mawdsley, *Selling Science*: 14, 21–22; Rivers, *Reflections on a Life*: 470–73, 482; Jane S. Smith, *Patenting the Sun: Polio and the Salk Vaccine* (New York: William Morrow and Co., 1990): 143; Heather Green Wooten, *The Polio Years in Texas: Battling a Terrifying Unknown* (College Station: Texas A&M University Press, 2009): 139.

14. Kluger, *Splendid Solution*: 146–47. Consult as well Wooten, *Polio Wars in Texas*: 139.
15. "Infantile Paralysis: Confidential Information Presented to Board of Managers on 4/24/52": 4. Stokes Papers, Folder 6: NFIP, APS.
16. Mawdsley, *Selling Science*: 99–100, 107, 109, 110, 113, and 115, respectively. See also pages 101–2, 104, 106–8, and Bert Hansen, *Picturing Medical Progress from Pasteur to Polio: A History of Mass Media Images and Popular Attitudes in America* (New Brunswick, NJ: Rutgers University Press, 2009): 245–46. Finally, refer to Mawdsley, "Fighting Polio": 210, 212–13, 215, 217.
17. Mawdsley, "Fighting Polio": 241, and 243, respectively. Further consult "For Mutual Newsreel, Wednesday, August 6, 1952, 7:45 p.m." Poliomyelitis (1951–1954), Folder 1: Poliomyelitis, 1951–1954, St. John to Melville, The Children's Hospital, May 19, 1952. Folder 8, and "A Discussion of Epidemiologic Problems of Mass Prophylaxis and Contact Prophylaxis of Poliomyelitis with Gamma Globulin," Stokes Papers, Folder 4: NFIP, APS. Secondary sources include Victor Cohn, *Four Billion Dimes* (Minneapolis: 1955): 98; Daniel and Robbins, *Polio*: 12–13; Tony Gould, *A Summer Plague: Polio and its Survivors* (New Haven: Yale University Press, 1995): 130–31; Klein, *Trial by Fury*: 60–62, 63, 70–71, 142; John R. Paul, *A History of Poliomyelitis* (New Haven: Yale University Press, 1971): 391, 393; Edith Powell Chappell and John F. Hume, *A Black Oasis: Tuskegee Institute's Fight Against Infantile Paralysis, 1941–1975* (Tuskegee, AL: Tuskegee University, 2008): 121–22; John Troan, *Passport to Adventure: Or, How a Typewriter from Santa Led to an Exciting Lifetime Journey* (Pittsburgh, PA: Geyer Printing, 2000): 201; Wooten, *Polio Wars in Texas*: 139–40.
18. Rivers, *Reflections on a Life in Medicine and Science*: 484–85.
19. David W. Rose, *Images of America: March of Dimes* (Charleston, SC: Arcadia Publishing, 2003): 62. Allen, *Vaccine*: 177.
20. Stokes Papers, Folder 9: NFIP, APS.
21. Weaver to Salk, November 18, 1952; Stokes to Salk, November 25, 1952; Salk to Stokes, December 1, 1952; Weaver to Stokes, December 4, 1952,

Stokes Papers, Folder 9: NFIP, APS. MOD; Salk and Sabin Polio Vaccine Records; Series 3: Salk Vaccine Gamma globulin activity, 1953–1958.

22. Paul, *A History of Poliomyelitis*, 393. See also "Polio: 1953," *Newsweek*, July 29, 953: 91. Salk File, Box 7, File Folder, Salk (Collection # 90/36/7), Jonas Salk Papers, UPITT; Klein, *Trial by Fury*, 71; and Troan, *Passport to Adventure*: 199.
23. "Panic Triumphant," *Newsweek*, September 7, 1953: 76, UPITT; Mawdsley, *Selling Science*: 122; Smith, *Patenting the Sun*: 250.
24. Mawdsley, *Selling Science*: 133.
25. US Department of Health, Education, and Welfare, "Gamma Globulin in the Prophylaxis of Poliomyelitis as Used in the United States, 1953." Report of the National Advisory Committee for the Evaluation of Gamma Globulin in the Prophylaxis of Poliomyelitis. Public Health Monograph No. 20 (Washington, DC: Government Printing Office, 1954): 21. Refer as well to pages iii, 1, 3, 9.
26. HEW, "Gamma Globulin in the Prophylaxis of Poliomyelitis": 21. Furthermore, consult Etheridge, *Sentinel for Health*: xv–xviii, 3, 67–70; Mawdsley, *Selling Science*: 135, 137–38.
27. Colgrove, *State of Immunity*: 112.
28. Mawdsley, "Fighting Polio": 254. See page 287 as well. In *Selling Science*, Mawdsley asserts the following: "The GG study served as a precedent for mass public experimentation and would be referenced by NFIP officials when they later considered the prospect of a large trial of the first polio vaccine" (93).
29. Gretchen Ann Reilly, "The Task Is a Political One: The Promotion of Fluoridation," in *Silent Victories*: 325, 329, and 330, respectively. Consult pages 324, 326, as well as Anne Taylor Kirschmann, "Making Friends for 'Pure' Homeopathy: Hahnemannians and the Twentieth-Century Preservation and Transformation of Homeopathy," in Robert D. Johnston, ed., *The Politics of Healing: Histories of Alternative Medicine in Twentieth-Century North America* (New York: Routledge, 2004): 34–35; Kluger, *Splendid Solution*: 210; Gretchen Ann Reilly, "'Not a So-Called Democracy': Anti-Fluoridationists and the Fight Over Drinking Water," in *The Politics of Healing*: 133, 136–38.
30. Reilly, "'Not A So-Called Democracy'": 134. See also page 136; Colgrove, *State of Immunity*: 124–26; Krischmann, "Making Friends for 'Pure' Homeopathy": 34.
31. Bert Hansen, "True-Adventure Comic Books and American Popular Culture in the 1940s: An Annotated Research Bibliography of the Medical Heroes," *International Journal of Comic Art*, 6 (Spring 2004): 117–18.

32. Russell Viner and Janet Goldman, "Children's Experiences of Illness," in Roger Cooter and John Pickstone, eds., *Medicine in the Twentieth Century* (Australia: Harwood Academic Publishers, 2000): 579 and 583, respectively. Further consult Colgrove, *State of Immunity*: 11–15; Bert Hansen, The Image and Advocacy of Public Health in American Caricature and Cartoons from 1860 to 1900," *American Journal of Public Health*, 87 (November 1997): 1804–5, and "Medical History for the Masses: How American Comic Books Celebrated Heroes of Medicine in the 1940s," *Bulletin of the History of Medicine*, 78 (2004): 180.

PART III

Ethical Authority?

CHAPTER 8

Mistakes and Misdeeds

The history of medicine, historian Roy Porter explains, is "extremely heterogeneous and subdivided, with numerous islands of knowledge linked by an erratic network of other temporary bridges. The picture is one of competition and cooperation, rivalry and symbiosis–universities, hospitals, research units, government funding agencies and schools of researchers interacting in ad hoc and unpredictable combinations."[1] Porter's generalization deftly characterizes four decades of polio research, one initially marked with little practical knowledge about viruses; any substantive advances would thus depend on the emerging field of virology. Regardless, many early researchers forged ahead with a false sense of certainty. This created serious risks for the children being tested.

Efforts to develop a safe and effective polio vaccine took place during the first half of the twentieth century, within the research traditions of their time, when the nature of human experimentation remained unresolved. On March 15, 1952, for instance, the *Lancet's* editors took a less than subtle jab at investigators using the term *volunteers*: "One of the reasons for the richness of the English language is that the meaning of some words is continually changing. Such a word is 'volunteer.' We may yet read a scientific journal that an experiment was carried out with twenty volunteer mice, and that twenty other mice volunteered as controls."[2] This prestigious medical journal raised a fundamental question: Who really acted as a volunteer in a scientific experiment? Jonas E. Salk, the investigator, and

R. J. Altenbaugh, *Vaccination in America*,
Palgrave Studies in the History of Science and Technology,
https://doi.org/10.1007/978-3-319-96349-5_8

the National Foundation for Infantile Paralysis (NFIP)/March of Dimes, which sponsored a series of local and national trials, found themselves in a formative period of medical ethics, a mere seven years after the release of the Nuremberg Code. What role did it play, if any, in this unprecedented mass, medical experiment?

Knowledge Is Power

Researchers employed divergent approaches in their quest to unlock the secrets of this enigmatic contagion. When viewed chronologically, some proved to be of limited value or misleading while others produced useful insights but were too often dismissed. What they did not know proved daunting and elusive. What they thought they knew was fraught with errors, resulting in flawed assumptions. Contradictions sowed confusion. In sum, serious problems plagued this entire "disease paradigm" before World War II.[3]

Ivan Wickman, a Swedish epidemiologist, methodically charted his country's 1905 polio epidemic and discovered three characteristics. First, it appeared to strike in irregular patterns. Yet, in spite of its randomness, the "search for a common source of infection led to the public school…." Second, it appeared to be a contagious disease. Austrian Karl Landsteiner confirmed this in 1908 when he successfully transferred poliovirus from a dead child's spinal cord to a monkey. Studies of Nebraska's 1909 and 1910 outbreaks corroborated this. Third, Wickman found this disease did not always result in paralysis. A 1910 study in St. Paul, Minnesota, arrived at the same conclusion.[4]

Simon Flexner, director of the Rockefeller Institute for Medical Research (RIMR), in New York City, appeared to successfully decipher this seemingly new and sudden disease. In 1909, he followed in Landsteiner's footsteps, demonstrating in the laboratory that it spread from one monkey to another. He also established the fact that this was not a bacterium but a virus. He, however, failed in his experiments to immunize monkeys. Nevertheless, "[t]o many at the time it appeared that Flexner, who had recently prepared a successful antiserum against … meningitis, was on the edge of conquering polio."[5] He and his associates published fifteen articles in the *Journal of the American Medical Association* between 1909 and 1913, establishing the Rockefeller Institute as the center of polio research in the USA, supplying poliovirus samples to researchers everywhere.[6]

Founded by John D. Rockefeller, RIMR was incorporated on June 14, 1901. New research centers in Europe, including the Koch and Pasteur institutes, had been successfully applying laboratory science to understanding diseases. Following their lead, the Rockefeller Institute became the first biomedical research center in the USA, publishing its own periodical, the *Journal of Experimental Medicine*. "[W]holly devoted to vivisectional medicine and science," with its own breeding farm, it spurred the inception of the Society for the Prevention of Abuse in Animal Experimentation in 1907, based in New York City, and the New York Anti-Vivisection Society in 1908.[7]

Flexner, based on his successful breakthrough against meningitis and the Institute's superbly equipped and well-staffed laboratories, confidently stated that infantile paralysis too would be controlled. The general public and medical community held high expectations, fully expecting him to find a cure, and quickly. On July 20, 1916, at the height of New York City's polio epidemic, the *New York Times* announced to its readers that Flexner was indeed close to a breakthrough.[8]

He had already committed Rockefeller Hospital to clinical investigations of polio. That facility, crucial to the Institute's mission, opened in 1910, a place where researchers could bridge the gap between science and bedside care, studying diseases both in the laboratory and as they manifested themselves in patients. It would serve as a model for dozens of subsequent clinical research centers. Flexner commissioned a study of the 1911 New York City epidemic based on 161 cases observed at that hospital, published the following year as a monograph, "A Clinical Study of Acute Poliomyelitis." Coauthored by Francis W. Peabody, George Draper, and Alphonse R. Duchey, it represented the first thorough and comprehensive study published in English: the cutting edge of polio research at that time.

It had serious limitations, however. John R. Paul, a pioneer polio researcher, asserts: "the reason for this lay in Flexner's rigid separation of the character of the work carried out in the Rockefeller Hospital as opposed to his own research laboratory in the institute." Paul continues his insightful analysis of Flexner's complex character: "He supported clinicians to the limit of his own beliefs, but he clung to the idea that physicians were supposed to practice their craft at the bedside–not in the experimental laboratory. They were not even allowed access to monkeys to perform clinical virological tests." As a result, the Peabody, Draper, and Duchey report appeared to be intellectually detached. It provided remarkably detailed descriptions of the patients' symptoms and analyses of their blood and spinal fluids, yet less than four percent of the report dealt with the "methods of treatment," which Paul generically summarizes as "bed rest, mild sedative, medication for the control of pain." Otherwise, doctors and nurses should treat it as "any other contagious disease" through quarantine and disinfection. Paul harshly concludes that this study "was not so novel or so important that it deserves a particularly high place in the history of poliomyelitis." Flexner's leadership in this field discouraged investigators from testing any experimental immunogen on humans. His cautious attitude reflected two concerns. First, an insufficient number of animal trials had been completed. Second, infantile paralysis maintained a low morbidity rate compared with other contagious diseases; in short, the real risks simply did not outweigh the possible gains. Flexner, for all intents and purposes, unilaterally imposed restrictions on human trials, maintaining this position until his retirement in 1935, about the same time as the disastrous Park-Brodie and Kolmer polio vaccine trials. In the end, Flexner's early research enthusiasm dissipated as his efforts to find a cure eluded him; this was due, in large part, to a "certain rigidity in clinging to ideas he considered sacrosanct."[9]

Wade H. Frost, of the US Public Health Service (USPHS), compiled data for outbreaks in 1910 in Mason City, Iowa, 1911 in Cincinnati, Ohio, and 1912 in Baltimore, Maryland, as well as Batavia and Buffalo, New York, employing "clinical observations, statistical analyses, and laboratory experiments on the sera he collected." He amassed mountains of evidence, created spot maps, compared urban and rural cases, checked meteorological patterns, accounted for sanitation conditions, and tracked insect infestations. His pioneer epidemiological study of poliomyelitis, concluded in 1913, asserted that it spread by human contact, primarily through the digestive system.[10]

In 1894, Charles S. Caverly, President of the Vermont State Board of Public Health, began to publish comprehensive, annual reports on outbreaks in that state. By 1911, he acknowledged the growing threat of polio, a "disease that is so obviously spreading and recurring in epidemic form…." At the same time, he saw promise because the "recent intense study of poliomyelitis has not been without results and promises soon to solve the mysteries surrounding its causation and spread," stressing the work of Landsteiner and Flexner.[11]

Finally, in 1931, Flexner wrote a retrospective, "Poliomyelitis (Infantile Paralysis)," summarizing the body of knowledge to that point, though it represented more than a simple compilation of information. While the piece echoed the authority of Flexner's many years of research experience, it contained numerous ambiguities and errors. Polio infections resulted in immunity, marked by the presence of antibodies. Based on this assumption, the blood of those previously infected could be processed to create a serum, reflecting a centuries-old belief. Blood serum originated in 1758 when Francis Howe, a Scottish physician searching for a measles cure, inoculated twelve children with blood from infected individuals. A similar approach with polio showed promise at the Rockefeller Institute. First used on monkeys in 1910, this "convalescent serum" began to be used on humans a year later. Not only did it exert "curative (therapeutic) properties," Flexner declared, "but [it appeared] to posses preventive powers as well." Yet, at the same time, he equivocated: "No assurance of absolute protection can, of course, be given, but by analogy with measles, benefit may be hoped for or even expected. Time and experience alone will make it possible to ascertain the value of this procedure."[12]

Flexner also erred, according to historian Margaret L. Grimshaw, about how this virus invaded the body. Based on his early findings from animal experiments, Flexner posited that polio entered through the nasal passages. On the other hand, Frost's data pointed to the digestive system. "In short, scientific specialization had discouraged the microbiologist and epidemiologist from seriously considering evidence in studies conducted outside their respective disciplines." Flexner's interpretation, because of his stature, dominated, but Frost's proved correct.

Other researchers, meanwhile, had begun to unlock the incredible complexity of this disease. In 1928, Grimshaw points out, W. Lloyd Aycock at the Harvard Infantile Paralysis Commission confirmed that passive immunization occurred independent of epidemics through an ongoing, subclinical process. The Yale Poliomyelitis Unit, led by John Paul and James D. Trask, based on data they accumulated between 1931 and 1932 during epidemics in the Northeast, confirmed Ivan Wickman's 1905 finding of abortive paralysis. By 1937, they demonstrated that immunization depended on its various strains; that is, "poliomyelitis was primarily a systemic illness caused by a family of antigenically different viruses rather than a single, homogeneous neurotropic virus." Moreover, they challenged the nasal route of infection.

This led Flexner to recruit Albert Sabin to confirm his olfactory passages theory. By 1940, after performing a series of systematic experiments, Sabin reported that no evidence existed to support it. This sparked a sea change. By 1942, "America's medical research community … generally accepted the view of poliomyelitis as an intestinal, often inapparent viral infection caused by a family of immunologically unique polioviruses.... Flexner graciously accepted the inevitable. Many of his ideas about the disease were abandoned and major medical textbooks were rewritten."[13]

Flexner and the Rockefeller Institute had so completely dominated the field that as the NFIP emerged in 1938 to pursue a vaccine, its leaders found themselves hard pressed to find any other expertise. An Australian researcher, visiting the USA in 1945, observed a high sense of "'frustration amongst most [American scientific] workers'" regarding polio research. "'It has proved very difficult,'" he continued, "'to put together a clear picture of how infection occurs in infantile paralysis, and even more difficult to provide any methods for effective administrative action against the disease.'"[14] Four decades of hypotheses and investigations failed to produce significant insights into this contagion. Nevertheless, this did not inhibit polio researchers from pursuing experiments on humans with their ill-conceived concoctions during that period.

Vulnerable Populations

Laboratory research did not become a routine part of the of the medical community's activities until the latter part of the nineteenth century. Not surprisingly, the American Medical Association's (AMA) Code of Ethics, approved unanimously in 1847 at its founding conference in Philadelphia, "did not explicitly address the use of human subjects in research. Like the Hippocratic Oath and the English physician Thomas Percival's 1803 code on which it was modeled, the AMA's code of ethics focused on the physician's responsibilities to patients at the bedside and in the consulting room."[15] They should "unite *tenderness* with *firmness*, and *condescension* with *authority*, as to inspire the minds of their patients with gratitude, respect and confidence." Furthermore, medical historian Susan E. Lederer notes, doctors needed to avoid dispensing negative information to patients, operating as the "minister of hope and comfort to the sick" in order to "smooth the bed of death." Concerning peers, it spelled out expected conduct: "There is no profession, from the members of which greater purity of character, and a higher standard of moral excellence are required than the medical…." This stood until 1902, when the only significant change appeared in the title, *Principles of Medical Ethics*. For patients, "[s]ecrecy and delicacy should be strictly observed." Such "benevolent deception" calmed patients. Depressing news about their frail condition, after all, would cause "mental anguish" and thereby hasten death. In all cases, these *Principles* stipulated, physicians should act with "dignity and honor."[16]

AMA leadership occasionally revisited these *Principles*, revising and amending them, but the "viewpoint on the patient-physician relationship remained largely unchanged."[17] The association did recommend guidelines regarding animal tests in 1909; most medical research institutions complied. However, seven years later when it attempted to enact a similar set of rules for humans, it failed. Questions arose over "therapeutic and nontherapeutic experiments…." and researchers debated the meaning of "patient consent," as Lederer outlines:

> When a pediatrician conducted tests of a new vaccine in a state-sponsored orphanage was permission needed? Who should authorize the children's participation? Was patient permission really necessary for such relatively benign procedures as urine testing or obtaining a small blood sample? … Providing the necessary information to enable the patient to make an informed decision, these investigators argued, would not only burden patients and their doctors but interfere with the vital progress of medical knowledge.

No major changes ensued. Fundamental issues reappeared in the 1930s. Foremost among them, how was consent established? Scientists avoided this issue altogether by using institutionalized children as test subjects, thus completely dispensing with parental permission. Another key question remained: What constituted risk? Researchers routinely dismissed any dangers children may face with "untested vaccines and sera."[18] In sum, the USA had no codified guidelines to protect human subjects until the 1970s. Meanwhile, the mistreatment of human subjects continued.

Both anti-vivisectionists and scientists, according to Lederer, resorted to the media to influence public opinion. The former used public displays of dissected animals, published pictures in magazines and pamphlets, and produced films. The American medical science community guarded itself by the careful selection of photographs. A 1938 issue of the popular photojournal *Life* magazine exploited images of children in distress, paralyzed by infantile paralysis, to cultivate pity. First, it illustrated how Philip Drinker had used cats to develop his artificial respirator, and second, it portrayed "four children suffering from infantile paralysis ... encased in iron lungs...." This manipulative, visual array played on the heartstrings of readers, effectively delivering the message that animal research saved children's lives. It went even further. Sensitive to cinematic projections of researchers, their work, and especially their use of animals and humans, they negotiated with major studios. And Hollywood complied. The film *Arrowsmith* (1931) depicted a young investigator struggling to reconcile the "goal of the scientist to advance knowledge with the responsibility of the physician to heal a patient." Testing his plague serum on a Caribbean island, he is torn between maintaining a control group, leaving them vulnerable or injecting all volunteers in the hope that he will successfully protect them. He chooses the latter never knowing if it actually worked: a tragically flawed decision assures scientific uncertainty. This theme of personal and professional struggle, an excruciating choice between experimentation or healing, permeated many of these pictures. Nonfiction movies, like *The Story of Louis Pasteur* (1936), *Yellow Jack* (1938), and *Dr. Ehrlich's Magic Bullet* (1940), went so far as to cast children to portray test subjects. All three taken together, Lederer summarizes, "celebrated heroic medical research and the explicit use of 'human guinea pigs' in the advance of medical knowledge." She adds that the enforcers of Hollywood's 1930 Hays Motion Production Code seemed to be more concerned about scenes depicting animals more so than human experimentation. Ultimately, the research community prevailed in its use of this new medium to propagate "ideas about medical research in the 1930s and 1940s."[19] From the beginning of the century to the development of flu vaccines and antibiotics in the 1940s, medical science seemed poised to solve all health threats. Concomitantly the "[c]ontrolled testing of experimental drugs in human beings" steadily grew.[20]

Notes and Sources

1. Roy Porter, *The Greatest Benefit to Mankind: A Medical History of Humanity* (New York: W. W. Norton Company, 1997): 534.
2. Anon. "Poliomyelitis: A New Approach," *Lancet* (March 15, 1952): 552.

3. Margaret L. Grimshaw, "Scientific Specialization and the Poliovirus Controversy in the Years Before World War II," *Bulletin of the History of Medicine*, 69 (Spring 1995): 44.
4. Ivan Wickman, *Acute Poliomyelitis* (1913; Rpt. New York: Johnson Reprint Corporation, 1970): 103–4. Refer to pages 110, 112–13, as well as Saul Benison, "Speculation and Experimentation in Early Poliomyelitis Research," *Clio Medica*, 10 (1975): 11.
5. Benison, "Speculation and Experimentation in Early Poliomyelitis Research": 1. See also pages 12–13, and George Draper, *Acute Poliomyelitis* (Philadelphia: P. Blakiston's Son and Co., 1927): 4; Simon Flexner and Paul A. Lewis, "Experimental Poliomyelitis in Monkeys," *Journal of the American Medical Association*, 54 (May 1910): 1780–82; "Cause of Infantile Paralysis a Germ: Animal Experiments Reveal What Even a Microscope Cannot Detect, Says Dr. Flexner," *New York Times*, March 12, 1911, Sunday Edition: 7.
6. John R. Paul, *A History of Poliomyelitis* (New Haven: Yale University Press, 1971): 115. Nobel prize winning medical scientist David Bodian, "Poliomyelitis and the Sources of Useful Knowledge," *Johns Hopkins Medical Journal*, 138 (April 1976), unequivocally states: "Experimental polio research" officially began in 1908 with the "first successful transmission of the disease to monkeys" (130).
7. Susan E. Lederer, *Subjected to Science: Human Experimentation in America Before the Second World War* (Baltimore: Johns Hopkins University Press, 1995): 77. Also consult *Scientific Reports of the Corporation and the Board of Scientific Directors of the Rockefeller Institute for Medical Research*, Vol. 1 (1909–11), RG 439, Board of Scientific Directors, Annual Scientific Reports, Rockefeller University Archives, RAC.
8. "Gain in Paralysis Due to a Holiday," *New York Times*, July 20, 1916: 11. Further check Bert Hansen, *Picturing Medical Progress from Pasteur to Polio: A History of Mass Media Images and Popular Attitudes in America* (New Brunswick, NJ: Rutgers University Press, 2009): 111, 114; Paul, *History of Poliomyelitis*: 105.
9. Paul, *History of Poliomyelitis*: 118–19, 121, 123, 124. Consult Francis W. Peabody, George Draper, and Alphonse A. Duchey, "A Clinical Study of Acute Poliomyelitis," *Monograph of The Rockefeller Institute for Medical Research*, no. 4, 1912, Reprints Vol. 61 (New York: RIMR, 1913); *Scientific Reports of the Corporation and the Board of Scientific Directors of the Rockefeller Institute for Medical Research*, Vol. 1 (1909–11), as well as Saul Benison, "Poliomyelitis and the Rockefeller Institute: Social Effects and Institutional Response," *Journal of the History of Medicine*, 29 (January 1974): 74–76, 84.

10. Paul, *History of Poliomyelitis*: 142. Further consult page 137. Paul was a member of the Yale Poliomyelitis Study Unit that later confirmed the existence of poliovirus in stool and sewerage samples (284, 286–89).
11. Robert W. Lovett, *Infantile Paralysis in Vermont, 1894–1922: A Memorial to Charles S. Caverly* (Burlington, VT: Vermont State Department of Health, 1924): 39–40. Also see page 69. This collection includes reprints of Caverly's annual reports published in a variety of professional journals and proceedings.
12. Simon Flexner, "Poliomyelitis (Infantile Paralysis)": 1–7, box 27, RG 210.3 Administration (Operations), Business Manager Files, Rockefeller Institute for Medical Research, Rockefeller University Archives, RAC. A handwritten date of August, 1931 appears on the first page. Paul, in his classic, *History of Poliomyelitis*, harshly criticizes Flexner's early mistakes, arguing that they seriously delayed the development of a way to counteract this disease (114–15). But Benison, "Poliomyelitis and the Rockefeller Institute," hints that Paul wrote from a presentist perspective (19–20). He asserts, instead, that given the limited technology and knowledge that Flexner was working with, he could not be faulted. Susan E. Lederer and Michael A. Grodin, "Historical Overview: Pediatric Experimentation," in Michael A. Grodin and Leonard H. Glantz, eds., *Children as Research Subjects: Science, Ethics, and Law* (New York: Oxford University Press, 1999), page 5, and Andrew Cliff, Peter Haggett, and Matthew Smallman-Raynor, *Measles: An Historical Geography of a Major Human Viral Disease, from Global Expansion to Local Retreat, 1840–1990* (Cambridge, MA: Blackwell Publishers, 1993), page 218, allude to early blood serum experiments. Finally, see Thomas M. Rivers, *Reflections on a Life in Medicine and Science* (Cambridge: Massachusetts Institute of Technology, 1967): 228.
13. Grimshaw, "Scientific Specialization and the Poliovirus Controversy": 55–56, 63, 65, respectively. Check pages 57–59, 62, 64, as well as Wickman, *Acute Poliomyelitis*: 110.
14. This Australian visitor is quoted by Bodian, "Poliomyelitis and the Sources of Useful Knowledge": 133.
15. Lederer, *Subjected to Science*: 97–98. Further check American Medical Association, *Code of Medical Ethics* (Chicago: AMA, 1847), http://www.ama.assn.org/ama/pub/about-ama/our-history-ama (Accessed January 24, 2014); Robert B. Baker, *Before Bioethics: A History of American Medical Ethics from the Colonial Period to the Bioethics Revolution* (New York: Oxford University Press, 2013): 102, 145, 147–49; Robert B. Baker, et al., eds., *The American Ethics Revolution: How the AMA's Code of Ethics Has Transformed Physicians' Relationships to Patients, Professionals, and Society* (Baltimore: Johns Hopkins University Press, 1999): xxv, xxvii–xxviii.

16. *Code of Medical Ethics of the American Medical Association* (Chicago: American Medical Association, 1847): 93, 94, 98, respectively. Emphasis is in original. *The Principles of Medical Ethics* is not paginated. Also check the timelines at this site. American Medical Association, http://www.ama-assn.org (Accessed October 2, 2007 and April 10, 2016). The "benevolent deception" quote by Lederer is on page 14, *Subjected to Science*. Finally, consult Baker, et al., *American Ethics Revolution*: xxx, as well as Susan E. Lederer, "Medical Ethics and the Media: Oaths, Codes, and Popular Culture," in *The American Ethics Revolution*: 98–99.
17. Faden and Beauchamp, *A History and Theory of Informed Consent*: 74. Refer to page 73, as well as American Medical Association, http://www.ama-assn.org; Baker, *Before Bioethics*: 214; Baker, et al., *American Ethics Revolution*: xxx–xxxi.
18. Lederer, *Subjected to Science*: 74, 107, respectively. See pages 104–5, as well as Roger Cooter, "The Ethical Body," in Roger Cooter and John Pickstone, eds., *Medicine in the Twentieth Century* (Australia: Harwood Academic Publishers, 2000): 459; Louis Lasagna, "Special Subjects in Human Experimentation," *Daedalus*, 98 (Spring 1969): 457; Susan E. Lederer, "Children as Guinea Pigs: Historical Perspective," *Accountability in Research*, 10 (2003): 8; Charles E. Rosenberg, "Codes Visible and Invisible: The Twentieth-Century Fate of a Nineteenth-Century Code," in Baker, et al., *American Ethics Revolution*: 210.
19. Susan E. Lederer, "Hollywood and Human Experimentation: Representing Medical Research in Popular Film," in Leslie J. Reagan, Nancy Tomes, and Paula A. Treichler, eds., *Medicine's Moving Pictures: Medicine, Health, and Bodies in American Film and Television* (Rochester, NY: University of Rochester Press, 2007): 294–95, 296, 301, respectively. Refer also to pages 282–84, 286, 288–89. For the Hays code, see, http://www.artsreformation.com/a001/hays-code.html (Accessed June 6, 2010).
20. Mariner, "AIDS Research and the Nuremberg Code": 288.

CHAPTER 9

Blood

Harry K. Beecher, an anesthesia and pain researcher at the Massachusetts General Hospital in Boston, published a 1966 article, "Ethics and Clinical Research," that reported the disturbing persistence of uninformed human subjects in American medical experiments. This represented a universal phenomenon, encompassing the armed services, hospital facilities, university laboratories, and public and private organizations. Since World War II, federal funds available for research had increased exponentially. Within this context, Beecher pointed to a tension between the advancement of "medical sciences" and "interests of the patient." He found it difficult to accurately assess the level of unethical human experimentation, but settled on a low range of 3–12%. Beecher substantiated his claim by summarizing twenty-two public and private research studies, a variety of tests conducted on adults and children, including two-day-old infants. The "patients" did not even know that they were participating in experiments, which consisted of withholding antibiotics for bacterial infections, inducing infections, injecting cancer cells, performing a variety of surgical procedures, and testing X-rays on human subjects. Sicker patients, more intense and longer hospitalization, and higher mortality rates resulted.[1]

A certain degree of risk was (and is) certainly involved in any medical research. However, existing histories of polio often imply a cliché-type of approach: *the end justified the means*; that is, Salk's polio vaccine worked. They dwell on the dispute between Jonas E. Salk's observed approach

R. J. Altenbaugh, *Vaccination in America*,
Palgrave Studies in the History of Science and Technology,
https://doi.org/10.1007/978-3-319-96349-5_9

versus Thomas Francis's double-blind protocol. Analyses of the national test thus founder on the process rather than those being tested. The research culture at that time remains virtually unexamined. This oversight is crucial because it suffers from a presentist perspective. John Paul's 1971 classic, *A History of Poliomyelitis*, is a case in point. His is an enigmatic analysis: On the one hand, he offers a damning appraisal of the experiment's design that used children on such a large scale; on the other, he justifies it because of the dire need for a vaccine and its subsequent success.[2] Indeed, the stakes could not have been higher. The success of this mass experiment on young children could be nothing less than miraculous; its failure could prove to be barbaric. The outcome remained unknown. And ethics played a vital role. But what was that context?

Four chronologically overlapping, ethical patterns provide the scaffolding to reconstruct the culture of medical research during the first half of the twentieth century in general and poliomyelitis in particular: natural experimentation; moral ethics; the Nuremberg Code; and *laissez faire*. They follow an approach known as, ethicist Gregory E. Pence's terms it, "case-based reasoning." Moral analysis is grounded on "paradigms or model cases." Similar experiences permit generalizations, allowing us to better grasp the fundamental meanings of context: "Each situation or case will present a unique array of people, interests, conflicting principles, incompatible role-duties, strong passions, and concerns about the larger good, about resources, about institutional policies, and about political consequences."[3]

Natural Experiments

Many researchers used an organic justification to carry out risky investigations on human beings of all races and ages: They passively watched how a disease ran its full course, even if it resulted in death, in order to identify all of the symptoms and the various stages of infection.[4] This operated as a unilateral determination, with the subject completely outside of the decision-making process, oblivious to the fact that he or she was involved in an experiment.

The syphilis and hepatitis observations at Tuskegee Institute and Willowbrook State School, respectively, serve as stark illustrations. The US Public Health Service (USPHS) launched the "Study of Syphilis in the Untreated Negro Male" in 1932, rooted in a racially based stereotype of sexual promiscuity. Researchers offered some 600 poor sharecroppers, residents of Macon County, Alabama, free care to facilitate the study. They used Tuskegee's medical center as their base of operations to reinforce the illusion of treatment while recording how syphilis ravaged their bodies. They also dispensed placebos, like "aspirin, iron tonic, and vitamins," to add to this deception. Public Health Service authorities further enticed them by providing free burial services, but solely to ensure that they could perform autopsies and gather additional data on the impact of unfettered syphilis. In 1936, these observers presented their early findings at the American Medical Association's (AMA) annual meeting. No one raised any objections to the study. The experiment proceeded. In 1958, "the USPHS awarded a certificate of appreciation, signed by the surgeon general and replete with a gold seal, to each infected man, along with twenty-five dollars–a dollar for each year of the study." The observations and dissections continued. Eleven years later, the Centers for Disease Control (CDC) endorsed the study, all the while that disease exacted a horrible price. By 1972, at least twenty-eight of the original subjects had died from syphilis. One lost his sight. About sixty of the subjects' wives and children suffered infections.[5]

The Willowbrook experiments exploited institutionalized children. New York built Willowbrook, as an institution for the "retarded," on Staten Island in the early 1940s. By 1956, it held 3600 inmates, about 600 over capacity. This figure jumped to 6000 in 1963; even with an expansion of the facilities, it remained grossly overpopulated, producing deplorable conditions. Because of inadequate staffing, patients lay naked in their own feces. The smell of urine and excrement permeated this dark and dreary building. Wailing flooded the hallways and rooms while the children suffered injuries from violent residents. These wounds compounded with unsanitary conditions produced open sores. Diseases, like measles, pneumonia, and shigella, besieged the children. Parasites abounded. And hepatitis infections became routine. This provided the perfect petri dish for Saul Krugman.

He began his pediatrics career at Bellevue Hospital and as a member of New York University's medical school in 1947, becoming the consulting physician for Willowbrook seven years later. His use of that institution's children involved three distinct stages spread over several years. The first focused on hepatitis. It struck Willowbrook's patients at a rate of 25:100 in 1955 compared to 25:1000 for New York state's general population. Krugman saw a basic research question: Would immune gamma globulin (GG) act as a prophylactic against hepatitis? In 1956, his research team injected "various doses" of GG into about 1800 newly admitted children while withholding it from another cohort of 1800. Those inoculated remained virtually unscathed, with an infection rate of 2:1000, but the control group attained a morbidity ratio of 23:1000. While GG provided a temporary reprieve, what would occur in the long run? Krugman and his colleagues took what they viewed as the next logical step and attempted to use it in conjunction with live virus to develop permanent immunity. They again split newly admitted patients, aged three through seven, into two groups and isolated them from that institution's general population. The first received GG injections and were fed live hepatitis virus, "obtained from the feces of Willowbrook hepatitis patients." The second set, likewise exposed to hepatitis, served as Krugman's control group and received no GG; in short, they had been systematically infected. At intervals of six, nine, and twelve months, the researchers fed the first group additional doses of the live virus. Krugman discovered an anomaly, however. Even infected children, who supposedly had developed a natural immunity, fell ill again to what appeared to be hepatitis. He would address this conundrum later.

His second investigation involved a prototype measles vaccine. Although this somewhat strayed from the natural experiment paradigm, it callously illustrated how easily children under state authority routinely became test subjects. John F. Enders and his Harvard University colleagues in 1960 developed a relatively untested immunogen. In an attempt to reduce measles outbreaks at Willowbrook, Krugman contacted Enders to employ it. It seemed to work, with those injected remaining disease free. Based on this trial, he proceeded to inoculate 90% of Willowbrook's residents by 1963 with this "experimental vaccine" before it had even been licensed.

Krugman's third project returned to the perplexing hepatitis question. His team purposefully infected groups of newly admitted children in order to observe their symptoms and relapse rates. The results led to the discovery of two hepatitis strains: A and B. Therefore, in his original hepatitis study, the boys did not experience a reversion but had contracted another virus type. Krugman received accolades from the medical community, most notably the editors of the *Journal of the American Medical Association* and *New England Journal of Medicine*, who touted his findings based on "'experiments in nature.'"

Neither Tuskegee nor Willowbrook focuses on polio research, but the context remains germane. First, they exemplify how scientists routinely exploited vulnerable and captive populations for their own purposes. Second, these incidences occurred concurrently with the polio vaccine trials, providing insight into the general nature of the broader ethical context. Third, they reveal how medical investigators rationalized ethical concerns. African-American males in the Tuskegee study already suffered from syphilis; scientists merely wanted to monitor the normal course of that disease. Willowbrook's children, because of overcrowding, poor hygiene, and under staffing, unavoidably fell prey to hepatitis. Thus, infecting newly admitted children only followed what would occur anyway. In their minds, these researchers had done nothing inappropriate, and they claimed they "could not alter the outcome," merely observing the inevitable course of events as they normally unfolded.[6]

However, they were not helpless bystanders. Known treatments and preventative measures existed; rather, social deprivation, i.e., "poverty, ignorance, filth, and institutional miseries," profoundly influenced these endeavors. They pursued these particular subjects because they represented "devalued" members of society. As such, scientists could easily and justifiably "manipulate" their consent. "In Tuskegee ... blacks were informed that they had 'bad blood,' not syphilis, and so were ignorant of the potential risks of contagion." At Willowbrook, the "consent form that parents signed to allow their children to be infected with the virus read as though their children were to receive a vaccine against the virus."[7]

Moral Ethics

Historian Sydney A. Halpern presents three reasons for the resilience of "moral tradition," which served as another, albeit informal, operational control over human experimentation during most of the twentieth century. First, expertise conferred authority on physicians, i.e., those who "command and decide," while patients remained subordinate and compliant. Second, investigators generally sought consent from healthy subjects, or if a test might cause undue discomfort. "But when the subject was a patient and the medical procedure was expected to yield therapeutic benefit, physician beneficence ruled and researchers did not consider consent to be mandatory." Experiments thus became conflated with treatments as the decision-making process fell to doctors alone. Third, the concept of "lesser harm" meant that clinical tests could be implemented if they caused less risk than the disease itself, and resulted in a possible, meliorative outcome.[8] This was especially true for early polio treatments.

In July 1916, New York City's Special Committee, appointed by the mayor to combat the raging epidemic, endorsed research in order to mitigate future, large-scale paralysis of young children, often triggered by this disease. These involved either "physiological treatments" or serum therapy. Samuel J. Meltzer, at the Rockefeller Institute for Medical Research (RIMR), used adrenaline to treat infected monkeys. He hypothesized that the "congested membranes of the cord could be restored to their normal physiological condition by [its] introduction ... into the spinal canal after lumbar puncture."[9] News of this approach quickly reached the public. On July 14, a *New York Times* front-page headline gave epidemic-weary readers hope: "Offers New Cure in Fight Against Infant Paralysis." At a special meeting at New York City's Academy of Medicine, where physicians gathered to share ideas and information about the ongoing outbreak, Meltzer urged intraspinal injections of adrenaline to prevent paralysis. The next day, the *Times* ran another front-page headline heralding the possibility of a "remedy." The following Sunday, July 16, Louis C. Ager, cooperating with that city's department of health, announced that he had field tested Meltzer's laboratory results. His sudden revelation of this experiment, which took place at the Kingston Avenue Hospital where he served as "physician in charge," failed to reveal the total number of children involved, but the *Times* quoted him as describing it as "'a reasonable trial.'" This ambiguity continued when he stated that no clear results had been observed and added that any reliable conclusions could not be drawn for two or three months. E. J. Bermingham, chief surgeon at the New York Throat, Nose, and Lung Hospital, championed this treatment, telling the *Times*: "'We feel that Dr. Meltzer's theory has been sustained.... [Each child] had these intra-spinal injections administered every six hours, day and night. We have had no ill effects, and the improvement in all cases has been so marked that all acute symptoms have subsided....'" He also used the *Times* to announce that "'Dr. Ager [had] used adrenalin in the Kingston Avenue Hospital in Brooklyn at the request and under the direction of Dr. Meltzer.'" The *Times* further disclosed that physicians at the Willard Parker Hospital had implemented this approach.[10] In the end, this treatment proved to be an utter failure. Meltzer gave adrenaline intraspinally to eighty-one children at Riverside Hospital; thirty-five died, a 45% mortality rate. He also injected adrenaline intramuscularly to an additional twenty-four young patients; fourteen of them died, a 67% death rate.[11] The total number of children used in this therapeutic trial at all of the hospitals remained unknown.

Another physiological treatment involved lumbar punctures. In one case, physicians withdrew spinal fluid from a patient and re-injected it "'subcutaneously or intramuscularly into the same patient.'" This procedure, city health department head Haven Emerson explained to the *Times*, would produce antibodies: "'the method is one of active immunization.'" In a similar, but more simplified, approach, doctors removed spinal fluid from patients' spinal cords in order to relieve, they believed, abnormal stress on it. As Emerson observed: "'... this has been followed by a striking improvement, possibly through the relief of excessive intraspinal pressure.'" Emerson concluded that "'physicians in charge of hospital patients are unable to say what value, if any, the foregoing treatments actually possess'" but they were, he stressed, "'based on sound scientific principles.'" In both approaches, virtually all those treated remained paralyzed.[12]

Most physicians and researchers though seemed to gravitate toward serum therapy to avoid paralysis. Simon Flexner, the Rockefeller Institute's director, and his associates conducted the earliest experiments. Based on a similarly successful approach he had used for meningitis and an apparent breakthrough by a French researcher on thirty human subjects who had contracted polio, Flexner pursued it with laboratory monkeys in 1910. He found that the blood of a recovered monkey injected into those infected seemed to prevent paralysis, a "passive serum protection" as he termed it. This research did not long remain in RIMR's laboratory. Several distinct trials occurred over the next twenty years.[13]

Serum preparation involved several stages. The blood was allowed to clot and then cooled in an "ice box" for twenty-four hours to facilitate the separation of the serum. It was then decanted, placed in a centrifuge to remove residual blood clots and red cells, treated with trikresol, a preservative, chilled for another twenty-four hours, and finally filtered. In some cases, when the serum was needed immediately, laboratory workers simply allowed it to separate for a few hours and then injected it into the patient.

William H. Park, as head of New York City's Bureau of Laboratories, at the department of health, had been regularly using child patients at

Willard Parker Hospital to test various experimental vaccines and sera, soon earning fame for his diphtheria breakthrough. That institution became the main site for the use of convalescent serum therapy. Abraham Zingher, Park's colleague, employed three types: horse blood; normal human blood, drawn from donors who never had polio; and blood provided by volunteers who had recovered from polio. Zingher and his team drew blood from patients at Willard Parker, Kingston Avenue, Riverside, and Queensboro hospitals to pave the way for a large trial comparing three groups injected intrathecally (spinal cord injection). Zingher injected three of them with the normal horse type, 34 with the normal human version, and 93 with immune human serum, and 12 with an "unspecified" substance. He gave 88, or 62%, of the total 142 children about 15 cc. each within twelve hours of being admitted to the hospital. After one injection, body temperatures dropped in 18 patients, rose in 54, and remained constant in 54. Zingher injected 36 children with a second dose within twenty-four hours. Temperatures dropped in two, increased for seven, and remained constant for 27. Responses appeared to be mixed.[14]

Nevertheless, the press trumpeted that trial and proudly published the names of the many donors. "Experiments ... indicated," the *New York Times* reported on August 9, 1916, "that the serum was more efficacious in preventing paralysis and in stopping it in its early stages than in curing it after full development." Consequently, ambulance attendants received blood serum to inject into children as they picked them up at their homes for transport to hospitals. Tests continued because, as the *Times* added, it had "been used in the cases of children in advanced states of the disease with encouraging results in some instances." That newspaper continued to monitor this development and, the following day, the front page described how doctors continued to draw blood from volunteers who had recovered from poliomyelitis: "Ten patients received the serum treatment yesterday, and two, a boy of five years and a girl of six, who were in an advanced stage of paralysis, showed striking signs of improvement. The other eight were not so badly paralyzed and the serum was used to stop the progress of the disease. Whether it was successful ... cannot be determined for several days." Results trickled in but no systematic data collection occurred; consequently, no scientific analysis resulted. Moreover, this effort created an unintended outcome. Mothers, it was reported, resisted bringing their ill children to hospitals for fear that doctors would drain them of their blood to make serum.[15]

Five days later, the *New York Times* proclaimed that a cure may be at hand: "A suggestion that the epidemic might be halted altogether and that children who had not yet contracted the disease might be immunized by injections of serum made from the blood of their parents, regardless of whether the parents have had poliomyelitis, was offered by Dr. Abraham Zingher of the Willard Parker Hospital." He recommended that two ounces be given to children younger than three and four ounces to those between three and six years. The *Times* further reported that Zingher expressed optimism about this new approach but, at the same time, expressed doubt about its effectiveness. He nevertheless declared it "harmless."[16]

This episode revealed serious flaws. First, only anecdotal reports of recovery served as evidence, most of it through newspaper accounts, proving nothing since those patients may have simply improved on their own. Nevertheless, this appeared sufficient for the general public; numerous adults, who had survived infection, queued to give blood. Seven volunteers alone donated fifty-four ounces at Parker Hospital, where physicians "administered" it to fourteen additional children. Second, the crude processing of "immune serum" raised other questions. For Zingher, the drawn blood should be left to stand in a "cold place for twenty-four hours, or until the serum has separated itself from the solid matter of the blood." Third, children received single or multiple doses, with little regard for the time period between them.[17] The experiment quickly unraveled since it lacked "uniformity and a failure to adhere to approved plans, not only at the various hospitals but particularly by some private physicians." They represented the weakest link. They did not follow a strict testing regimen, John R. Paul reflected fifty-five years later, believing the patient's welfare superseded the need to "assemble a suitable control group." Paul adds that "by the time the trial had progressed very far it had shrunk to a comparison of two groups, not three, and these were of unequal size (119 versus 43). In any event a comparison between the two would have been meaningless. It was a situation to be repeated over and over again."[18]

Herman Schwarz, in a second attempt, found the use of convalescent blood serum problematic. In a report to the New York Academy of Medicine in November 1916, he revealed that he had tried it in both hospital settings and private practice. He injected twenty-one infected children; nine, or 43%, recovered without paralysis. An additional twenty-one patients received no serum whatsoever and seventeen, or 81%, remained free from paralysis. Schwarz, as a result, did not see sufficient evidence justifying this approach. Moreover, because no standard, blood serum formula existed, it would prove impossible to duplicate it under similar circumstances, raising questions about its reliability as a therapy.[19]

Researchers Harold L. Amoss and Alan M. Chesney, working under the auspices of the Rockefeller Institute, oversaw a third field trial in the immediate outskirts of the city, across the Hudson River at Westchester Isolation Hospital. They preferred intravenous injections, so more of it could be introduced into patients, treating a total of twenty-six with "human serum from recovered and convalescent cases of the disease." Children aged ten years or younger comprised 23 of the subjects, of whom 21 were younger than five. The largest subgroup consisted of seven children aged two years or younger. The dosages ranged from 5 to 80 cc. Twelve patients already had some degree of paralysis when injected. One of them died while two experienced more paralysis and nine completely recovered. Of the remaining 14 nonparalytic cases, two died, two developed some degree of paralysis, and 10, or 71%, showed no signs of it. Amoss and Chesney concluded that using this approach produced "favorable results."[20] In sum, trials in New York City and its suburbs produced inconsistent outcomes.

Serotherapy forged ahead, as Francis W. Peabody conducted a fourth experiment for the Harvard Infantile Paralysis Commission during the 1916 epidemic. The research team, working in close cooperation with the Massachusetts State Department of Health, focused on Boston and implemented its plan through family physicians. They injected 10 cc. of serum into each patient, decided when to give subsequent doses, and recorded the data within a 51-patient cohort, 27 only received one treatment, 16 had 2, 6 had 3, and 2 had 4. These additional jabs occurred on an irregular basis, sometime within twelve and twenty-four hours of the previous ones. Thirty-five children, or 69%, recovered without paralysis, 11, or 21%, lived but were paralyzed, and 5, or 10%, died. The Harvard Commission likewise treated 60 paralyzed patients and found that "8, or 12 percent, showed definite rapid improvement, while 33, or 51 percent, did not improve much after the injection of the serum, or became worse." Twenty-one, or 32%, of these children died. Finally, no results were reported in three cases, or 5%. Since no control group existed, Peabody could not report any valid comparisons. This did not stop him from making inferences from his, the Zingher, and the Amos-Chesney experiments. In Zingher's case, 44, or 82%, of 54 preparalytic cases recovered without paralysis. Amos and Chesney treated 14 nonparalytic patients, and 10, or 71%, recovered without paralysis, 2 became paralyzed, and 2 died. The absence of legitimate control groups did not hinder Peabody from arriving at this dubious conclusion: "... there is apparently general agreement among those who have used the immune serum as to its harmlessness, and as to the fact that in certain, possibly in numerous, instances its administration is beneficial."[21]

A fifth endeavor unfolded during the 1927 Massachusetts's epidemic. Boston doctors W. Lloyd Aycock and Eliot H. Luther, working in conjunction with the Harvard Infantile Paralysis Commission, saw this as an "opportunity for further testing of the use of convalescent serum under fairly uniform conditions." They injected 106 preparalytic patients both intrathecally (spinal) and intravenously on either the first, second, third, or fourth day after diagnosis as the possibility of paralysis increased with each day's delay. One, or 0.9%, died. They also left 482 cases untreated to serve as a control group; it experienced a one percent fatality rate. While 19% of those treated developed significant paralysis, 63% of the latter cohort developed it. Aycock and Luther concluded that the use of blood serum lowered both mortality and paralysis rates. And when paralysis did occur, injected patients experienced milder manifestations.[22]

This study, in particular, and convalescent serum, in general, received high-profile endorsements and widespread support. Simon Flexner saw it as validation of his earlier work and projected a twofold use for it: as a prophylactic and a therapy. New York's Governor Franklin D. Roosevelt not only donated a pint of blood but appealed for volunteers. Finally, convalescent stations across the country supplied it, like the Deutsch Serum Center at the Michael Reese Hospital distributed it in the Chicago area.

A sixth set of trials unfolded during the Northeast's 1931 epidemic, addressing the need for uniform dosage and the use of a control group. It concentrated on the outbreaks in Connecticut and Brooklyn, New York. It further narrowed its scope to one hospital in each setting and included a total of 82 preparalytic patients. W. Lloyd Aycock and his associates at the Harvard Infantile Paralysis Commission oversaw the Connecticut cases while William Park and his colleagues conducted the other in Brooklyn. In each locale, they limited the serum to 41 patients. They injected each with 60 cc. intravenously and an additional 40 cc. intraspinally, spacing the two doses eight to twelve hours apart.

Forty-two, or 51%, of the total 82 patients developed paralysis. Two deaths occurred but only in the treated cohort. John Paul noted, in retrospect, both "well-controlled experiments yielded discouraging results,"

Physicians employed convalescent serum for the last time during the 1934 Los Angeles epidemic. Edwin W. Schultz and L. P. Gebhardt, two Stanford University researchers, published the results: It was wholly ineffective, formally closing the door on this line of investigation. After two decades of experimentation with convalescent serum, the best John Paul could muster "was that apparently it did little harm to the patient." Research could now move in "other directions, because so much progress in poliomyelitis investigations had been sorely held up for many years over controversies on the use of serum."[23]

These episodes expose the limitations of the moral tradition. Because of inadequate knowledge about polio, because of the absence of oversight or standard protocols, this approach proved wholly unproductive. Relying on unchallenged hypotheses and perfunctory laboratory experiments, it also wrought mistreatment. Researchers' determinations, based on the respect accorded to their expertise, resulted in young patients remaining outside of the decision-making process. As a result, "[h]uman experimentation continued to be left to the discretion of individual investigators."[24]

Nuremberg Code

During World War II, Allied and Axis powers alike exploited medical science to achieve military superiority. All of the warring parties were guilty of abuses to one degree or another. In the USA, the Committee on Medical Research (CMR) approved projects involving human subjects. Initially, only volunteers who signed waivers could be eligible. "Some 600 research proposals, many of them involving humans," amounted to an unheard of sum of $25 million. The mass mobilization of civilian resources for military purposes and the concomitant use of selective service compulsion overlapped with medical investigations. The "common understanding that experimentation required the agreement of the subjects–however casual the request or general the approval–was often superseded by a sense of urgency that overrode the issue of consent." The CMR organized research at agricultural stations, pharmaceutical companies, and universities to address health problems, like dysentery, influenza, malaria, and sexually transmitted diseases that might impair the fighting ability of the military. This also included wartime deprivations, such as surviving exposure in icy water after a plane crash or ship sinking. Most important, World War I's influenza pandemic haunted military planners, since it had drained vital manpower from combat situations. Vaccine tests, largely under the supervision of Jonas E. Salk, who worked as a research assistant for Thomas Francis, Jr., at the University of Michigan, involved 20,174 civilian and military subjects. As medical historian David J. Rothman explains,

> ... the record of human experimentation during World War II constitutes a curious mixture of high-handedness and forethought. The research into dysentery, malaria, and influenza revealed a pervasive disregard of the rights of subjects–a willingness to experiment on the mentally retarded, the mentally ill, prisoners, ward patients, soldiers, and medical students without concern for obtaining consent. Yet, research into survival under hardship conditions and into gonorrhea was marked by formal and carefully considered protocols that informed potential subjects about the risks of participation.

These latter protocols, Rothman concludes, "contradict blanket assertions that in the 1940s and 1950s [American] investigators were working in an ethical vacuum."[25]

The Nazi medical experiments, usually performed by highly credentialed doctors and scientists on concentration camp prisoners, sparked the Doctors' Trial, one of thirteen proceedings in Nuremberg. "From late 1946 to the middle of 1947, twenty-three Nazi German officials, twenty of them physicians, were tried for complicity in medical experiments." However, the legal defense team, with significant "effectiveness," attempted to "turn the proceedings into a trial of the Allies' wartime medical research, as well as that of the Third Reich."[26] The endless supply of "'human material'" available at American penal institutions, they contended, allowed countless experiments involving animal blood transfusions, gonorrhea, and malaria.[27]

Life magazine described one example in a June 4, 1945, article titled "Prison Malaria." Researchers tasked with finding a cure chose three penal institutions: the US Penitentiary (in Atlanta), Illinois State Penitentiary, and New Jersey State Reformatory. The magazine skillfully depicted the infection process with large black-and-white photographs, reporting that 800 prisoners "volunteered to be infected with malaria so medical men can study the disease." One caption read "Army doctors expose patients to infected mosquitoes," placing a glass jar filled with them and covered in gauze on convicts' arms and abdomens. Other pictures showed deadly ill subjects suffering from malaria's symptoms, with a description stating: "Some of the prisoners are allowed to progress considerably before they are treated with drugs."[28] These wanton actions backed the Allies into an ethical corner, one awkwardly involving a double standard.

Because of *Life*'s coverage, this experiment had become public knowledge, and the medical community had to extricate itself from its shadow. The AMA assigned Andrew C. Ivy, a research physiologist at the University of Illinois medical school, as its representative and prosecution advisor. He departed from Chicago for Europe on July 18, 1946, stopping in Washington, D.C., to consult with War Department officials. He knew all too well that this trial struck too close to home.

Ivy was present, in January 1947, to hear the testimony about the use of American convicts in the malaria trial. The defense's argument cast the Allies as hypocrites. Ivy had been given the responsibility of formulating a "'pragmatic instrument' ... to label the German experiments as 'unethical' and those of the Allies as 'ethical.'" He returned to Illinois and contacted Dwight H. Green, that state's governor, about forming a committee to review the Illinois State Penitentiary's malaria project. On March 13, the governor invited six individuals to participate as members of what became known as the Green Committee. In April, Ivy contacted them about their charge but never convened a meeting, though he unilaterally drafted a report, "Outline of Principles and Rules of Experimentation on Human Subjects," on its behalf.[29]

Ivy returned to Nuremberg in June 1947 to rebut the defense's case equating American wartime experiments with those of Germany. He presented the Green Committee's report as evidence, asserting that it "had carefully considered and approved the Stateville research." It contained three general points to guarantee safety: voluntary consent, prior animal tests, and professional, medical oversight. Ivy deliberately omitted the fact that the committee had never met and flatly denied under oath that its formation was connected with this trial. He effectively undermined the defense's contention "that [American] experimentation on prisoners was ... unethical." But Ivy's actions were clearly duplicitous. "Ivy's stance can be seen as a symptom of a broader refusal among U.S. medical scientists to draw lessons about their own actions from the

Nuremberg Medical Trial."[30] It, in fact, granted them tacit approval to continue their unfettered use of human subjects.

The Nuremberg justices convicted fifteen of the twenty-three defendants and, on August 20, 1947, issued sentences ranging from ten years of incarceration to death. The principles encapsulated in the Nuremberg Code evolved from the trial's proceedings as the judges sought guidelines by which to write their opinion. It also was intended to inform future researchers: Subjects had to be volunteers with full and informed consent; studies had to benefit society; successful animal tests had to precede human trials; experiments had to avoid any undue mental and physical harm or be life threatening; risks had to be predicated on benefit; only scientific personnel conducted them; subjects had the option to withdraw at any time; and investigators had to terminate the project if any participants suffered injuries, disabilities, or death. These regulations attempted to inject a humane element in the experimental process, ensuring that individuals would participate as informed volunteers with appropriate safeguards. The Nuremberg Code therefore protected subjects in two different ways. First, it granted them direct control; that is, it gave them the "absolute right to refuse to be subjects and the absolute right to terminate their participation at any time." Second, it protected their welfare by instructing researchers "what things may not be done and what things must be done regardless of the subject's consent."[31] In sum, the Nuremberg Code "constitutes the first authoritative, and surely the most stringent, pronouncement on the rights of research subjects."[32]

It influenced numerous discussions of and proposals by international commissions forging human rights initiatives in the "general field of medical ethics and, more specifically in the field of human experimentation." The World Medical Association (WMA), founded in 1947, developed "professional ethical codes and guidelines," resulting in the 1954 Declaration of Geneva. The WMA would continue to refine its stance. Nevertheless, although the Nuremberg Code condemned state-sponsored medical investigations, private projects remained in a labyrinthian world. It likewise had negligible impact on subsequent, state-sanctioned research in the USA. Because scientists had become heroes—through their discoveries, media's depictions of them, and wartime breakthroughs—the American public, more often than not, trusted them. But such confidence proved groundless. In a 1962 article, "Human Guinea Pigs," Maurice H. Pappworth exposed the work of thousands of British researchers, who proved cavalier in their approach to using humans as test subjects, including children. Pappworth further detailed similar American attitudes and abuses; that is, ethical decisions continued to operate in a unilateral manner as scientists alone evaluated the level of risk to human subjects.[33]

In 1947, as medical historian Harriet Washington points out, the Manhattan Project's blue-ribbon panel, the Medical Board of Review, released a three-part research protocol sanctioned by the US Advisory Committee on Biology and Medicine. First, all such research had to have the reasonable hope of therapeutic benefits. Second, participants had to give their consent in writing. Third, they had to be informed about any potential dangers. While on the surface, this "1947 policy demonstrates that such abusive experiments were as morally unacceptable in their time as they are in ours, sweeping protections of the AEC policy were not widely distributed and scientists routinely flouted their own policy." Even in 1953, when Secretary of Defense Charles Wilson, in what would become known as the Wilson memo, consciously adopted the Nuremberg Code, American scientists continued their experiments as before, unfazed and undeterred, completely disregarding this consideration.[34]

Experiments on civilians and military troops focusing on atomic, biological, and chemical weaponry dominated America's post-Nuremberg agenda. The Atomic Energy Commission (AEC), Central Intelligence Agency, and US Army funded studies carried out by university researchers. The USA and its allies quickly recruited and hired German doctors who had been indicted but escaped prosecution in Nuremberg. This became known as operation Paperclip. Because of Cold War fears, intelligence and military communities sanitized, if not totally buried, their war crimes, permitting them to enter the country. According to medical ethicist Jonathan Moreno, "[i]f the Paperclip story is not mainly a medical ethics story, it surely has implications for governmental attitudes toward the importance of ethical standards in medicine." From 1945 to 1947, researchers assisted nuclear scientists, who were developing America's atomic arsenal, to approximate conditions for soldiers exposed to nuclear fallout during wartime conditions. They injected eighteen hospital patients with minute amounts of plutonium without their knowledge or consent. In 1950, physicians, funded by the defense department, began a two-decade total-body irradiation experiment on hundreds of hospitalized cancer patients, rationalizing them as potentially therapeutic. In reality, these widespread tests, in Bethesda, Maryland,

Houston, and New York City, merely "substituted patients for troops in radiation experiments…." A further objectification of human bodies occurred between 1953 and 1957 when the AEC funded a study at Massachusetts General Hospital involving the injection of uranium into eleven terminally ill, generally comatose, brain cancer patients. A similar story unfolded between 1962 and 1972 at Cincinnati's General Hospital. Researchers, based at the College of Medicine, University of Cincinnati, conducted radiation experiments sponsored by the US Department of Defense. Eighty-seven cancer patients, fifty-six of whom were African-American, underwent full-body irradiation in a specially constructed lead-lined room in that hospital's basement. None of the patients realized they had been exposed for experimental purposes; for the first five years, none even saw a consent form. Physicians had simply told them they were being treated for cancer. Few of these cases had been diagnosed terminal, but all suffered mental confusion and hallucinations, severe nausea, horrible pain, and vomiting: Twenty-one died. A well-known and highly regarded project, research-team members published articles in leading medical journals and delivered ten reports to the defense department. AEC-sponsored experiments also irradiated male prisoners' testicles. Even "[a]fter Nuremberg, America was the only Western country that maintained an extensive program of prison experiments." The Army, moreover, conducted open-air tests in crowded airports, ports, and subways in numerous, major cities; researchers released biological agents to determine dispersion rates and vulnerability to enemy attacks. Finally, "[a]pproximately 6,700 human subjects were used by the government in experiments with psychoactive chemicals, most pervasively lysergic acid diethylamide (LSD), but also in private contract research with universities and chemical companies, other agents were used, including morphine, Demerol, Seconal, mescaline, atropine, psilocybin, and Benzedrine."[35]

A stunning example of radiation absorption tests involved children residing at the Walter E. Fernald State School for the Feebleminded. Founded by Boston's Samuel Gridley Howe in 1848 and expanded and relocated to Waltham in 1887, it represented the first institution to serve children with mental disabilities in the USA. During the early decades of the twentieth century, it housed boys and girls labeled "retarded," because of their low test scores on intelligence tests, and deemed unsuitable for reproduction by eugenics advocates; in the most severe cases, judges ordered them sterilized. By 1949, the already overcrowded Fernald facility hosted 1900 residents; this expanded to 2032 in 1952 and 2242 two years later, including children disabled by infantile paralysis. They "were often segregated as mentally deficient. (The so-called feeble-mindedness and even insanity that accompany other forms of paresis had long been associated in the popular mind with poliomyelitis paralysis.) And polios were often housed in the same institutions as people with mental diseases." Fernald State Hospital in Waltham hosted a large number since it had one of the "best pools for therapeutic purposes in New England ..." Nevertheless, because of overcrowding and neglect, Fernald was a filthy institution: The staff daily hosed urine and feces off the floors. Roaches and rodents infested the facility, and its buildings stood in disrepair. A single outdoor shower served the entire population. And the staff regularly abused patients.[36]

According to Michael D'Antonio's account, Clemens E. Benda, a psychiatrist and neuropathologist, arrived at Fernald in 1947 and oversaw several medical tests, among them a seemingly benign nutrition study that began in 1949 and lasted six years. Funded by Quaker Oats corporation, to determine calcium absorption, and under the auspices of the Massachusetts Institute of Technology (MIT), Benda sent a note home to the parents of several dozen boys seeking permission to place them on a special, enriched diet of various cereals as well as iron and vitamin supplements. He would also draw blood and collect their urine and feces for testing, but downplayed it as an experiment. All parents granted permission when they could be found. When neither parents nor guardians could be located, the state acted as official surrogate.

To facilitate cooperation, Benda and his MIT colleagues called them "smart boys," recruiting them for the newly formed "Science Club." These bored, isolated, and mistreated boys enthusiastically agreed to participate after hearing such rare praise and given special privileges, like attending Red Sox games and visiting MIT to celebrate Christmas. There, D'Antonio adds, scientists hosted a holiday party where they received Mickey Mouse wristwatches from Santa Claus; the next year Santa gave them Hopalong Cassidy mugs.

Benda quarantined these boys within Fernald to strictly control their diets. He did this because the hot oatmeal and farina they ingested had been laced with radio active calcium. In truth, this medical experiment not only involved Fernald's doctors, Quaker Oats, and MIT, but also Harvard University and the AEC, which supplied the radioisotopes. No one ever informed the boys, or their parents or guardians, that they would be ingesting a toxin. This callousness appeared typical, D'Antonio insists. Fred Boyce, one of the Science Club's members, recalled how "Fernald's doctors had considered children at the institution their 'personal property' and disregarded their civil rights." In addition to the tests described above, Fernald's doctors even "pierced the sternums of patients to withdraw samples of bone marrow."

Two key points grow out of this episode. First, Fernald did not represent an isolated radiation study: "more that 23,000 Americans had taken part in more than 1,400 different experiments." Like the children at Fernald, they had unknowingly participated. This "posed legal and ethical problems and that as early as 1947 efforts were made to hide the radiation research to avoid scandal."[37] Anemia investigations used children and pregnant women. Such "nutrition research with radioiron focused on healthy human subjects"; in 1945, doctors from Vanderbilt University's medical school "administered trace doses of iron-59 in lemonade" to "189 children from two Nashville school districts" to detect iron ingestion. Over the next two years, they also "administered oral doses of radioactive iron to over 800 pregnant women to track its absorption," many of them, poor and white. Researchers then drew blood from them and their umbilical cords to determine if iron had been assimilated in both mothers and their fetuses.[38]

Second, according to D'Antonio, wards of the state commonly became the objects of research, including Jonas Salk's use of children housed at Pennsylvania's Polk State School. Thus, "there was nothing remarkable about [Fernald's] oatmeal experiment." Even as late as 1994, Constantine Maletskos, one of MIT's investigators in the Fernald study, still saw captives as the best approach: "'... in all of these experiments you have control of the subjects. You can't just let them walk around; you have to collect 100 percent of their excretions....'"[39]

The Nuremberg Code, invoking the key principles of choice and protection, failed to move human subjects from the margins. "The least that can be said is that if the Second World War was good for medical ethics, it took an unconscionable amount of time for it to be realized."[40] In the interim, scientists continued to be somewhat supercilious in their treatment of humans—particularly children. Western scientists largely ignored the Code because, first, they reasoned, it applied only to Nazis not to them and, second, they balked at the notion of any central control over their work.

Notes and Sources

1. Henry K. Beecher, "Ethics and Clinical Research," *New England Journal of Medicine* 274 (June 1966): 1355 and 1356, respectively. Consult pages 1354–60, as well as Lawrence K. Altman, *Who Goes First? The Story of Self-Experimentation in Medicine* (Berkeley: University of California Press, 1998): 18–19; Ruth R. Faden and Tom L. Beauchamp, *A History and Theory of Informed Consent* (New York: Oxford University Press, 1986): 157–60; Stephen R. Graubard, "Preface," *Daedalus* 98 (Spring 1969): v–viii; Jonathan D. Moreno, *Undue Risk: Secret State Experiments on Humans* (New York: Routledge, 2001): 239–42; and Lainie Friedman Ross, *Children in Medical Research: Access Versus Protection* (New York: Oxford University Press, 2006): 12. On the one hand, J. N. Hays, *The Burdens of Disease: Epidemics and Human Response in Western History* (New Brunswick: Rutgers University Press, 1998), uses the term, "powerless populations" (162). This however implies a lack of human agency. Vulnerable populations represents a better term. While it certainly denotes an exploited group, it does not necessarily connote complacency or passiveness. Susan E. Lederer, *Subjected to Science: Human Experimentation in America Before the Second World War* (Baltimore: Johns Hopkins University Press, 1995), uses the label "vulnerable populations" (142). For examples of groups exploited in experiments, refer to Robert B. Baker, *Before Bioethics: A History of American Medical Ethics from the Colonial Period to the Bioethics Revolution* (New York: Oxford University Press, 2013); Allan M. Brandt, "Polio, Politics, Publicity, and Duplicity: Ethical Aspects in the Development of the Salk Vaccine," *International Journal of Health Services* 8 (1978): 264; Sharla M. Fett, *Working Cures: Healing, Health, and Power on Southern Slave Plantations* (Chapel Hill: University of North Carolina Press, 2002): 151–52; Vanessa Northington Gamble, "Under the Shadow of Tuskegee: African Americans and Health Care," *American Journal of Public Health* 87 (November 1997): 1774; Gerald L. Geison, "Pasteur's Work on Rabies: Reexamining the Ethical Issues," *Hastings Center Report* 8 (April 1978): 27, 29–30; Evelynn Maxine Hammonds, *Childhood's Deadly Scourge: The Campaign to Control Diphtheria in New York City, 1880–1930* (Baltimore: Johns Hopkins University Press, 1999): 188; Stephen C. Kenny, "'I Can Do the Child No Good:' Dr. Sims and the Enslaved Infants of Montgomery, Alabama," *Social History of Medicine* 20 (August 2007): 226–29; Susan E. Lederer and Michael A. Grodin, "Historical Overview: Pediatric Experimentation," in Michael A. Grodin and Leonard H. Glantz, eds., *Children as Research Subjects: Science, Ethics, and Law* (New York: Oxford University Press, 1994): 7; Timothy

R. Murphy, "The Ethics of Research with Children," *Virtual Mentor: Ethics Journal of the American Medical Association* 5 (August 2003): 1–3; and Harriet A. Washington, *Medical Apartheid: The Dark History of Medical Experimentation on Black Americans from Colonial Times to the Present* (New York: Doubleday Books, 2006). Finally, the AMA's positions may be gleaned at http://www.ama-assn.org/ama/pub/category/print/10817.html (Accessed April 29, 2004).

2. John R. Paul, *A History of Poliomyelitis* (New Haven: Yale University Press, 1971): 410. Histories of polio dwell on the research design, downplaying the ethical context. Some of these include Paul's venerable *History of Poliomyelitis*, Chapter 40, and Pulitzer Prize winning David M. Oshinsky, *Polio: An American Story* (New York: Oxford University Press, 2005), Chapter 11.
3. Gregory E. Pence, *Classic Cases in Medical Ethics*, 4th ed. (Boston: McGraw-Hill, 2004): 24–25; Robert B. Baker, "The American Medical Ethics Revolution," in Robert B. Baker, et al., eds., *The American Ethics Revolution: How the AMA's Code of Ethics Has Transformed Physicians' Relationships to Patients, Professionals, and Society* (Baltimore: Johns Hopkins University Press, 1999): 17–51. Baker points out a historiographic discrepancy. Some see the 1847 Code of Ethics as nothing more than rules of "professional etiquette that parroted the language of an English physician, Thomas Percival…." This represents an "institutional myth." Baker instead argues that Isaac Hays and John Bell, authors of the 1847 Code, "transformed and transcended Percival's ideas…." (19), synthesizing Percival's "ethics of conduct" with Jacksonian "egalitarian ethics," resulting in a "social contract" (36). Thus, Baker draws a straight line from Percival to modern bioethics standards and implementation.
4. See David J. Rothman, "Were Tuskegee and Willowbrook 'Studies in Nature?'" *Hastings Center Report* 12 (April 1982): 5–7, for a detailed explanation this practice. Also refer to Sydney A. Halpern, *Lesser Harms: The Morality of Risk in Medical Research* (Chicago: University of Chicago Press, 2004): 7–8.
5. Washington, *Medical Apartheid*: 165–66. See pages 137, 157, 160, 162–63, 181, as well as Faden and Beauchamp, *A History and Theory of Informed Consent*: 165–67; Pence, *Classic Cases in Medical Ethics*: 283–85, 294; Susan M. Reverby, "Rethinking the Tuskegee Syphilis Study: Nurse Rivers, Silence and the Meaning of Treatment," *Nursing History Review* 7 (1999): 3–28; and Rothman, "Were Tuskegee and Willowbrook 'Studies in Nature?'" 5.
6. David J. Rothman and Sheila M. Rothman, *The Willowbrook Wars: Bringing the Mentally Disabled into the Community* (1984; Rpt. New Brunswick, NJ: Transaction Press, 2005): 23, 261–62, and 265,

respectively. Refer to pages 16, 17, 20–21, 22, 25, 261–64, 267, as well as Faden and Beauchamp, *History of Informed Consent*: 163–64; Joel D. Howell and Rodney A. Hayward, "Writing Willowbrook, Reading Willowbrook: The Recounting of a Medical Experiment," in Jordan Goodman, Anthony McElligott, and Lara Marks, eds., *Useful Bodies: Humans in the Service of Medical Science in the Twentieth Century* (Baltimore: Johns Hopkins University Press, 2003): 191; Louis Lasagna, "Special Subjects in Human Experimentation," *Daedalus*, 98 (Spring 1969): 458; Rothman, "Were Tuskegee and Willowbrook 'Studies in Nature?'" 6; and Washington, *Medical Apartheid*: 13. According to Lainie Friedman Ross, *Children in Medical Research: Access versus Protection* (New York: Oxford University Press, 2006), Beecher references Krugman's experiments at Willowbrook without explicitly identifying them (12).

7. Rothman, "Were Tuskegee and Willowbrook 'Studies in Nature?'" 7. Further check Howell and Hayward, "Writing Willowbrook, Reading Willowbrook": 195, 197, 200, 206–7. Finally, consult Rothman and Rothman, *Willowbrook Wars*. The entire cover letter is reprinted on page 266.
8. For the first and third quotes, see Halpern, *Lesser Harms*: 3–4. Further consult pages 11–13. For the second, refer to Faden and Beauchamp, *History and Theory of Informed Consent*: 61–62. Also check pages 12–13, 15. For related studies, see Lawrence K. Altman, *Who Goes First? The Story of Self-Experimentation in Medicine* (Berkeley: University of California Press, 1998); Baker, *Before Bioethics*: 73; and John R. Wilson, *Margin of Safety: A Doctor Tells the Controversial Inside Story of the Development of the Polio Vaccine* (Garden City, NY: Doubleday & Co., 1963): 23–24.
9. Haven Emerson, *The Epidemic of Poliomyelitis (Infantile Paralysis) in New York City in 1916: Based on the Official Reports of the Bureau of the Department of Health* (1917. Rpt. New York: Arno Press, 1977): 245. Also consult page 254, and Rockefeller Foundation Minutes, November 14, 1916: 112, folder 275, "Infantile Paralysis, 1916–1918," box 25, RG 1.1, Rockefeller Foundation Archives, RAC. Finally, refer to "Fewer Fatal Cases on Paralysis Rolls," *New York Times*, July 24, 1916: 16, as well as International Committee for the Study of Infantile Paralysis, *Poliomyelitis: A Survey* (Baltimore: Williams and Wilkins Co., 1932): 221–22; Naomi Rogers, *Dirt and Disease: Polio Before FDR* (New Brunswick, NJ: Rutgers University Press, 1996): 97.
10. "Offers New Cure in Fight Against Infant Paralysis," *New York Times*, July 14, 1916: 1 and 5; "31 Die of Paralysis," *New York Times*, July 15, 1916: 1 and 16; "Paralysis Deaths Following Off," *New York Times*, July 17, 1916: 1 and 4.

11. Bermingham is quoted in "Scientists to Study New York Paralysis," *New York Times*, July 28, 1916: 5. Emerson is quoted in "Schools to be Shut Till Epidemic Ends," *New York Times*, August 10, 1916, Section I: 1 and 5. Also see Emerson, *The Epidemic of Poliomyelitis*: 254.
12. "Schools to be Shut Till Epidemic Ends": 1 and 5; "Suggests Serum for All Children," *New York Times*, August 15, 1916: 18; Emerson, *The Epidemic of Poliomyelitis*: 245, 254.
13. Simon Flexner and Paul A. Lewis, "Experimental Poliomyelitis in Monkeys," *Journal of the American Medical Association* 54 (May 28, 1910): 1781. See also Paul, *History of Poliomyelitis*: 190–91.
14. Emerson, *Epidemic of Poliomyelitis*: 248–49, 251. See pages 245, 249–50, 266–67, as well as Evelynn Maxine Hammonds, *Childhood's Deadly Scourge: The Campaign to Control Diphtheria in New York City, 1880–1930* (Baltimore: Johns Hopkins University Press, 1999): 180; Paul, *History of Poliomyelitis*: 191; Rogers, *Dirt and Disease*: 96, 97, 99.
15. "Paralysis Cripples Glad to Aid Others," *New York Times*, August 9, 1916: 3. For brief historical background, refer to Rogers, *Dirt and Disease*: 101–103.
16. "Schools to be Shut Till Epidemic Ends": 1 and 5; "Suggests Serum for All Children": 18.
17. "Suggests Serum for All Children": 18.
18. Paul, *History of Poliomyelitis*: 193 and 191, respectively.
19. Herman Schwarz, *Proceedings of the New York Academy of Medicine* 33 (November 1916): 859–60. Further consult Paul, *History of Poliomyelitis*: 191. John F. Landon and Lawrence W. Smith, *Poliomyelitis: A Handbook for Physicians and Medical Students* (New York: Macmillan Company, 1934), point to at least 15 experiments or statistical analyses, spanning the period 1916–1931, that reached no conclusive results (189–90). I include five of the most prominent ones.
20. Harold L. Amoss and Alan M. Chesney, "A Report on the Serum Treatment of Twenty-Six Cases of Epidemic Poliomyelitis," *Journal of Experimental Medicine*, 25 (February 1917): 581 and 608, respectively. See pages 585, 586–95, 596, 598, 599, and Paul, *History of Poliomyelitis*: 194, 195.
21. Francis W. Peabody, "A Report of the Harvard Infantile Commission on the Diagnosis and Treatment of Acute Cases of the Disease During 1916," *New England Journal of Medicine* 176 (May 1917): 639, and 642, respectively. Further check pages 637–38; George Draper, Peabody's colleague, writing that same year stated, in *Acute Poliomyelitis* (Philadelphia: P. Blakiston's Son & Co., 1927), that the serum's effectiveness remained "open to doubt" but he supported its continued use (pages 97 and 101). The absence of any discussion of consent or patient

rights remained completely absent (87–103). Also consult Paul, *History of Poliomyelitis*: 193; Anne Finger, *Elegy for A Disease: A Personal and Cultural History of Polio* (New York: St. Martin's Press, 2006): 83.

22. W. Lloyd Aycock and Eliot H. Luther, "Preparalytic Poliomyelitis: Observations of One Hundred and Six Cases in which Convalescent Serum Was Used," *Journal of the American Medical Association* 91 (August 1928): 387. Also see pages 389, 390, 391, 392, 393.
23. Paul, *History of Poliomyelitis*: 198 and 210, respectively. Also refer to pages 196–97, and especially Franklin D. Roosevelt, "Wanted: Enlistment for a Crusade" and "Rockefeller Institute Sends Dr. Rhoads to Secure Blood for Serum," *Polio Chronicle* 1, no. 1 (July 1931): 1 and 3; Leroy W. Hubbard, "Serum," *Polio Chronicle* 1, no. 2 (August 1931): 1 and 3; Benjamin Kramer,"Serum as Used for Anterior Poliomyelitis," *Polio Chronicle* 2, no. 7 (February 1933): 1 and 4; and Philip Lewin, "Infantile Paralysis," *Polio Chronicle* 3, no. 7 (February 1934): 2, WSA. Finally refer to Roland H. Berg, *Polio and Its Problems* (Philadelphia: J. B. Lippincott, 1948): 64, 71–72; S. D. Kramer, W. Lloyd Aycock, Charles I. Solomon, and C. L. Thenebe, "Convalescent Serum Therapy in Preparalytic Poliomyelitis," *New England Journal of Medicine* 206 (March 1932): 432–35; Edwin W. Schultz and L. P. Gebhardt, "On the Problem of Immunization Against Poliomyelitis," *California and Western Medicine* 43 (August 1935): 111–12.
24. Halpern, *Lesser Harms*: 124. Further consult pages 7–8, 63–65, 68, 109–110, as well as Wendy K. Mariner, "AIDS Research and the Nuremberg Code," in George J. Annas and Michael A. Grodin, eds., *The Nazi Doctors and the Nuremberg Code: Human Rights in Human Experimentation* (New York: Oxford University Press, 1992): 288.
25. David J. Rothman, *Strangers at the Bedside: A History of How Law and Bioethics Transformed Medical Decision Making* (New York: Basic Books, 1991): 47–48, and 42–43, respectively. Also see pages 30–35, 39, 41, 49–50. For additional analyses, consult George J. Annas, "The Nuremberg Code in U.S. Courts: Ethics versus Expediency," in Annas and Grodin, *The Nazi Doctors and the Nuremberg Code*: 201–22; Baker, *Before Bioethics*: 255–63; Jenny Hazelgrove, "The Old Faith and the New Science: The Nuremberg Code and Human Experimentation Ethics in Britain, 1946–73," *Social History of Medicine* 15 (April 2002): 109–35; Jonathan D. Moreno, *Undue Risk: Secret State Experiments on Humans* (New York: Routledge, 2001): 15–16, 20, 22, 25–27, 30, 33–35, 40–43; Ulf Schmidt, *Justice at Nuremberg: Leo Alexander and the Nazi Doctors' Trial* (New York: Palgrave Macmillan, 2004): 215.
26. Moreno, *Undue Risk*: 54–55. Consult pages 15 and 67, as well as Lasagna, "Special Subjects in Human Experimentation": 449–51;

Jordan Goodman, Anthony McElligott, and Lara Marks, eds., *Useful Bodies: Humans in the Service of Medical Science in the Twentieth Century* (Baltimore: Johns Hopkins University Press, 2003): 6–7.

27. Allen M. Hornblum, "They Were Cheap and Available: Prisoners as Research Subjects in Twentieth-Century America," *British Medical Journal* 315 (November 1997): 1437–41. http://www.bmj.com/cgi/content/full/315/7120/1437? (Accessed April 19, 2007). Also see Jon M. Harkness, "Nuremberg and the Issue of Wartime Experiments on U.S. Prisoners: The Green Committee," *Journal of the American Medical Association* 276 (November 27, 1996): 1672–73.
28. "Prison Malaria: Convicts Expose themselves to Disease So Doctors Can Study It," *Life* 4 (June 4, 1945): 43–44, 46, respectively.
29. Schmidt, *Justice at Nuremberg*: 136. Refer to pages 3 and 134–35, as well as Harkness, "Nuremberg and the Issue of Wartime Experiments on U.S. Prisoners": 1673; Hornblum, "They Were Cheap and Available": 1437–41.
30. Harkness, "Nuremberg and the Issue of Wartime Experiments on U.S. Prisoners": 1673–74 and 1675, respectively. Harkness further states: "The Green Committee report arose from the Nuremberg Medical Trial because Ivy refused to concede even a remote moral similarity between the experimental atrocities committed in Nazi concentration camps and the medical tests that had been carried out on U.S. prisoners during the war." Also check Hornblum, "They Were Cheap and Available": 1437–41; Schmidt, *Justice at Nuremberg*: 135–39, 234–35.
31. Leonard H. Glantz, "The Influence of the Nuremberg Code on U.S. Statutes and Regulations," in Annas and Grodin, eds., *Nazi Doctors and the Nuremberg Code*: 183–84. Also consult Paul B. Beeson, Philip K. Bondy, Richard C. Donnelly, and John E. Smith, "Panel Discussion: Moral Issues in Clinical Research," *Yale Journal of Biology and Medicine* 36 (June 1964): 455–56; Alan Beyerchen, "Ethics and Science," in J. L. Heilbron, ed., *The Oxford Companion to the History of Modern Science* (New York: Oxford University Press, 2003): 275; Robert N. Proctor, "Nazi Doctors, Racial Medicine, and Human Experimentation," in Annas and Grodin, eds., *Nazi Doctors and the Nuremberg Code*: 26; and Schmidt, *Justice at Nuremberg*: 3, 248. Finally, Annas and Grodin, eds., *Nazi Doctors and the Nuremberg Code*, and Roy Porter, *The Greatest Benefit to Mankind: A Medical History of Humanity* (New York: Norton, 1997), include all ten points of the Nuremberg Code on pages 2 and 651, respectively. Authorship remains a highly debated point among historians as Schmidt so effectively points out (246–53). See also pages 259–60.
32. Harkness, "Nuremberg and the Issue of Wartime Experiments on U.S. Prisoners": 1675. Further check Jay Katz, "The Consent Principle of

the Nuremberg Code: Its Significance Then and Now," in Annas and Grodin, eds., *Nazi Doctors and the Nuremberg Code*: 227.

33. Sharon Perley, Sev S. Fluss, Zbigniew Bankowski, and Françoise Simon, "The Nuremberg Code: An International Overview," in Annas and Grodin, eds., *Nazi Doctors and the Nuremberg Code*: 155 and 154, respectively. Also see pages 152–53, as well as Goodman, McElligott, and Marks, *Useful Bodies*: 8–9; Maurice H. Pappworth, "Human Guinea Pigs: A Warning," *The Twentieth Century*, 171 (Autumn 1962): 66–75; and Ross, *Children in Medical Research*: 12, 15–16. Further consult Hazelgrove, "The Old Faith and the New Science," 118–19, 122, 123–24, 125, 132. Hazelgrove completely enumerates the Nuremberg Code in Appendix A of the article (133–34).
34. Washington, *Medical Apartheid*: 230 and 232, respectively. Refer also to pages 216–17, 218, 231, 233–36, and Martha Stephens, *The Treatment: The Story of Those Who Died in the Cincinnati Radiation Tests* (Durham: Duke University Press, 2002): 257.
35. Moreno, *Undue Risk*: 101, 211, 230, and 251, respectively. Further consult pages 10, 92–98, 120–29, 132–33, 136–37, 167–68, 210–13, 226–29, 233–34. Operation Paperclip is examined in detail in Chapter 4, "Deals with Devils." Also see Gilbert Whittemore and Miriam Boleyn-Fitzgerald, "Injecting Comatose Patients with Uranium: America's Overlapping Wars Against Communism and Cancer in the 1950s," in Goodman, McElligott, and Marks, eds., *Useful Bodies*: 166–67; David S. Jones and Robert L. Martenson, "Human Radiation Experiments and the Formation of Medical Physics at the University of California, San Francisco and Berkeley, 1937–1962," in Goodman, McElligott, and Marks, eds., *Useful Bodies*: 84; Lasagna, "Special Subjects in Human Experimentation": 449. Especially refer to Stephens, *The Treatment*: xix, 4, 17, 49, 53, 155, 157, 173, 178–79, 208, 257. The number of subjects varies somewhat. See Appendix I, "Table of Cincinnati Radiations," for Stephens's explanation for this apparent discrepancy.
36. Marc Shell, *Polio and Its Aftermath: The Paralysis of Culture* (Cambridge: Harvard University Press, 2005): 102. Moreover, check Michael D'Antonio, *The State Boys Rebellion* (New York: Simon and Schuster, 2004): 5, 6, 12–13, 18, 52, 80; Moreno, *Undue Risk*: 217; Margaret Winzer, *The History of Special Education: From Isolation to Integration* (Washington, DC: Gallaudet University Press, 1993): 132, 140, 181–84.
37. D'Antonio, *The State Boys Rebellion*: 238, 56, 240, 249, and 262, respectively. Moreover, see pages 53, 55–56, 57–58, 239, 243, 269, 277. The entire "consent" letter is reproduced on page 242. Finally, consult Angela N. H. Creager, *Life Atomic: A History of Radioisotopes in Science and Medicine* (Chicago: University of Chicago Press, 2013): 292–97.

38. Creager, *Life Atomic*: 280, 284, and 261, respectively. Further explore pages 281, 287, 291–92.
39. D'Antonio, *The State Boys Rebellion*: 244. Maletskos is quoted on page 244. Annas, "The Nuremberg Code in U.S. Courts," agrees with D'Antonio: In the early 1950s, "it was seen as perfectly appropriate to test new polio vaccines on institutionalized, mentally retarded children. Utilitarianism was the ethic of the day" (204). Refer also to Schmidt, *Justice at Nuremberg*: 276; J. David Smith and Alison L. Mitchell, "Sacrifices for the Miracle: The Polio Vaccine Research and Children with Mental Retardation," *Mental Retardation* 39 (October 2001): 408.
40. Roger Cooter, "The Ethical Body," in Roger Cooter and John Pickstone, eds., *Medicine in the Twentieth Century* (Amsterdam: Harwood Academic Publishers, 2000): 460. Also consult Hazelgrove, "The Old Faith and the New Science": 117, 140; Schmidt, *Justice at Nuremberg*: 264–65.

CHAPTER 10

A Moral Compass?

On September 4, 1952, Pope Pius XII addressed an international group of 427 doctors and scientists about the use of humans in experiments. Two contextual phenomena effected this speech. The first grew out of Nazi atrocities conducted in the name of science during World War II. The second involved the numerous and profound medical breakthroughs during and after the war. This included an ill-fated polio trial the year before. Pius pointed to two linchpins in guiding the research process. He condemned "'strong' paternalism." Physicians and investigators had to always consider "'moral possibilities and obligations.'" This meant that they had to seek the "'free and informed consent of the patient.'" Pius also charged "'committees for ethical research'" to monitor the safety of human subjects, to avoid any mistreatment. Individual rights, not the pursuit of knowledge, must prevail. The Pope's position seemed unequivocal.[1]

Private sources funded polio research from the very beginning. "Human testing of serum preparations," according to sociologist Sydney Halpern, "began in earnest during the 1916 epidemic and continued through the early 1930s." The Rockefeller Foundation—separate from the Rockefeller Institute for Medical Research—began its support during the 1916 epidemic, while New York City philanthropist Jeremiah Milbank established the Milbank Foundation in 1924. His "grantees produced eighty scientific papers in a period of four years," and, through a $280,000 grant, he underwrote the International Committee

R. J. Altenbaugh, *Vaccination in America*,
Palgrave Studies in the History of Science and Technology,
https://doi.org/10.1007/978-3-319-96349-5_10

for the Study of Infantile Paralysis four years later. Finally, the National Committee for the Celebration of the President's Birthday sponsored several experiments during the 1930s.

Inconsistent treatment of human subjects and uneven oversight characterized philanthropically sponsored research. Halpern, focusing on two separate influenza-virus tests between 1935 and 1937, points out these weaknesses. The Rockefeller Institute closely monitored the work of Thomas Francis, Jr. He thoroughly and methodically implemented animal trials, used consent forms for human tests, and followed a structured, experimental approach with his subjects. On the other hand, the various organizations that sponsored Joseph Stokes, Jr., of Philadelphia's Children's Hospital, delegated responsibility to him alone. He did not have to adhere to any organizational restrictions as he administered his immunogen at the New Jersey State Colony in New Lisbon, a "state correctional facility for boys, a home for epileptics, and two other institutions for the mentally disabled."[2]

Laissez-Faire

Generally speaking, conscience alone guided researchers' use of human subjects. This attitude originated, philosopher Robert B. Baker asserts, "as advice to individual [American Medical Association] members and did not presuppose any organizational responsibility to society…."[3] Medical historian Allan M. Brandt contends that this mindset failed to keep pace with the ever-expanding world of research, pointing to three nagging problems. First, the use of human subjects, from institutionalized children with mental disabilities to "parent-volunteered" children, failed to follow prescribed guidelines. Second, pharmaceutical production of vaccines proceeded without oversight. Third, the federal government largely abrogated its responsibility to oversee major medical advancements. The absence of regulations for philanthropic organizations and researchers proved wholly inadequate as numerous polio experiments unfolded.[4]

William H. Park, who had produced an immunizing agent for diphtheria and prototyped mass testing and immunization of school children, chaired Milbank's committee studying polio. In 1932, it published *Poliomyelitis: A Survey*, the most comprehensive report about this disease at that time. Park, employed at New York City's department of health and as a lecturer at New York University at age seventy-one, began to assume a "position of leadership in the poliomyelitis field, and the new responsibilities bestowed upon him by the Milbank Committee gave him a degree of confidence about the disease almost equal to that of Dr. [Simon] Flexner."[5]

In 1933, Park enlisted Maurice Brodie, a young Canadian researcher working at New York University's medical school, to work on a polio vaccine. Brodie met with Thomas M. Rivers, a pioneering virologist, at the Rockefeller Institute, to explain how he had combined formalin with an emulsion of infected monkey's spinal cord to develop a killed-virus solution. He had injected it into twenty rhesus monkeys and believed he had produced antibodies. Rivers had doubts about Brodie's claim but, out of respect for Park, remained mum. Park secured funding from the Milbank ($5000), New York ($3000), and Rockefeller ($3000) foundations. To inaugurate its research agenda, the President's Birthday Ball Commission (later, the March of Dimes) awarded him $4000. Park, Paul de Kruif writes, "assured our advisory committee that this vaccine was ready for human test."[6]

Brodie and Park began their experiment by using it themselves and on six health department "volunteers." Three of them received one dose, two had two injections, and only one underwent three. They all developed antibodies. Brodie and Park, however, did not know if they had been previously exposed to the virus—and perhaps had acquired a natural immunity.[7] Satisfied with its safety, in July 1934, Brodie tried it on "twelve children in a New York asylum," ranging in ages from one to six. Antibodies in their blood appeared to increase, once again denoting immunity. They injected seventeen additional children, using various doses and number of shots. Their blood also revealed antibodies. At a Rotary Club luncheon in New York City, on January 25, 1935, Park announced that the vaccine was 85 percent effective. Following experiments on 500 California and 100 New York City children, he confidently declared, "'if given two or three times a year, [it] offers definite protection against poliomyelitis.'"[8]

Based on these preliminary results, Park and Brodie sought and received permission from the US Public Health Service (USPHS) to administer their vaccine in North Carolina and Virginia. They hoped to carry out a trial using observed controls. Injections began in Greensboro in May of 1935. Private physicians administered the entire experiment, conducting all of the jabs in their offices. They gave two doses to 422 children while 36 received only one, for a total of 458.

Another 686 acted as controls. Because these physicians varied in the method and injection sites, because they proved erratic in their follow-up visits and reporting, USPHS experts reported that the entire enterprise had been botched.[9]

Regardless, public optimism swelled. On August 8, 1935, the *New York Times* hailed it as a godsend, announcing "Dr. Brodie in his laboratory at the New York University Medical College now produces about 5,000 doses a week." The tone of this article implied safety and efficacy. The chair of the Fall River, Massachusetts, health board felt enough confidence in it to plead with Brodie for some to stem a polio outbreak. However, after Park and Brodie had used their immunogen on between 8000 and 9000 children, serious illnesses began to be reported. In North Carolina, inoculated participants developed abscesses and experienced fevers, headaches, leg cramps, and nausea, while others suffered from encephalitis. The presence of monkey nervous tissue had been the cause.[10] Their experiment screeched to a halt.

John A. Kolmer, a pathologist at Temple University Medical School, with a grant from the Research Institute of Cutaneous Medicine, a consortium of Philadelphia hospitals and medical schools, had also been working on a polio vaccine. Like Albert B. Sabin some twenty-five years later, he used a live-virus approach. He administered it to forty-two rhesus monkeys. None became ill. Believing it safe, in July of 1934 he injected it into himself and Anna M. Rule, a laboratory assistant, before he gave it to his two young children, aged eleven and fifteen. With no apparent deleterious effects, he added twenty-three children, between eight months and fifteen years at Temple University's hospital in Philadelphia, all "with written consent of the parents." His research team observed no ill effects and found that an overwhelming majority of them showed antibodies where none previously existed, or an increase in antibodies where they had been present.[11] Brimming with confidence, he enlarged the trial: 446 additional individuals—319 of them were children aged six months to fifteen years—in the Philadelphia area. Beginning in April of 1935, the Institute of Cutaneous Medicine "distributed vaccine to 582 physicians in 36 states, including Canada, mostly in … epidemic areas, and the William S. Merrell [Pharmaceutical] Company of Cincinnati" distributed additional doses to "137 physicians to the same areas." In total, Kolmer ultimately experimented, directly or indirectly, on 10,725 individuals. Ninety percent were fifteen years of age or younger.

On November 19, 1935, both Brodie and Kolmer formally reported their findings at the Annual Meeting of the Southern Branch, American Public Health Association, in St. Louis. Kolmer, in particular, revealed that reactions, such as abscesses, fever, headache, and muscle stiffness, had occurred. Worst of all, ten children had developed polio. However, he dismissed these because they had received two doses instead of three. What he described in reality was a medical catastrophe but seemed oblivious, relegating these infections to statistical anomalies that fail to either "prove definitely or disprove that they were caused by the vaccine."[12] Attendees uniformly condemned both vaccines after Brodie and Kolmer read their papers. As Rivers described it, the most dramatic moment occurred when James P. Leake of the Public Health Service "presented … clinical evidence to the effect that the John Kolmer live-virus vaccine caused several deaths in children and then point-blank accused Kolmer of being a murderer."

Both the Park-Brodie and Kolmer research teams pursued vaccines based on incomplete and inaccurate information. They thought that only one strain of this virus existed. Also, because they could not grow it outside of a live host, they used brain and spinal cord tissue from monkeys infected with polio; this precluded a completely sterile solution. Moreover, the Park-Brodie vaccine, Rivers points out, "was made in the most incredibly sloppy manner." Finally, the selection of the right test animal, even sometimes depending on the season of the year, affected the progress of a medical experiment. What is safe for animals may harm humans.[13] "Now the public had two things to fear: the disease itself, and false promises of vaccines."[14]

The scientific community expressed doubts before human trials were even launched. Simon Flexner, harboring early suspicions about the Park-Brodie vaccine, asked Peter Olitsky, a Rockefeller Institute scientist, and Albert Sabin, his assistant, to replicate it: They failed. Flexner cautioned Park, Bodie, and Kolmer, suggesting extended laboratory tests before experimenting on humans but could do nothing to stop them. The research establishment directed its anger less at Park, out of respect for his storied career, and more at lesser-known Brodie and Kolmer. Park strongly objected to this condemnation and promptly retired while New York University fired Brodie. Kolmer never recanted and continued to work at Temple's School of Medicine. The National Committee for the Celebration of the President's Birthday became the final casualty: "Responsible scientists were disgusted with the Birthday

Ball Commission for underwriting the Brodie-Park gamble and creating an atmosphere conducive to the Kolmer tragedy."[15]

Meanwhile, other approaches unfolded. On October 4, 1935, the front page of the *New York Times* beamed, "Serum Held a Cure of Poliomyelitis in Its Early Stage." Edward C. Rosenow, of the Mayo Clinic, University of Minnesota, believed "'poliomyelitis streptococcus' normally lived in the tonsils, but sometimes invaded the bloodstream, where it homed in on the spinal cord and caused paralysis." He claimed to have developed a vaccine and demonstrated it by injecting sixty infected children in Louisville, Kentucky. Researchers criticized Rosenow's fundamental notion connecting streptococcus to polio. Another endeavor, based largely on Flexner's 1910 findings, operated on the theory that the poliovirus invaded the body through nasal passages; that is, children breathed in droplets containing this virus circulating in the air or swallowed it after they stuck their fingers in their mouths. From there, it passed into the brain and ultimately attacked the spinal cord. This idea gained considerable traction in the medical community and remained unchallenged for twenty-five years.[16] In retrospect, John R. Paul wrote in 1971 that such "prevailing interpretations, emanating in large part from the Rockefeller Institute, were accepted as the authoritative word on the human disease, and it was a bold man indeed who offered contrary opinions."[17] Applying a chemical shield into the olfactory canal to protect the mucous membrane thus became popular.

Paul de Kruif, on behalf of the Birthday Ball, approved funding for two of the better-known experiments. In 1934, the Public Health Service's Charles Armstrong squirted alum nasal spray into monkeys and humans with apparent success. Alabama's state health authorities requested that Armstrong employ his mixture to halt an epidemic there. They had crafted a careful field trial, using it both on adults and on children, but in practice it went awry. Although 4600 recorded persons received this treatment through "general practitioners," unknown thousands, some self-medicated, had also used it. This anarchy destroyed any chance of reliable and valid results. De Kruif remained undeterred and backed Edwin W. Schultz, a Stanford University microbiologist, with a $12,000 grant. Schultz had used zinc sulfate on monkeys and firmly claimed it achieved better results. He also believed that a temporary loss of smell would indicate protection. Finally, he felt that proper application required professionals with background in nasal anatomy.

His opportunity for a field trial arose when a severe epidemic struck Toronto, Ontario, during the summer of 1937. That city's health officials agreed to host Schultz's experiment. Newspapers advertised it and embedded permission forms in printed announcements. Some 7000 parents responded. By late August, Schultz's team had sprayed 5000 children, aged three to ten. The second application occurred ten to twelve days later, with 95% of the children showing up to receive it. Another 6300 children served as a control group. However, this trial's integrity became jeopardized because private physicians had administered the spray to another 1000 children, who were not counted in the final sample. Moreover, temporary but serious side effects appeared, among them fever, severe pain between the eyes, nausea, neck stiffness, and vomiting. Worse yet, many children permanently lost their sense of smell. A statistical analysis revealed that the test group recorded eleven cases of infection, while the control group reported nineteen. In sum, the zinc sulfate spray "had not significantly altered the attack rate of the disease," a less than a one percent difference. This was not the answer and by 1939, researchers had grown disenchanted with the whole notion of infection through the olfactory pathway.[18]

Collectively, these tragic failures not only gave researchers pause, but they also proved to be embarrassing for Paul de Kruif and the National Committee for the Celebration of the President's Birthday. The Park-Brodie disaster created a public relations nightmare. The National Committee compounded this when it subsidized the purported development of nasal spray prophylactics. According to Thomas Rivers, a member of the National Committee, it had other, more promising, funding choices than William Park. W. Lloyd Aycock, of Harvard Medical School, and David Kramer, of the Long Island Medical School, "had a long experience working with polio through their association with the Harvard Infantile Paralysis Commission, and knew this virus from the clinical as well as the experimental side." The National Foundation for Infantile Paralysis—the National Committee's successor—eventually gave Kramer a grant; he succeeded in developing a killed-virus vaccine in mice but no one took his breakthrough seriously. As Rivers summarizes "… after the debacle of the Park-Brodie and Kolmer vaccines of 1935, most virologists exhibited a great deal of skepticism when anybody started to discuss the possibilities of making a vaccine."[19] All such developments paused, as a result.

Reboot

To protect itself from any future funding imbroglios, Basil O'Connor put into place three elements to make the National Foundation for Infantile Paralysis (NFIP) less risk averse. First, it maintained "scientific expertise" through an active advisory panel, one that made recommendations and provided supervision. Second, it utilized a clear mission, namely disease eradication. The Foundation's links to Franklin D. Roosevelt required that care be taken toward grantees to avoid embarrassing the office of the president of the USA. Moreover, the use of "charitable donations for operating expenses" relied on popular support and any dangerous experiments could seriously undermine it. Third, it relied on "experience and learning" based on prior failures, particularly the Park-Brodie calamity.[20]

With these criteria in place, the largest medical experiment in human history involving children would proceed. Some doubt lingered though. Thomas Rivers, who chaired the NFIP's Committee on Research, vividly recalls the ethical context in the USA during the early 1950s: "… there was no settled code or standard to which one could refer regarding such experiments." It appeared to be a fluid period. He alludes to the Nuremberg Code as ambiguous and, at times, "contradictory." Rivers, in sum, stresses "that the general question of human experimentation was one of the key questions that had to be considered in extending Dr. [Jonas] Salk's tests on the scale of a field trial."

The National Foundation and Salk cobbled an experimental protocol that, whether consciously or not, seemed to incorporate aspects of the Nuremberg Code but only loosely so, and at certain times and places. Some federal oversight occurred, but only in a limited and informal manner. Moreover, the much overlooked and unique feature of this story is the indispensable role of the public schools operating as an extension of the medical laboratory. Both the NFIP and Salk had to get it right. Since the public education system had no policy concerning school-based research, the onus rested on them. The subsequent use of these facilities as mass vaccination centers underscores their singular and essential character.[21]

A Delicate Balance

Nothing could be accomplished, however, unless scientists unlocked polio's secret. This began in 1937 when researchers at the Yale Poliomyelitis Study Unit, headed by John R. Paul, discovered two poliovirus strains, corroborating earlier findings of Australian and French scientists. World War II's demands on medical expertise and resources delayed progress, but in 1946 National Foundation officials deemed it a priority. Harry M. Weaver, appointed as research director in 1946, initiated "round-table discussions whereby NFIP-supported scientists came together to exchange ideas and think through problems...." At the first meeting, in 1947, "researchers agreed that before developing a human vaccine, they had to determine the number of different poliovirus types." A truly effective vaccine would have to confer immunity on all types. The typing project had begun. Over a two-year span, they oversaw a cooperative effort that initially involved 100 virus samples but soon expanded to 196, studied by researchers at the universities of Kansas, Southern California, Utah, and Pittsburgh. They successfully identified three members or serotypes. Type I seemed to be the most prevalent followed in order by Type II and Type III.[22] Jonas E. Salk's work on this project in Pittsburgh stood out.

Born in 1914 in New York City, son of Russian–Jewish immigrants, Salk grew up witnessing classmates wearing braces, bequeathed by the 1916 poliomyelitis epidemic, and horse-drawn carriages clopping down the streets piled with coffins, victims of the 1917 influenza pandemic. He qualified for admissions to a prestigious, classical high school and entered City College of New York. Inspired by Louis Pasteur's biography, Salk chose medical research over practice, attending New York University's medical school, and publishing his first paper before completing his degree. He worked with Thomas Francis, Jr., during his last year of study, focusing on his mentor's pet project, influenza. Their lives became inextricably intertwined.

Salk's mentor also came from a humble background, born into a Welsh immigrant family in 1900. His father worked in the mills of New Castle, Pennsylvania, a steel and foundry town located northwest of Pittsburgh. Francis completed high school followed by solid academic achievements at Allegheny College, in northwestern Pennsylvania, and entered Yale University's medical school in 1921. Following his internship, he spent ten years at the Rockefeller Institute for Medical Research. There he encountered Thomas Rivers, who encouraged him to focus on influenza research. This proved to be fortuitous advice leading Francis to become a vaccine pioneer. Medical canon embraced Edward Jenner's and Louis Pasteur's success with live-virus immunogen, but Francis deviated from this path by pursuing a dead-virus approach. In 1938, he accepted the chair position in the department of epidemiology at New York University's medical school. He continued his influenza research, using ultraviolet light to inactivate it at first but switched to formaldehyde. Salk not only became one of his students but a "junior colleague."[23]

Not long after Francis accepted the chair position at the University of Michigan's epidemiology department, his research became a significant part of the war effort. After the death of 45,000 American troops during World War I because of the great flu pandemic, military mobilization following Pearl Harbor focused on "protecting them with every vaccine available."[24] The "secretary of war formed the Commission on Influenza with Francis as Director." Working for the Armed Forces Epidemiological Board, he continued to refine his vaccine. According to biographer Charlotte DeCroes Jacobs, Jonas Salk, funded by an NFIP fellowship, joined Francis at Michigan in April of 1942.

The first sizeable field trials occurred in late 1942 when Salk and two other researchers inoculated "8,000 psychiatric patients" at Eloise Hospital for the Insane. No one experienced any adverse reactions, and 85 percent of them developed antibodies. Safety was assured. In order to measure effectiveness, Salk staged a blind-control group at Ypsilanti State Hospital, where he injected "200 male residents half with vaccine, half with saline.... In May of 1943," Jacobs continues, "two weeks after the last inoculation, they exposed inmates to influenza, spraying mist made from dried, infected mouse lung tissue into their nostrils." It proved successful: "just 16 percent of those inoculated with vaccine became ill whereas almost half of the control subjects who had received the placebo contracted the flu." With a larger trial needed to confirm efficacy, Francis directed his research team to implement a double-blind protocol using 15,000 army enlistees. The blood they drew brimmed with antibodies. "[O]nly 2 percent of the vaccinated group" fell ill during that flu season. When the surgeon general "decided to vaccinate the entire army ... Salk volunteered to consult with seven manufacturers on procedures and production."

Both Francis and Salk received considerable recognition for their work, with the former receiving the "Medal of Freedom and elected to the National Academy of Sciences."[25] Salk's "ability to design the vaccine trials and to carry them through to completion was of great help to Francis." This experience not only proved indispensable when it came time to design "his own experimental trials of the inactivated poliovirus vaccine" but also raised his stock as a virologist.[26] Moreover, that war's immunization program greatly boosted the drive to do the same for the American public. The "military campaign against flu would be transformed into a general attack on viruses and would lead to the conquest of polio and other viral diseases after the war." Salk, chaffing as a subordinate, accepted a position as director of the newly formed Virus Research Laboratory at the University of Pittsburgh. He moved there in August of 1947 where he found a wholly inadequate research facility located in the basement of Municipal Hospital, housing that city's contagious patients.[27]

Salk's original agenda focused on the influenza vaccine. However, when Harry Weaver visited the rising, young virologist in December 1947, Salk's research agenda took a dramatic turn. Weaver invited him to participate in the Foundation's typing program and arranged for a National Foundation grant of $148,000 to subsidize Salk's expenses for a year, beginning in the late summer of 1948. The Foundation's Committee on Typing, consisting of David Bodian, of Johns Hopkins University, Thomas Francis, and Albert Sabin, at the University of Cincinnati, among others, oversaw the entire operation. Salk announced the results of that project on September 3–7, 1951, at the Second International Poliomyelitis Congress held in Copenhagen, Denmark. During his return to the USA, Salk developed a close shipboard relationship with Basil O'Connor, another passenger on the Queen Mary. Although they knew of each other, they had never formally interacted. This long transatlantic passage proved propitious: Salk proposed the notion of pursuing a dead-virus vaccine to O'Connor, leaving a deep and lasting impression.[28]

Laboratory Results

By 1951, all of the pieces had fallen into place. John Enders's research team, at Harvard University, subsidized with NFIP money, had discovered how to grow poliovirus in non-nervous tissue cultures, an essential tool to provide an ample source of safe (i.e., free of animal contaminants) virus for study and development. Furthermore, David Bodian, at Johns Hopkins, and Dorothy Horstmann, at Yale, had confirmed that poliovirus used the bloodstream to reach the nervous system, confirming that a vaccine can confer immunity. Finally, the National Foundation had awarded Jonas Salk a three-year, $220,000 subsidy permitting the Virus Research Laboratory to expand, now covering three floors of Municipal

Hospital with over fifty supremely qualified staff members. Based at that hospital, they witnessed the impact of poliomyelitis on children's lives on a daily basis: It was their reality, not some distant or abstract consequence, as a nurse who worked there recalls: "One year the ambulances literally lined up outside the place.... It was an atmosphere of grief, terror, and helpless rage. It was horrible."[29]

Salk's experimental sequence consisted of five distinct stages: laboratory, preliminary, personal, local, and national trials. It spread outward geographically with each step. Although more methodical and thorough than previous polio experiments and many contemporary research efforts involving children, it still represented a hazardous undertaking at many different levels.

He and his research team first tested the experimental, dead-virus vaccine on mice and by the end of 1951 had ably immunized monkeys.

Salk delivered a paper, "Studies with Non-Infectious Poliomyelitis Virus Vaccines," at the International Poliomyelitis Congress, held in Rome, Italy, during the first week of September 1954, announcing this breakthrough. With a statistical flourish, he described the various stages of this research. The first involved laboratory experiments on tissue samples. These focused on the correct formalin ratio (to inactivate the virus), preparation temperature, and pH level. The second part followed a twofold operation, expanding production and pursuing more ambitious tests. Salk selected two different pharmaceutical companies to manufacture batches based on his specifications. His staff reviewed these to assure consistency. They injected these into laboratory monkeys and drew blood to test for antibodies. They used these results to adjust the potencies of each of the three polio strains as well as tentatively determine the number of injections needed to induce the most antigens. The third level involved human subjects.[30]

Trial and Error

The National Foundation's Committee on Immunization, reconstituted from the Committee on Typing, met in Hershey, Pennsylvania, on March 17, 1951, to consider vaccine experiments on humans. Attendees knew all too well about the ill-fated Park-Brodie and John Kolmer human trials twenty years earlier. They also were aware of current efforts.

Isabel Morgan, at Johns Hopkins, "demonstrated beyond a shadow of a doubt that she had been able to immunize rhesus monkeys with formalin-inactivated viruses of all three basic immunological types to a point where it was impossible to bring down such animals by the sensitive routes." Howard A. Howe, her colleague and a virologist, took these findings a step further.[31] Believing these results could be replicated with humans, Howe administered it to six children with mental disabilities at the Rosewood Training School in Owings Mills, Maryland. "He described [them] as 'low grade idiots or imbeciles with congenital hydrocephalus, microcephaly, or cerebral palsy.'" They developed antibodies against all three types of the virus. While this certainly represented a significant finding, Howe's vaccine contained monkey spinal-cord material and was deemed unsuitable for larger human trials.[32] Finally, scientists at the Centers for Disease Control (CDC) had worked on an attenuated polio vaccine, successfully testing it on laboratory monkeys and going as far as self-administration. However, since the National Foundation's looming national trial of Salk's immunogen overshadowed their efforts, the CDC's "Montgomery lab was asked to stop working on the oral vaccine altogether. Its strains of the attenuated virus ... were sent to Albert Sabin and others working in the field. It was a signal contribution" as historian Elizabeth W. Etheridge notes, "by CDC to the oral poliovirus program."[33]

Misdeeds

Others pushed the ethical envelope. Hilary Koprowski and Harold Cox at the Lederle Division of the American Cyanamid Company, a large pharmaceutical manufacturer, located in Pearl River, New York, also experimented with an attenuated virus. Koprowski, a Polish-trained medical scientist, had escaped from Nazi-occupied Poland and fled to Brazil where he worked on yellow fever for the Rockefeller Foundation. Cox, a vaccinologist and a former employee of the USPHS, lured Koprowski to Lederle in 1944. Together they improved the rabies vaccine, a highly profitable product for that company. Koprowski and Cox also began work on a poliovirus immunogen. This represented an unprecedented commercial venture into vaccine development, and Lederle laboratories intended to win this race. Koprowski, using a simple Waring blender, pureed the spinal cord and brain tissue of cotton rats to produce a Type II live-virus mixture, testing it on chimpanzees. In January of 1950, Cox, Koprowski, and several coworkers ingested it; they all developed antibodies.[34]

George A. Jervis, director of research at Letchworth Village, opened in 1912 to house children with mental disabilities and located in the Hudson River Valley, contacted Koprowski about using his vaccine on residents there. Jervis feared their throwing of feces may cause an outbreak. On February 27, 1950, Koprowski and his assistant fed a cubic centimeter of the gray vaccine, immersing it in chocolate milk, to a "nonimmune human [male] volunteer." Blood work, fourteen days later, revealed antibodies. They continued to observe him for six weeks. With no adverse signs, they fed the mixture to another boy who also developed antibodies. They monitored him for six weeks. With two successful outcomes, Koprowski gave it to eighteen other "volunteers," isolating them to minimize contamination. None displayed signs of illness, and all had antibodies in their blood.[35] Koprowski did not seek permission from the New York State Department of Health to conduct these experiments because he knew that he would be refused. In a 1956 article in the *Journal of the American Medical Association*, he vaguely refers to a pediatrician obtaining parental permissions for him, but doubt exists whether he actually acquired them. Cox only learned about the Letchworth trial after it had been completed.[36] When Koprowski conferred with Thomas Rivers at the Rockefeller Institute following the Letchworth experiment, Rivers expressed concern because it cast serious doubt on the basic notion of voluntary subjects. Nevertheless, in retrospect, Rivers readily admits that using "mentally defective children as test subjects" in general and "what Koprowski wanted to do in particular was not unusual–you might even say that it was standard practice."[37]

Koprowski delivered a ten-minute report at the National Foundation's March 1951 meeting of the Immunization Committee in Hershey, Pennsylvania. Albert Sabin, another attenuated-vaccine proponent, rebuked him for feeding live poliovirus to children. Another committee member inquired if Koprowski was aware that he may be vulnerable to a lawsuit filed by the Society for the Prevention of Cruelty to Children. Koprowski remained undeterred and began research into a Type I vaccine, though he and Cox went their separate ways in 1952.[38]

In July of that year, Koprowski sought and obtained official permission from California state authorities to continue tests of his oral vaccine on sixty-one children with mental disabilities residing at the Sonoma State Hospital. According to his biographer, Roger Vaughn, Koprowski obtained parental permission this time. He fed both Type I and Type II attenuated poliovirus to them; fifty-two of them developed antibodies.

However, Koprowski discovered that live virus existed in six of the tested children's feces. His research team mixed these children with eight non-immunized children for three hours a day to determine if the disease would be transmitted from the feces as they played together on a "plastic mat." The hospital staff washed that mat "to remove gross soils" but never disinfected it. Three of the eight children contracted polio, revealing that the live-virus approach could cause secondary infections, a hidden and serious danger. Koprowski continued his trials at New Jersey's Clinton Farms female prison. It housed not only women accused of crimes but also their children from unauthorized assignations. Koprowski and his assistants tested them and found no antibodies. The permissions needed for this experiment, from blood tests to dosing, seemed hopeless at first because many of the mothers were younger than twenty-one; nevertheless, that state's attorney general waived the age restriction, blithely disposing of any legal or moral barriers. Meanwhile, Lederle's executives began to drag their feet concerning the development of an attenuated vaccine, thus allowing Albert Sabin to take the lead and gain notoriety. Koprowski forged ahead anyway. But after the initial successes of Jonas Salk's dead-virus version in the USA, and the presence of millions of inoculated children, Koprowski had to turn elsewhere for unprotected test subjects. He persuaded Northern Ireland's medical community to conduct a large-scale trial in Belfast, but additional laboratory tests there revealed Koprowski's mixture to be potentially contagious. Having failed there, he made hasty arrangements in the Belgian Congo (now the Democratic Republic of the Congo), and squirted it into the mouths of 250,000 children. The absence of sophisticated laboratory facilities there to analyze the outcome left the entire enterprise incomplete.[39]

Koprowski's rush to enlist human subjects without regard for their safety sparked an international outcry. The Nuremberg Code unequivocally banned the use of children for medical tests. Great Britain, which held jurisdiction over Northern Ireland, officially prohibited such practices and Pope Pius XII "condemned experiments" that violated human rights.[40]

Hilary Koprowski left Lederle to become director of the Wistar Institute in Philadelphia while Harold Cox continued to pursue his research on Type I and Type III poliovirus. In 1959, Lederle arranged large-scale trials in Florida and Minnesota, providing 520,000 free doses. In the former case, "90 percent [of the Dade County] school population received ... the vaccine." Many became ill and, worse yet, six suffered paralysis. Like Koprowski, Cox had to go overseas to find non-immunized subjects. With the support of the Pan-American Health Organization, he pursued human trials in Colombia, Nicaragua, and Uruguay in 1958. Those conclusions, as with Koprowski's African experiments, proved useless. Lederle dumped Cox's work and pursued production of Albert Sabin's vaccine which, by that time, proved successful.[41] In the interim, Salk and the National Foundation forged ahead with its inactivated-virus approach.

Notes and Sources

1. Gonzalo-Herranz Rodríguez, "1952 Address of Pope Pius XII to the Medical Community," *L'Osservatore Romano English Edition,* July 30, 2003: 4. www.ewtn.com/library/ISSUES/BIORSP12.HTM (Accessed July 19, 2011). According to Thomas M. Rivers, *Reflections on a Life in Medicine and Science.* An Oral History Prepared by Saul Benison (Cambridge: Massachusetts Institute of Technology, 1967), the group that Pope Pius XII addressed was the First International Congress on the Histopathology of the Nervous System (498). Also see Tony Gould's reference to the Pope's address in, *A Summer Plague: Polio and Its Survivors* (New Haven: Yale University Press, 1995): 126.
2. Sidney A. Halpern, *Lesser Harms: The Morality of Risk in Medical Research* (Chicago: University of Chicago Press, 2004): 44–45 and 76, respectively. Also refer to pages 73–75.
3. Robert B. Baker, *Before Bioethics: A History of American Medical Ethics from the Colonial Period to the Bioethics Revolution* (New York: Oxford University Press, 2013): 229. See also pages 225–27.
4. Allan M. Brandt, "Polio, Politics, Publicity, and Duplicity: Ethical Aspects in the Development of the Salk Vaccine," *International Journal of Health Services* 8 (1978): 268. Focusing on the 1954 trial, Halpern, *Lesser Harms,* corroborates Brandt. Especially check pages 64–65. Finally, consult Baker, *Before Bioethics*: 17.
5. John R. Paul, *History of Poliomyelitis* (New Haven: Yale University Press, 1971): 302–3. Also check MOD; "History of the National Foundation for Infantile Paralysis, Volume III: The Medical Research Program of NFIP" (umpublished): 1–2. International Committee for the Study of Infantile Paralysis, *Poliomyelitis: A Survey* (Baltimore: Williams & Wilkins Co., 1932): v, ix, x; Wade W. Oliver, *The Man Who Lived for Tomorrow: A Biography of William Hallock Park.* New York: E.P. Dutton & Co., 1941): 434; Richard Carter, *The Gentle Legions: National Voluntary Health Organizations in America* (New Brunswick, NJ: Transaction

Publishers, 1992): 114; "Dr. Park, 71, Guards City's Health Still," *New York Times*, December 31, 1934: 15.

6. Paul de Kruif, *The Sweeping Wind: A Memoir* (New York: Harcourt, Brace and World, Inc., 1962): 179. Refer also to pages 177–78. See, especially, "History of the National Foundation for Infantile Paralysis": 4–5. Evelynn Maxine Hammonds, *Childhood's Deadly Scourge: The Campaign to Control Diphtheria in New York City, 1880–1930* (Baltimore: Johns Hopkins University Press, 1999), provides biographical and professional background for Park as does Oliver, *The Man Who Lived for Tomorrow*: 111–12. Further consult Arthur Allen, *Vaccine: The Controversial Story of Medicine's Greatest Lifesaver* (New York: W. W. Norton Company, 2007): 172; Roland H. Berg, *Polio and Its Problems* (Philadelphia: J. B. Lippincott, 1948): 59; Maurice Brodie, "Active Immunization in Monkeys Against Poliomyelitis with Germicidally Inactivated Virus," *Science* 79 (29 June 1934): 594–95; Carter, *Gentle Legions*, 115; Aaron E. Klein, *Trial by Fury: The Polio Vaccine Controversy* (New York: Charles Scribner's Sons, 1972): 20; Rivers, *Reflections on a Life in Medicine and Science*: 185; John Troan, *Passport to Adventure: Or, How a Typewriter from Santa Led to an Exciting Lifetime Journey* (Pittsburgh, PA: Geyer Printing, 2000): 206–7; and Heather Green Wooten, *The Polio Years in Texas: Battling a Terrifying Unknown* (College Station: Texas A&M University Press, 2009): 131. Finally, see "Paralysis Vaccine Is 85% Successful," *New York Times*, January 25, 1935: 23.
7. Maurice Brodie, "Active Immunization of Children Against Poliomyelitis with Formalin Inactivated Virus Suspension," *Proceedings of the Society for Experimental Biology and Medicine* 32 (November 1934): 301. See also pages 300 and 302.
8. "Paralysis Vaccine Is 85% Successful." Further consult Brodie, "Active Immunization of Children": 301–2; Maurice Brodie and William H. Park, "Active Immunization Against Poliomyelitis," *Journal of the American Medical Association* 105 (5 October 1935): 1090; Berg, *Polio and Its Problems*: 53; and John R. Wilson, *Margin of Safety: A Doctor Tells the Controversial Inside Story of the Development of the Polio Vaccine* (Garden City, NY: Doubleday & Co., 1963): 49, 53–54.
9. Rivers, *Reflections on a Life in Medicine and Science*: 185. Refer to page 187, as well as Maurice Brodie and William H. Park, "Active Immunization Against Poliomyelitis," *New York State Journal of Medicine* 35 (15 August 1935): 816–17; Maurice Brodie and William H. Park. "Active Immunization Against Poliomyelitis," *American Journal of Public Health* 26 (February 1936): 119, 123–24. For other descriptions of the Park-Brodie experiment, see Allen, *Vaccine*: 172–73; Lawrence R. Altman, *Who Goes First? The Story of Self-Experimentation in Medicine*

(Berkeley: University of California Press, 1998): 126; Berg, *Polio and Its Problems*: 54–55; Carter, *Gentle Legions*: 116; Victor Cohn, *Four Billion Dimes* (Minneapolis: 1955): 47–48; Thomas M. Daniel and Frederick C. Robbins, eds,. *Polio* (Rochester, NY: University of Rochester Press, 1997): 14; A. G. Gilliam and R. H. Onstott, "Results of Field Studies with Poliomyelitis Vaccine," *American Journal of Public Health*, 26 (February 1936): 113–15; Tony Gould, *A Summer Plague: Polio and Its Survivors* (New Haven: Yale University Press, 1995): 61, 66, 125; Halpern, *Lesser Harms*: 48, 50; Klein, *Trial by Fury*: 20–21; Susan E. Lederer, *Subjected to Science: Human Experimentation in America before the Second World War* (Baltimore: Johns Hopkins University Press, 1995): 108; David M. Oshinsky, *Polio: An American Story* (New York: Oxford University Press, 2005): 57; Paul, *History of Poliomyelitis*: 254–55, 261; Naomi Rogers, *Dirt and Disease: Polio Before FDR* (New Brunswick, NJ: Rutgers University Press, 1996): 171; and Edmund J. Sass, with George Gottfried and Anthony Sorem, *Polio's Legacy: An Oral History* (Lanham, MD: University Press of America, 1996): 8.

10. "200 Secretly Get Paralysis Vaccine," *New York Times*, August 24, 1935: 17. For the *Times*'s entire coverage during these months, see "Anti-Paralysis Serum Reported a Success In Wholesale Tests Made in California," *New York Times*, February 3, 1935: 21; "City Sends Serum to North Carolina," *New York Times*, July 7, 1935: 18; "Boston Awaits Serum," *New York Times*, August 31, 1935, Social News Section: 6; "The Week in Science," *New York Times*, September 1, 1935, Aviation-Automobiles-Science-Travel-Resorts-Steamships Section: 5. Moreover, check Brodie and Park, "Active Immunization Against Poliomyelitis": 123–24; Gilliam and Onstott, "Results of Field Studies with Poliomyelitis Vaccine": 116.
11. John A. Kolmer, George F. Klugh, and Anna M. Rule, "A Successful Method for Vaccination Against Acute Anterior Poliomyelitis," *Journal of the American Medical Association* 104 (9 February 1935): 457. See pages 458 and 460, as well as John A. Kolmer and Anna M. Rule, "Concerning Vaccination of Monkeys Against Acute Anterior Poliomyelitis," *Journal of Immunology* 26 (June 1934): 507, 509; John A. Kolmer, "Susceptibility and Immunity in Relation to Vaccination in Acute Anterior Poliomyelitis," *Journal of the American Medical Association* 105 (December 1935):1960–62. Secondary sources include Allen, *Vaccine*: 173; Altman, *Who Goes First?* 127; Berg, *Polio and Its Problems*: 53–54; Halpern, *Lesser Harms*: 49–50; Carter, *Gentle Legions*: 116; Daniel and Robbins, *Polio*: 14; Lederer, *Subjected to Science*: 107; Oshinsky, *Polio*: 57; Paul, *History of Poliomyelitis*: 257–58; and Wilson, *Margin of Safety*: 52–54. Further consult "The Week in Science," *New York Times*, Aviation-Automobiles-Science-Travel-Resorts-Steamships Section: 5.

12. John A. Kolmer, "Vaccination Against Acute Anterior Poliomyelitis," *American Journal of Public Health* 26 (February 1936): 128 and 132, respectively. Consult pages 129–31, 135, as well as Berg, *Polio and Its Problems*: 55; Paul, *History of Poliomyelitis*: 258.
13. Rivers, *Reflections on a Life in Medicine and Science*: 189 and 185, respectively. Refer to pages 186–88, 190, as well as Brodie and Park, "Active Immunization Against Poliomyelitis": 119–25; Henry Vaughan, C. H. Halliday, John F. Kessel, Maurice Brodie, and James P. Leake, "Discussion of Poliomyelitis Papers," *American Journal of Public Health* 26 (February 1936): 148; "Questions Safety of Paralysis Vaccine," *New York Times*, October 9, 1935: 11; "Infantile Paralysis Vaccines," *New York Times*, October 11, 1935: 24; "Flexner Rejects Paralysis Vaccine," *New York Times*, November 1, 1935: 23; "Dr. Brodie Upholds Paralysis Vaccine," *New York Times*, November 3, 1935: 28. For secondary treatments, consult Allen, *Vaccine*: 173; Altman, *Who Goes First?*: 298–99; Berg, *Polio and Its Problems*: 54–55, 59; Allan Chase, *Magic Shots: A Human and Scientific Account of the Long and Continuing Struggle to Eradicate Infectious Disease by Vaccination* (New York: William Morrow & Co., 1982): 280–83; Daniel and Robbins, *Polio*: 14; Klein, *Trial by Fury*: 21; Lederer, *Subjected to Science*: 107–8; Paul A. Offit, *The Cutter Incident: How America's First Polio Vaccine Led to the Growing Vaccine Crisis* (New Haven: Yale University Press, 2005): 16–17; Oshinsky, *Polio*: 57; and Paul, *History of Poliomyelitis*: 258–59. Halpern, *Lesser Harms*, oddly minimizes the Park-Brodie and Kolmer episodes, treating them as aberrations (48–50, 54, 61, 64–65).
14. Nina Gilden Seavey, Jane S. Smith, and Paul Wagner, *A Paralyzing Fear: The Triumph Over Polio in America* (New York: TV Books, 1998): 164–65. Berg, *Polio and Its Problems*, states a similar outcome, describing it as a "spectre" for "many years" (59). Refer especially to James P. Leake, "Poliomyelitis Following Vaccination Against this Disease," *Journal of American Medical Association*, 105 (December 28, 1935): 2152.
15. Carter, *Gentle Legions*: 116. John A. Kolmer, "Active Immunization Against Acute Anterior Poliomyelitis with Ricinoleated Vaccine," *Journal of Immunology*, 32 (May 1937): 343, 355–56. For additional background, consult Allen, *Vaccine*: 173; Daniel and Robbins, *Polio*: 14; Klein, *Trial by Fury*: 21; Leake, "Poliomyelitis Following Vaccination": 2152; Lederer, *Subjected to Science*: 107–8; Charles L. Mee, *A Nearly Normal Life* (Boston: Little, Brown & Co., 1999), 54–55; Oshinsky, *Polio*: 57; Rivers, *Reflections on a Life in Medicine and Science*: 186–88, 190.
16. Gareth Williams, *Paralyzed with Fear: The Story of Polio* (New York: Palgrave Macmillan, 2013): 73. "Serum Held a Cure of Poliomyelitis in

Its Early Stage," *New York Times*, October 4, 1935: 1; "Experts Skeptical on Paralysis 'Cure,'" *New York Times*, October 5, 1935: 17; "Doubts Paralysis 'Cure,'" *New York Times*, October 6, 1935: 45. Rosenow two decades previous had claimed this connection and announced a possible vaccine. For this, refer to "Paralysis Germ Probably Found," *New York Times*, October 22, 1916: Section 1: 19; "Remedy for Infantile Paralysis," *New York Times*, November 18, 1917, Sunday Magazine Section: 3; Flexner and Lewis, "Experimental Poliomyelitis in Monkeys": 1782.

17. Paul, *History of Poliomyelitis*: 241. Also see Rivers, *Reflections on a Life in Medicine and Science*: 191–93.
18. Berg, *Polio and Its Problems*: 38–39, 43, respectively. Also check pages 35, 40–42. See especially MOD; Medical Programs Records; Series 8: Grants and Appropriations; Approved Grants, 1938–1941. Carter, *Gentle Legions* (117–18) provides fine background on these nasal spray experiments, as does Paul, *History of Poliomyelitis* (241). In the latter, further consult pages 191, 238, 240, 244, 247–48, 250, 385, 389. For additional information, refer to Allen, *Vaccine*: 172; Saul Benison "Speculation and Experimentation in Early Poliomyelitis Research," *Clio Medica* 10 (1975): 17; Cohn, *Four Billion Dimes*: 49; Gould, *A Summer Plague*: 69; Klein, *Trial by Fury*: 34; and Oshinsky, *Polio*: 124; Rivers, *Reflections on a Life in Medicine and Science*: 191–92.
19. Rivers, *Reflections on a Life in Medicine and Science*, 190–91 and 421, respectively. Refer also to Allen, *Vaccine*: 172; Cohn, *Four Billion Dimes*: 49; Liza Dawson, "The Salk Polio Vaccine Trial of 1954: Risks, Randomization and Public Involvement." *Clinical Trials* 1 (2004): 124; and Gould, *A Summer Plague*: 69.
20. Halpern, *Lesser Harms*: 44–45, 73–76, and 80–88, respectively. Also consult Carter, *Gentle Legions*: 114.
21. Rivers, *Reflections on a Life in Medicine and Science*: 498–99. Especially consult Lainie Friedman Ross, *Children in Medical Research: Access versus Protection* (New York: Oxford University Press, 2006): 154.
22. Charlotte DeCroes Jacobs, *Jonas Salk: A Life* (New York: Oxford University Press, 2015): 75. Also see pages 73, 76, and Box 7, File Folder, International Poliomyelitis Congress (Collection #90/36/7), Jonas Salk Papers, UPITT; Jonas E. Salk, et al., "The Use of Adjutant to Facilitate Studies on the Immunologic Classification of Poliomyelitis Viruses," *American Journal of Hygiene* 54 (September 1951): 157–73. Finally, consult Stuart Blume, *Immunization: How Vaccines Became Controversial* (London: Reaktion Books, 2017): 77.
23. John R. Paul, "Thomas Francis, Jr., July 15, 1900–October 1, 1969," *Biographical Memoirs*, 44 (1974): 70. Refer to pages 57, 62–65, 67–68,

71, as well as Allen, *Vaccine*: 162, 179; Tony Gould, *A Summer Plague: Polio and Its Survivors* (New Haven: Yale University Press, 1995): 128–29; Jacobs, *Jonas Salk*: 14, 17, 20, 23–24, 37–38; Oshinsky, *Polio* Oxford University: 96–99, 102–3; and Jane S. Smith, *Patenting the Sun: Polio and the Salk Vaccine* (New York: William Morrow and Co., 1990): 172.

24. Allen, *Vaccine*: 137. Also check Offit, *Cutter Incident*: 30.
25. Jacobs, *Jonas Salk*: 32–33, 43–44, 45–46, and 47, respectively. Of all of the existing biographies, Jacobs most humanizes Salk. Moreover, unlike other biographers and medical historians, she points out Salk's pioneering work on both flu and polio vaccines. Further consult Jeffrey Kluger, *Splendid Solution: Jonas Salk and the Conquest of Polio* (New York: G. P. Putnum's, 2004): 64, 66; Paul A. Offit, *Deadly Choices: How the Anti-Vaccine Movement Threatens Us All* (New York: Basic Books, 2011): 59; and Oshinsky, *Polio*: 100–1, 104–5; Paul, "Thomas Francis, Jr.": 74.
26. Paul, "Thomas Francis, Jr.": 74. Refer again to Jacobs, *Jonas Salk*: 47–48.
27. Allen, *Vaccine*: 137, 142–59. See also Chase, *Magic Shots*: 275; Jacobs, *Jonas Salk*: 44, 50–52, 54–57; Kluger, *Splendid Solution*: 75, 77; and Offit, *Deadly Choices*: 54.
28. News Release from the Second International Poliomyelitis Conference of the International Poliomyelitis Congress–8/28/51 and News Release from the Second International Poliomyelitis Conference of the International Poliomyelitis Congress–9/5/51 (Copenhagen), Box 7, File Folder, International Poliomyelitis Congress (Collection #90/36/7), Jonas Salk Papers, UPITT. Further consult Allen, *Vaccine*: 179–80; Blume, *Immunization*: 78; Jacobs, *Jonas Salk*: 60–61, 64, 79, 81, 89–90, 93; Offit, *Cutter Incident*: 28; and Williams, *Paralyzed with Fear*: 136, 196.
29. This nurse is quoted by Dawson, "The Salk Polio Vaccine Trial of 1954: Risks": 123. I also use it in Richard J. Altenbaugh, *Last Children's Plague: Poliomyelitis, Disability, and Twentieth-Century American Culture* (New York: Palgrave Macmillan: 2015): 24. Also refer to MOD; Medical Programs Records; Series 9: Human Resources; Biographical Data, Dr. Harry M. Weaver, 1953. "Chronology of Events in Salk Research for Polio Vaccine," Jonas Salk Papers, Box 7, File Folder, Salk Polio Research: Chronology (Collection #90/36/7), UPITT; MOD; Medical Programs Records; Series 8: Grants and Appropriations; Approved Grants, 1950–51 and Approved Grants, 1952–53; Jonas E. Salk, et al., "Studies in Human Subjects on Active Immunization Against Poliomyelitis. Part. I. Preliminary Report of Experiments in Progress," *Journal of the American Medical Association*, 151 (March 28, 1953):

1081–82. Finally, consult Allen, *Vaccine*: 179–80; Blume, *Immunization*: 76; Victor Cohn, *Four Billion Dimes* (Minneapolis: 1955): 95–97; Dorothy M. Horstmann,, Robert W. McCollum, and Anne D. Mascola, "Viremia in Human Poliomyelitis." *Journal of Experimental Medicine*, 99 (April 1954): 355–69; Jacobs, *Jonas Salk*: 96; Klein, *Trial by Fury*: 63; Kluger, *Splendid Solution*: 121; "Obituary: Polio Pioneers Joseph Melnick, Dorothy M. Horstmann Die," *Washington Post*, January 22, 2001: B-6. http://pitt.idm.oclc.org/login?url=http://search.proquest.com/docview/409060435?accountid=14709 (Accessed November 7, 2015); Oshinsky, *Polio*: 107, 109, 150, 123–24, 153; Smith, *Patenting the Sun*: 116. See especially Paul, *History of Poliomyelitis*: 493–94. As a member of the NFIP's immunization committee, he provides an insider's look of that organization's trappings.

30. Salk, et al., "Studies in Human Subjects on Active Immunization against Poliomyelitis, Part I": 1082, 1085. The copy I reference is housed in the Jonas Salk Papers, Box 8, File Folder, Written Papers (Collection #90/36/7), UPITT. *News Release of the Third International Poliomyelitis Conference*, 9/6–9/10, 1954 (Rome, Italy), Jonas Salk Papers, Box 7, File Folder, International Poliomyelitis Congress (Collection #90/36/7), UPITT. Moreover, see Jacobs, *Jonas Salk*: 104–5; Klein, *Trial by Fury*: 48, 69; Rivers, *Reflections on a Life in Medicine and Science*: 458; and J. David Smith and Alison L. Mitchell, "Sacrifices for the Miracle: The Polio Vaccine Research and Children with Mental Retardation," *Mental Retardation* 39 (October 2001): 406.
31. Rivers, *Reflections on a Life in Medicine and Science*: 456–58. Also check Chase, *Magic Shots*: 295; Klein, *Trial by Fury*: 48, 69. MOD; Medical Programs Records; Series 5. Committees. Memo from Dr. Kumm to Dr. Van Riper, July 11, 1955.
32. See Salk, et al., "Studies in Human Subjects on Active Immunization against Poliomyelitis, Part I": 1082. Please see Chase, *Magic Shots*: 295; Jacobs, *Jonas Salk*: 78.
33. Elizabeth W. Etheridge, *Sentinel for Health: A History of the Centers for Disease Control* (Berkeley: University of California Press, 1992): 88. Refer also to pages 70–71.
34. Hilary Koprowski, George A. Jervis, and Thomas W. Norton, "Immune Responses in Human Volunteers Upon Oral Administration of a Rodent-Adapted Strain of Poliomyelitis Virus," *American Journal of Hygiene* 55 (1952): 108–9; Wilson, *Margin of Safety*: 148–49. See as well pages 136–38, 142, 145. Wilson seems to be the only writer to grapple, to any extent, with the ethics of human subjects research regarding polio, but much of it, written in 1963, appears to be a rationalization at best and

glib at worst (148–54). Also see Chase, *Magic Shots*: 303; Klein, *Trial by Fury*: 59; and Williams, *Paralyzed with Fear*: 222–23.

35. Koprowski, et al., "Immune Responses in Human Volunteers": 109–9. Further check Leonard Kriegel, *The Long Walk Home: An Adventure in Survival* (New York: Appleton-Century, 1964): 93; Wilson, *Margin of Safety*: 149, 154.
36. Hilary Koprowski, Thomas W. Norton, George A. Jervis, Thomas L. Nelson, David L. Chadwick, Doris J. Nelsen, and Karl F. Meyer, "Clinical Investigation on Attenuated Strains of Poliomyelitis Virus," *Journal of the American Medical Association* 160 (March 17, 1956): 955. Also refer to Allen, *Vaccine*: 183; Altman, *Who Goes First?*: 114–15; Anne Finger, *Elegy for A Disease: A Personal and Cultural History of Polio* (New York: St. Martin's Press, 2006): 106; Klein, *Trial by Fury*: 59, 60, 127–31; Oshinsky, *Polio*: 133, 135; Smith and Mitchell, "Sacrifices for the Miracle": 406–7; Roger Vaughn, *Listen to the Music: The Life of Hilary Koprowski* (Oxford, MD: Roger Vaughn, 1999): 11–16, 47–48; Williams, *Paralyzed with Fear*: 226; and Wilson, *Margin of Safety*: 145. Especially see Rivers, *Reflections on a Life in Medicine and Science*: 462, 465–66, 468, and consult his footnote on page 463.
37. See the footnote on page 466 in Rivers, *Reflections on a Life in Medicine and Science*. Refer to page 463, as well as Klein, *Trial by Fury*: 68–69, 131; Smith and Mitchell, "Sacrifices for the Miracle": 407; and Williams, *Paralyzed with Fear*: 224
38. Vaughn, *Listen to the Music*: 11–16, 47–48. Further consult Allen, *Vaccine*: 184; Klein, *Trial by Fury*: 68–69, 131, 133; Sheryl Persson, *Smallpox, Syphilis and Salvation: Medical Breakthroughs that Changed the World* (Wolloesbi, NSW: Exisle Publishing Limited, 2009); Rivers, *Reflections on a Life in Medicine and Science*: 461–62; Williams, *Paralyzed with Fear*: 224; and Wilson, *Margin of Safety*: 188.
39. Smith and Mitchell, "Sacrifices for the Miracle": 407. Moreover, see Allen, *Vaccine*: 18; Etheridge, *Sentinel for Health*: 92; Klein, *Trial by Fury*: 134–35; Vaughn, *Listen to the Music*: 11–16, 47–48; and Wilson, *Margin of Safety*: 156, 163, 165–66, 185.
40. Gould, *A Summer Plague*: 126–27. Gould's comment on the U.S. position seems to dismiss this significant issue in a matter-of-fact tone. Refer also to Klein, *Trial by Fury*: 136–37; Wilson, *Margin of Safety*: 169–70.
41. Etheridge, *Sentinel for Health*: 92–93. Further consult Allen, *Vaccine*: 206; Klein, *Trial by Fury*: 141, 147; Williams, *Paralyzed with Fear*: 228, 239, 247.

CHAPTER 11

A Problematic Process

Jonas E. Salk formulated two goals for human trials: "Although the primary objective of these studies was to determine whether or not the preparations selected on the basis of their safety for animals still retained antigenic capacity, another purpose was to obtain information in answer to the question of safety for human subjects."[1] However, he faced a serious bureaucratic obstacle. Both Salk and Harry M. Weaver, research director for the National Foundation for Infantile Paralysis (NFIP) "knew the Committee on Immunization would block such a study, and agreed Salk should conduct it sub rosa."[2]

Risky Business

Salk began with children housed at the Polk State School, about seventy-five miles northeast of Pittsburgh, and D. T. Watson Home for Crippled Children, in the Pittsburgh suburbs. In January 1952, Salk wrote Gale Walker, Polk's superintendent, requesting permission to test his vaccine. Walker consented but had to, first, consult with the state's attorney general and, second, contact parents and guardians. Selecting Polk School, housing 3400 children with mental deficiencies, with a mean age of twelve, seemed unremarkable. In "postwar America, many medical researchers turned to state institutions for human subjects to use in experiments."[3] Salk, however, encountered an unforeseen problem. Polk's staff had routinely sanctioned medical experiments, but this all

R. J. Altenbaugh, *Vaccination in America*,
Palgrave Studies in the History of Science and Technology,
https://doi.org/10.1007/978-3-319-96349-5_11

changed in 1944 when Pennsylvania's "attorney general had intervened to stop a major vaccine trial at a state facility, claiming the government could not allow patients to be used as 'guinea pigs' in a project where 'many might suffer serious side effects' and 'some might even die.'" However, Gale Walker sought "consent for Salk using a loophole that forbade commercial medical research but permitted it for humanitarian reasons."[4]

While Salk awaited the outcome of that process, he embarked on trials at the Watson Home. His research team wanted to begin with these children because they had already been disabled by polio, arthritis, or "congenital deformities."[5] Therefore, this would represent an opportunity to minimize the possibilities of infections and concomitant physical disabilities with this untried vaccine. For this, he turned to Jessie Wright, Watson's medical director. As a leading physical therapist, instructor, and innovator, she devised a universal Paralysis Record Sheet to evaluate treatment progress and supplied institutions, like the Gonzales Warm Springs Foundation for Crippled Children, with trained administrators and therapists. She also worked closely with several highly placed National Foundation officials, including Basil O'Connor. Seeking advice and assistance, Harry Weaver visited Wright at Watson as early as 1947. Catherine Worthington, the National Foundation's Director of Technical Education, sought her counsel. Further, the National Foundation subsidized equipment for the care of Watson's patients, and O'Connor personally invited Wright to participate in the twentieth anniversary of the founding of the Georgia Warm Springs rehabilitation facility. Finally, Wright served on the Foundation's Committee on Research for the Prevention and Treatment of After-Effects, listing her officially as Medical Director, D. T. Watson School of Physical Therapy and Consultant on Acute Poliomyelitis, Pittsburgh Department of Health and Municipal Hospital. By this time, the "Watson facility was rivaled only by the legendary Warm Springs in Georgia for the good name it had earned in the polio community." It was no coincidence, therefore, that Jonas Salk turned to her to pursue his testing process. He submitted his request in the spring of 1952, and Wright consulted with the board of directors, who approved, and sent Salk a list of fifty-two candidates.[6]

Meanwhile, Pennsylvania's attorney general expressed "grave reservations" in his response to Gale Walker's inquiry about Salk's proposed experiments at Polk School but offered a caveat: He would approve if those researchers could demonstrate it may be of some to value the residents. On the one hand, benefits for Watson's children seemed opaque since Salk merely wanted to jab them "with a type of polio they already had and to which they were ... already immune. The initial injections would ... avail the children nothing at all." On the other hand, inoculations against the other polio types would make them immune to all of them, reducing their risk of becoming ill again. Polk's patients could likewise be protected. "With the state grudgingly willing to allow the work to go ahead, Salk, Walker, and Wright, along with the lawyers for Polk, Watson, the National Foundation for Infantile Paralysis, and the University of Pittsburgh, had to draft and redraft a waiver letter to the parents of the volunteers would be required to sign." In the end, the National Foundation prepared that document.[7]

Human trials officially began on May 23, 1952, when Salk went to Polk State School to conduct "pre-bleedings." Three weeks later he drove to the Watson Home to meet with parents and draw blood samples. Lucille Cochran, Nursing Superintendent and Administrator, "undertook the project of talking with parents and explaining that although we believed in the vaccine, the actual effect on human beings could not be known until it was tried." In addition to working with the staff and parents who supported this experiment, she maintained its secrecy.[8] "The parents of every youngster at the Watson Home," according to *Pittsburgh Press* science correspondent John Troan, and Salk confidante, "signed consent forms, and Salk personally *inoculated* all of them" on July 2, 1952, with a mixture that matched their particular virus type. Seventeen-year-old Bill Kirkpatrick, a former football player at a local high school who resided there to rehabilitate his paralyzed legs due to a polio attack in the fall of 1951, volunteered for this initial test. His parents hesitated at first but eventually assented to his compelling plea to do something to help other children.[9] Robert Nix, Watson's chief pediatrician, who had first met Salk at New York University, where he was working with Thomas Francis, Jr., recalls Salk's attitude: "When he first did his vaccine on the patients in the D. T. Watson Home he was so concerned, because this was the first time the had done it on many live people."[10] Salk drove out to that facility every evening for several days to check on those children.

Salk's research team chose forty-seven residents who had no "detectable" antibodies and another fifty-one who served as a "control for booster." The injections produced antigens in the former and successfully increased them in the latter.[11] Nevertheless, from the beginning, according to his contemporary John R. Paul, Salk had used an inordinate number of children: 100 subjects for his first dead-virus trial seemed "large," marking a "bold and prodigious step."[12]

These preliminary data led Salk to other determinants. First, the injection intervals became increasingly vital as they moved from mice to monkeys to humans. The "antigenic effectiveness of many of the vaccine preparations with which we have been concerned this past several years," Salk wrote, "would never have been recognized if direct experimentation in man [sic] had been deferred...." Second, he emphasized the need for additional trials, though he admitted that many issues remained unresolved. Salk and his staff had used—and would continue to use—a complex calculus of vaccine potency and injection sequence. Still, he expressed confidence: "... it does appear from data presented, as well as from additional studies now under way, that by suitable manipulation of the dose of vaccine, and the intervals between inoculations, it should be possible with relatively few injections, properly spaced, of course, to provide long term immunity."[13] Third, Salk's Watson tests had demonstrated that if polio antibodies already existed, then the immunogen would increase them. But he did not know what would occur if he injected children without disabilities and who had none of the three antibodies, i.e., blank slates. Could antibodies develop for all three virus strains? Moreover, the duration of this induced immunity remained unknown. Would one injection be sufficient to stimulate permanent immunity? Or would it need two? More? The next step proved especially risky because he and his colleagues had to answer the most crucial question: In the process of producing antibodies, would this experimental vaccine cause paralysis?

To answer these questions, on November 11, 1952, Jonas Salk introduced the experimental immunogen into sixty-three male residents at Polk State School. Forty-four of them were aged four through twelve. He and his assistants wanted to see if it would produce the same results as natural immunization: They succeeded. Antibodies appeared in their blood, and none became ill. Safety seemed to be assured. John Paul, in retrospect, points to the collective outcome of the Watson and Polk tests as absolutely critical:

> Had the experiments gone wrong at this point there might have been a tremendous outcry. Some would have called it unnecessarily hasty to

> use so many subjects all at once. Some would have said that Salk, having received his training in the trial use of vaccines in 'captive' military populations, was not the man to be trying his experiments on juvenile or adult civilians, were they crippled or otherwise. And others would have called it a crime to subject helpless children and adults to this sort of experimentation.

Of course, these trials had been conducted without public knowledge. Because Salk prevailed, most historical accounts have downplayed the potential for catastrophe.[14]

Salk unveiled his findings at the January 23, 1953, meeting of the National Foundation's Committee on Immunization in Hershey, Pennsylvania. On the surface, this group met there because it represented a central, travel location for its members; in reality, it was a journalistic backwater, keeping that meeting's agenda under wraps. Committee members included John Enders, John Paul, Thomas M. Rivers, Albert B, Sabin, Joseph Smadel (Walter Reed Army Medical Center), and Harry Weaver. Basil O'Connor also invited William Workman, director of the National Institutes of Health's (NIH) Laboratory of Biologics Control. Rivers, Weaver, and O'Connor knew about the Watson and Polk experiments, committee members did not. After hearing this news, the "group sat speechless," but matters grew heated as they peppered Salk with questions. The following day, Sabin challenged Salk's selection of polio strains and criticized his use of mineral oil adjutant to boost the vaccine's effectiveness. Committee members eventually shifted their attention to additional human tests, but could not decide if the design should focus on "safety tests or a controlled trial to determine whether the vaccine could prevent paralytic polio." The majority of them, including Salk, "considered a large trial premature." Weaver and O'Connor, the lone exceptions, favored an ambitious experiment, though they "didn't know how the public would view this."[15]

Committee members decided to focus on safety before embarking on a full-scale trial and recommended Salk test an additional 500–600 children in Allegheny County. That spring he and his team proceeded to administer immunogen to children, aged four through twelve, in Pittsburgh's northwestern suburbs, using the Watson Home as his base of operations. Salk again tapped Watson's Robert Nix, who recalls that Salk

> wanted to determine how much vaccine to give, what vehicle to put it in, whether it be saline, oil, or whatever, and at what intervals the vaccine should be given to give the best antibody response....
>
> He came to me and said, "Bob can you get me five hundred families I'd like the children, the parents, the grandparents, and everybody."
>
> I said, "Sure, I'll get you five hundred."
>
> And [e]very Saturday and Sunday, we would stick these people with a needle, take blood, then give them an injection of the vaccine. He had them come back at two, four, six, and eight weeks.... There was lots of action and crying with the small ones.

Word quietly spread. Friends and relatives of these same families phoned Nix and pleaded to become part of the test group. Fear of contracting this disease and the possibility of becoming paralyzed consumed them as Nix recalls: "... some of the grandparents who had a fair amount of money got their children to come down from New England or wherever they were, to get included in the study.... [T]hey were so thankful that they were lucky enough to be included...."[16] The whole affair assumed a party atmosphere, as historian Jane S. Smith describes it: "Whole families trooped off to be vaccinated together, and brought picnics and spent the rest of the day on the grounds of the Watson estate, enjoying the mild spring weather.... None of them talked much about what they were doing, and no reporters were on hand. It was serene, bucolic, secret, and secure."[17]

The permission process replicated that used during the initial trial. A one-page "Form of Consent," distributed to parents and guardians of children residing at Watson, delineated four points. First, in unambiguous language, this would constitute a medical "test," one necessary for the "advancement of science only" with "no commercial or profit motive." Second, each child would be injected with the "poliomyelitis vaccine." However, the inclusion of the term *vaccine* appears to contradict the first stipulation, since it implies unequivocal safety and efficacy. This becomes even more confusing when the third point assures parents that the experimental antigen is safe, having been administered "without subsequent harm" to laboratory animals. Fourth, blood would be drawn from these children. A cover letter from Salk, dated May 15, 1953, accompanied this form and elaborated the experiment's intentions, determining the dose required and the "number of inoculations necessary." Only then, Salk stressed, "can we say that we <u>have</u> a vaccine that can be used with assurance in protecting against polio." He closes this missive by providing both his office and home phone numbers. Finally, Salk attached a "Reaction Sheet" for parents to note side effects or "any other symptoms" they may observe throughout this trial, indicating the date and type.[18]

Behind the scenes, Lucille Cochran once again proved indispensable. She recruited candidates by sending packages of consent forms, with explanatory cover letters, to families, small-scale care facilities, and physicians. She also posted a total of 1550 to five local elementary schools, four public and one parochial, as well as Sewickley Academy, a private school located in the same Pittsburgh suburb as the Watson Home. On October 9, 1953, Salk reported that "637 children and adults ... had been inoculated in the early trials." Pre-trial blood screenings indicated that 60% of them had no antibodies to any of polio's three strains, 30% had them for one type, and 10% to two variants.[19]

One slight irregularity occurred during this phase. As Salk's team began these injections, they discovered they had brought an insufficient amount of the experimental immunogen. Two "different inocula were used," Salk later noted, "because the absorbed viral vaccine was inadvertently left in the laboratory and the aqueous vaccine was therefore substituted at the last minute." The team proceeded to give a total of three injections of the two solutions, alone or in combination, at one-week intervals. It also did not matter whether the children received an intradermal or intramuscular injection; they displayed antibodies to all three of virus's strains during "post-vaccination" screening.[20]

While this unfolded, Salk tempered any false hopes about the first human trials at Watson and Polk as he concluded in an article published in the March 28, 1953, issue of the *Journal of the American Medical Association*: "Although the results obtained in these studies can be regarded as encouraging, they should not be interpreted to indicate that a practical vaccine is at hand." He further cautioned,

> Because of the great importance of safety factors in studies of this kind, it must be remembered that considerable time is required for the preparation and study of each new batch of experimental vaccine before human inoculations can be considered. It is this consideration, above all else, that imposed a limitation in the speed with which this work can be extended. Within these intractable limits every effort is being made to acquire the

> necessary information that will permit the logical *progression* of these studies into larger numbers of individuals in specially selected groups.[21]

Salk projected a careful plan to gradually expand the sample population in methodical steps. But this is not what happened.

Basil O'Connor decided to expedite the national field test. For this, he needed a committee, biographer Charlotte DeCroes Jacobs writes, that dealt less with theoretical issues and more with practical matters, not scientists but public health experts. In April 1953, Thomas Rivers created the Vaccine Advisory Committee (VAC), consisting of David Price, federal assistant surgeon general, Norman Topping of NIH, and Ernest Stebbins from Johns Hopkins University. With a single stroke, O'Connor could now avoid the cautious and often contentious immunization committee. Salk appeared conspicuously absent from this new group and began to feel the entire enterprise slipping out of his control. With momentum building for a definitive test, underlying tensions within the National Foundation's organizational structure became apparent.

At the VAC's first meeting, on May 25, 1953. Weaver, Rivers, and O'Connor favored an observed-control group but other committee members preferred a randomized, placebo-control group. Further, according to Jacobs, "[i]n June of 1953, when Salk read in the newspaper that the NFIP was planning a [national] field trial to begin in the fall, he was stunned." He felt he should have been chosen to oversee it. He sought and received assurances about autonomy from O'Connor. Weaver, meanwhile, forged ahead, hiring Joseph A. Bell, an NIH epidemiologist and vaccine expert, as the National Foundation's scientific director to supervise and evaluate that experiment. Bell's proposed protocol involved three components. First, he wanted a double-blind control; that is, one group would receive the experimental immunogen while the second group would be injected with a placebo. This would give the trial credibility since a "randomized, placebo-controlled trial" represented the "gold standard for proving efficacy of a new vaccine." Second, he favored the flu vaccine, instead of a saline solution, as the placebo. By doing so, the placebo cohort, while not protected against polio, would at least receive immunization against another virus. Third, Bell refused to use any vaccine that contained mineral oil adjutant, opting instead for an "aqueous (water-based) vaccine for the field trial." Bell feared that the adjutant had been responsible for mild reactions when used in other vaccines and might taint the evaluation process.[22] This methodical, experimental design clashed with O'Connor's observed-control preference. Salk disagreed with Bell about the mineral adjutant. Salk and O'Connor viewed this entire process through the lens of a humanitarian agenda, "'a proposed immunizing agent as yet untested.'"[23] If they proceeded with Bell's plan, they believed too many children receiving the placebo would be susceptible to infection, hospitalization, paralysis, and even death from infantile paralysis.

While these deliberations ensued, Salk continued with human trials. In June 1953, he visited Pittsburgh's Industrial Home for Crippled Children as a luncheon guest with a related, but somewhat different, research agenda. That institution had opened in 1902, located in the city's East End, to patients ranging from age three to twelve, only admitting Pennsylvania residents, preferably from the western part of that state. It operated as a comprehensive treatment center, offering educational, medical, and therapeutic services. Instruction included academics, through that city's board of education, covering all grades, as well as "useful skills," such as cooking and sewing for girls and woodworking and shoe-making for boys. Its facilities and staff grew over the decades to encompass a brace-making shop, dormitory, gymnasium, playground, solarium, and a heated swimming pool, as well as an orthopedic surgeon, pediatricians, physical therapists, psychologists, and social workers. Beginning in 1916, it accommodated an increasing number of children disabled by polio; by 1954, they accounted for thirty-one of its thirty-eight new admissions.[24] Salk wanted to confront claims that because the "virus [used] for vaccine is propagated in cultures of monkey kidney tissue ... [it] may have some damaging effect upon [human] kidneys." The Industrial Home's board approved this endeavor. In November of 1953, Salk's staff chose fifteen children, after obtaining written parental consent, conducted physical exams, x-rayed their kidneys, and processed urine specimens, both before and after receiving antigen injections. They finished the following March and published the results in an August 1955 article. In it, Salk and his coauthors state, "this question has been [the] subject of continued study, *in a variety of ways*, and there are no indications thus far of any harmful effect upon the kidneys, either in experimental animals or in man." That article omits any mention of the Industrial Home, the number of children involved, and the procedures used.[25]

Meanwhile, tensions within the NFIP arose over the national trial. Hart Van Riper, that organization's medical director, and Basil O'Connor grew disenchanted with Harry Weaver's overzealousness and insubordination. Weaver responded with a four-page letter to O'Connor on August 29, 1953, enumerating concerns over "validity testing [of the] poliomyelitis vaccine." The Foundation, to this point, had functioned primarily as a fund-raising organization, he argued, but it now had to profoundly alter its fiscal allocations, reorganize internal operations, and expand field and office staff in order to oversee a massive field trial; that is, it had to embark on an institutional sea change. He further expressed deep frustration about his recommendations being stymied. Unable to resolve differences with Van Riper, and having rubbed too many staff members the wrong way, Weaver submitted his resignation on September 1. Van Riper replaced Weaver with Henry W. Kumm, a member of the Rockefeller Foundation's International Health Division. Joseph Bell, also weary from internecine battles over the research protocol and personality differences, resigned at the end of October 1953.[26]

O'Connor, in search of a "disinterested third party" to oversee the national field trial, in December, turned to Thomas Francis. Infantile paralysis was not new to him. While at the University of Michigan, in 1941, Francis participated in an epidemiological study of that disease in Akron, Ohio. He accepted O'Connor's offer, but only with the following stipulations, all of which echoed Bell's conditions. First, he insisted on a double-blind study, except in states where O'Connor had already secured permission through observed controls. O'Connor had used this enticement to win the cooperation of thirty-three state health departments. Salk supported it because he wanted to immunize as many children as possible, as soon as possible. Francis ultimately struck a bargain with them, persuading Salk, his former protégée, that a traditional and rigorous scientific experiment required a double-blind method; Salk relinquished this point. Recognizing O'Connor's political bind, Francis further agreed to retain thirty-three states as observed-control sites but insisted that O'Connor add eleven more for a control-group approach. Two assessment protocols would thus be implemented. Second, the National Foundation agreed to "give Francis's laboratory grants of several hundred thousand dollars a year throughout the field trial period, and Van Riper made it clear that they were willing to commit another million dollars to the separate establishment of the Vaccine Evaluation Center" at Michigan, officially established in February of 1954. Francis could then operate without any NFIP brass hovering over his shoulders. Third, Francis would keep all findings secret until the completion of the review process, forestalling any premature press releases by O'Connor.

Finally, National Foundation leaders alone assumed full authority and approved the protocol, locales, and number and ages of the subjects.[27]

Henry Kumm organized a meeting of the immunization committee at Detroit's Henry Ford Hospital, on October 24, 1954, where "Salk … present[ed] his most recent studies on the use of an inactivated vaccine." John Paul, who attended that meeting, recalls a serious "confrontation…. The trial with the Salk vaccine was by this time much further along than many of the immunization committee members realized. And it was immediately made apparent to the members that they were to have a passive, not active, further role in the plans."[28]

A Missing Step?

As a general rule, scientists experimented on themselves, prior to human trials, to catalogue reactions, if any, and ensure safety. This includes Friedrich Wilhelm Adam Serturner, a German who pioneered the use of morphine in 1803; William T. G. Morton, an American who demonstrated ether as an anesthetic in 1846; William Murrell, a British doctor who tested nitroglycerin for arteriosclerosis in 1879; and August Bier, a German who developed spinal anesthesia in 1886, among many others. However, self-experimentation did not guarantee success, as we saw with William H. Park, Maurice Brodie, and John A. Kolmer who self-tested themselves with polio vaccines but failed in 1935 field trials with children.

In spite of the context and this tradition, an apparent anomaly arose with the development of the Salk vaccine. According to Dan Wegemer, one of Salk's assistants, all of that lab's scientists and workers received jabs in 1953 as part of the protocol. However, not every member of the laboratory staff was an adult or a professional. "A high-school student who had a part-time job in the laboratory recalls the low level of ceremony attached to the process; after she had worked there for a few days, Dr. Salk came through the lab, told her to roll up her sleeve, and vaccinated her as though he were stamping a crate of fruit." Salk injected himself, his wife, and three sons, Darrell, Jonathan, and Peter, as well as the children of the National Foundation's Hart Van Riper and Harry Weaver on May 16, 1953. This took place after the Watson and Polk trials. No reason is given for these apparent divergences from the self-experimentation sequence.[29] From this point on, the number of children used in field trials steadily grew.

Notes and Sources

1. Jonas E. Salk, et al., "Studies in Human Subjects on Active Immunization against Poliomyelitis. Part I. A Preliminary Report of Experiments in Progress," *Journal of the American Medical Association*, 151 (March 28, 1953): 1088. Also see pages 1082–83.
2. Charlotte DeCroes Jacobs, *Jonas Salk: A Life* (New York: Oxford University Press, 2015): 106–7. According to Jeffrey Kluger, *Splendid Solution: Jonas Salk and the Conquest of Polio* (New York: G. P. Putnum's, 2004), on February 26, 1952, Weaver organized a meeting of the immunization committee at New York City's Waldorf-Astoria Hotel. It proved contentious. Rivers recommended an intermediate trial of 25,000 children in the Pittsburgh and Allegheny County area, but it adjourned without a firm resolution (197–98).
3. Michael D'Antonio, *The State Boys Rebellion* (New York: Simon and Schuster, 2004): 55. See also Kluger, *Splendid Solution*: 167–70; Charles L. Mee, *A Nearly Normal Life* (Boston: Little, Brown & Co., 1999): 149; Edmund J. Sass, George Gottfried, and Anthony Sorem, *Polio's Legacy: An Oral History* (Lanham, MD: University Press of America, 1996): 12; Jane S. Smith, *Patenting the Sun: Polio and the Salk Vaccine* (New York: William Morrow and Co., 1990): 137–39; John Troan, *Passport to Adventure: Or, How a Typewriter from Santa Led to an Exciting Lifetime Journey* (Pittsburgh, PA: Geyer Printing, 2000): 196. For background on Polk and Watson, refer to Richard J. Altenbaugh, *The Last Children's Plague: Poliomyelitis, Disability, and Twentieth-Century American Culture* (New York: Palgrave Macmillan, 2015), pages 81, 86–87.
4. David M. Oshinsky, *Polio: An American Story* (New York: Oxford University Press, 2005): 157–58. Also see Thomas M. Rivers, *Reflections on a Life in Medicine and Science.* (Cambridge: Massachusetts Institute of Technology, 1967): 494.
5. Susan Jacobs, "Watson Institute: Needs Have Changed, but Focus Is Still on Children with Disabilities," *Pittsburgh Post-Gazette*, January 19, 2002, Section N, 1–2. Further consult "The Watson Institute: Our History," http://www.thewatsoninstitute.org/ (Accessed August 5, 2011); Jonas E. Salk, et al., "Studies in Human Subjects on Active Immunization Against Poliomyelitis. Part. I. Preliminary Report of Experiments in Progress," *Journal of the American Medical Association*, 151 (March 28, 1953): 1086. Also see pages 1082–83, 1088; Nix in Nina Gilden Seavey, Jane S. Smith, and Paul Wagner, eds., *A Paralyzing Fear: The Triumph Over Polio in America* (New York: TV Books, 1998): 219–20.
6. Kluger, *Splendid Solution*: 170–71. Further consult Jessie Wright Papers, 1925–1970, Record Group 7/1, and "Poliomyelitis Vaccine: History,"

Box No. 1, File Folder No. 4, Salk Vaccine, Record Group 42, WIA. Also see Tony Gould, *A Summer Plague: Polio and its Survivors* (New Haven: Yale University Press, 1995): 132; Rivers, *Reflections on a Life in Medicine and Science*: 494; David W. Rose, *Images of America: March of Dimes* (Charleston, SC: Arcadia Publishing, 2003): 9.

7. Kluger, *Splendid Solution*: 174–75. Moreover, consult Smith, *Patenting the Sun*: 138–39; Troan, *Passport to Adventure*: 191–92; "Poliomyelitis Vaccine History," Box No. 1, File Folders No. 4 and 13, Salk Vaccine, Record Group 42, WIA.
8. "Poliomyelitis Vaccine History," Box No. 1, File Folder No. 4. Especially check File Folders 7 and 13, Salk Vaccine, Record Group 42 and Lucille Cochran Papers, Record Group 6/2, WIA, as well as Gould, *A Summer Plague*: 132.
9. Troan, *Passport to Adventure*: 193–94. (Emphasis is mine.) See page 196, and Smith, *Patenting the Sun*: 139.
10. Nix in Seavey, Smith, and Wagner, *A Paralyzing Fear*: 222. Refer to page 219, as well as Jacobs, *Jonas Salk*: 108–10; Kluger, *Splendid Solution*: 175–80.
11. Salk, et al., "Studies in Human Subjects on Active Immunization against Poliomyelitis, Part I": 1087. See moreover Gould, *A Summer Plague*: 132; Kluger, *Splendid Solution*: 178; David Smith and Alison L. Mitchell, "Sacrifices for the Miracle: The Polio Vaccine Research and Children with Mental Retardation," *Mental Retardation*, 39 (October 2001): 406; Troan, *Passport to Adventure*: 182.
12. John R. Paul, *A History of Poliomyelitis* (New Haven: Yale University Press, 1971): 417.
13. *News Release of the Third International Poliomyelitis Conference*, 9/6-9/10, 1954 (Rome, Italy), Jonas Salk Papers, Box 7, File Folder, International Poliomyelitis Congress (Collection # 90/36/7), 6, 7, and 9, respectively, UPITT. Salk attached numerous graphs and tables to this paper. Also consult Jonas E. Salk, et al. "Formaldehyde Treatment and Safety Testing of ExperimentalPoliomyelitis Vaccines," *American Journal of Public Health and the Nations Health*, 44 (May 1954): 563, 569–70.
14. Paul, *History of Poliomyelitis*: 418. Further consult Salk, et al., "Studies in Human Subjects on Active Immunization against Poliomyelitis, Part I": 1086–87; Mee, *A Nearly Normal Life*: 149; Smith, *Patenting the Sun*: 143.
15. Jacobs, *Jonas Salk*: 113, 114, 116, and 117, respectively. Refer to page 110, as well as Gould, *A Summer Plague*: 133; Oshinsky, *Polio*: 160; Paul, *History of Poliomyelitis*: 419; Rivers, *Reflections on a Life in Science and Medicine*: 488, 495–97, 499–500; Seavey, Smith, and Wagner, *A*

Paralyzing Fear: 182; Smith and Mitchell, "Sacrifices for the Miracle," 406; Smith, *Patenting the Sun*: 144, 248, 251.

16. Nix in Seavey, Smith, and Wagner, *A Paralyzing Fear*: 221. Also refer to Paul, *History of Poliomyelitis*: 419–20; Rivers, *Reflections on a Life in Science and Medicine*: 500; Smith, *Patenting the Sun*: 187–88, 205; Troan, *Passport to Adventure*: 208. Salk, et al., "Studies in Human Subjects on Active Immunization against Poliomyelitis, Part I": 1098.
17. Smith, *Patenting the Sun*: 189. Jonas E. Salk, et al., "Studies in Human Subjects on Active Immunization Against Poliomyelitis. Part II. A Practical Means for Inducing and Maintaining Antibody Formation," *American Journal of Public Health*, 44 (August 1954): 994–1009.
18. "Form of Consent" (with cover letter and reaction sheet), Box No. 1, File Folder No. 6, Salk Vaccine, RG 42, WIA. Emphasis is Salk's. No forms from the first test are in the Watson Archives. Since the existing blank consent forms are undated, I presumed that these served as the official documents for both trials. File Folder No. 9 contains lists of patients' names and those of family members. Italicized *vaccine* is mine.
19. "Chronology of Events in Salk Research for Polio Vaccine," Jonas Salk Papers, Box 7, File Folder, Salk Polio Research: Chronology (Collection # 90/36/7). Refer to Cochran's correspondence contained in Box No. 1, Folder No. 2, Salk Vaccine, RG 42, WIA, and Jonas E. Salk, "Recent Studies on Immunization Against Poliomyelitis," *Pediatrics*, 12 (November 1953): 474. Also check Gould, *A Summer Plague*: 136. Finally, consult Jacobs, *Jonas Salk*: 124.
20. Salk, "Recent Studies on Immunization Against Poliomyelitis": 476.
21. Salk, et al., "Studies in Human Subjects on Active Immunization against Poliomyelitis, Part I": 1098. Emphasis is mine. MOD: Medical Program Records; Series 2: Conferences and Meetings Records, 1938–2002; Committees; Minutes: "Summary of Meeting on Application of the Vaccine with Mr. O'Connor," February 11, 1954. See also Victor Cohn, *Four Billion Dimes* (Minneapolis: 1955): 104.
22. Jacobs, *Jonas Salk*: 121, 122 and 124–25, respectively. See pages 96–97, 121, 123, as well as Salk, et al., "The Use of Adjutant to Facilitate Studies on the Immunologic Classification of Poliomyelitis Viruses": 158, 161. Further check Kluger, *Splendid Solution*: 223; Oshinsky, *Polio*: 176–78, 180; Paul, *History of Poliomyelitis*: 411, 422–23; Smith, *Patenting the Sun*: 192, 194–95, 198.
23. Oshinsky, *Polio*: 182. Also refer to pages 176–78, 180–81. Especially consult Chapter 11, titled "The Rivals," for a thorough description of these conflicts. Moreover, see Rivers, *Reflections of Life in Science and Medicine*: 508–12; Smith, *Patenting the Sun*: 198, 203.

24. *Memories and Miracles: The Children's Institute* (Pittsburgh: Children's Institute, 2016): 5, CIA. See pages 10, 12–14, 18, 21, as well as MOD, Chapter Administration, Pennsylvania, Allegheny Chapter: 11; Jacobs, *Jonas Salk*: 124; Kluger, *Splendid Solution*: 215; Edith Reeves Solenberger, *Public School Classes for Crippled Children*, Department of the Interior, Bureau of Education, Bulletin No. 10 (Washington, DC: Government Printing Office, 1918): 184.
25. Salk, et al., "Studies in Human Subjects on Active Immunization Against Poliomyelitis, Part II": 1007. Emphasis is mine. The endnote for this statement reads "studies to be published" (1009). Further consult Board Minutes (1953–1954), no pagination, and Annual Report, 1954–1955, no pagination, CIA.
26. MOD; Medical Programs Records; Series 5: Committees; Weaver to O'Connor, August 29, 1953 and Weaver to O'Connor, September 1, 1953. For background on Kumm, consult Henry W. Kumm Collection, Alan Mason Chesney Medical Archives, Johns Hopkins Medical Institutions. http://www.medicalarchives.jhmi.edu/papers/kumm.html (Accessed August 11, 2011); MOD; Medical Records Programs; Series 9: Human Resources; Biographical Data, Henry W. Kumm, M.D., 1951–1958 and Van Riper to O'Connor, September 16, 1953; Series 5: Committees; Vaccine Advisory Committee, 1951–1960. Further check Kluger, *Splendid Solution*: 223–24; Paul, *History of Polio*: 423; Smith, *Patenting the Sun*: 197–98.
27. Smith, *Patenting the Sun*: 225 and 228, respectively. Also refer to pages 198, 204–5, 226. See especially MOD, Conferences and Meetings Records, 1938 2002, Series 2: Committees, "Meeting of Advisory Group on Evaluation, 11 January 1954," and "Minutes of the Meeting of the Advisory Committee on Technical Aspects of Poliomyelitis Field Trials," January 30–31, 1954. Further check "Memorandum No. 2," *Operational Memoranda: Vaccine Field Trial, 1954*. Box No. 14, BHL. No pagination. Much has been written about this episode in secondary sources. Examine, for example, Roland H. Berg, *Polio and Its Problems* (Philadelphia: J. B. Lippincott, 1948): 90; Allan M. Brandt, "Polio, Politics, Publicity, and Duplicity: Ethical Aspects in the Development of the Salk Vaccine," *International Journal of Health Services*, 8 (1978): 264; Liza Dawson, "The Salk Polio Vaccine Trial of 1954: Risks, Randomization and Public Involvement," *Clinical Trials*, 1 (2004): 128; Gould, *A Summer Plague*: 139–40; Jacobs, *Jonas Salk*: 126, 136; Aaron E. Klein, *Trial by Fury: The Polio Vaccine Controversy* (New York: Charles Scribner's Sons, 1972): 90–92, 96–97; Kluger, *Splendid Solution*: 242–43; Marcia. Meldrum, "'A Calculated Risk:' The Salk Polio Vaccine Field

Trials of 1954," *British Medical Journal*, 317 (October 31, 1998): 1223–36; Arnold S. Monto, "Francis Field Trial of Inactivated Poliomyelitis Vaccine: Background and Lessons for Today," *Epidemiological Reviews*, 21 (1999): 12–13; Paul A. Offit, *The Cutter Incident: How America's First Polio Vaccine Led to the Growing Vaccine Crisis* (New Haven: Yale University Press, 2005): 184–85; Oshinsky, *Polio*: 181, 184–85; Paul, *History of Polio*: 422–23; John R. Paul, "Thomas Francis, Jr., July 15, 1900-October 1, 1969," *Biographical Memoirs*, 44 (1974): 76–78; Paul, *History of Polio*: 423.

28. Paul, *History of Poliomyelitis*, 423.
29. Wegemer in Seavey, Smith, and Wagner, *A Paralyzing Fear*: 191–92, 196. Chronology of Events in Salk Research for Polio Vaccine, Box 7, File Folder, Salk Polio Research: Chronology (Collection # 90/36/7), Jonas Salk Papers, UPITT; MOD; Photography Collection. Secondary sources appear somewhat unclear about when or how Salk injected himself and his family. See Cohn, *Four Billion Dimes*: 104; Klein, *Trial by Fury*: 86; Kluger, *Splendid Solution*: 2, 203–4; Oshinsky, *Polio*: 153; Rivers, *Reflections on a Life in Science and Medicine*: 545; Smith, *Patenting the Sun*: 137; Troan, *Passport to Adventure*: 186. However, Offit, *Cutter Incident*, appears to be the accurate account. Finally, see Donna Salk in Seavey, Smith, and Wagner, *A Paralyzing Fear*: 202. For self-experimentation, refer to Lawrence K. Altman, *Who Goes First: The Story of Self-Experimentation in Medicine* (Berkeley: University of California Press, 1998): 53–63, 72–73, 89–90, 95–96, 127–28. For Altman's treatment of Salk, see page 129.

CHAPTER 12

School Days

In 1953, as the National Foundation for Infantile Paralysis (NFIP) pushed ahead with plans to experiment on tens of thousands of schoolchildren, the public and private sectors absented themselves from any oversight of human subjects. Prior to the 1960s, legal professor William J. Curran states, "there was little 'law' in the United States concerning medical research." No federal or state statutes existed to "regulate organizations or investigators in their research methods, their areas of research, or the use of subjects or patients in such work." Curran adds that "no reported court actions involving liability issues or criminal action against research organizations or personnel" had occurred. The Nuremberg Code received only nominal mention at the National Conference on the Legal Environment of Medical Science held in Chicago in 1958. Moreover, in a study published in 1961, Louis C. Welt, at the University of North Carolina, found through the use of a national questionnaire that, out of sixty-six total responses from medical schools, "only eight departments had a procedural document, and only twenty-four had or favored the establishment of a committee to review research protocols."[1]

The Steel City

In November 1953, Basil O'Connor committed the NFIP to purchase 27 million doses of experimental antigen, amounting to $9 million, for mass immunization from Cutter, Eli Lilly, Parke-Davis, Pittman-Moore,

R. J. Altenbaugh, *Vaccination in America*,
Palgrave Studies in the History of Science and Technology,
https://doi.org/10.1007/978-3-319-96349-5_12

and Wyeth pharmaceutical companies. Further, Eli Lilly and Parke-Davis would subsidize the production of polio vaccine for the national field test. "O'Connor was gambling that … trial would show the vaccine worked. But even if it didn't work and wasn't later sold, companies would still be paid. O'Connor and the National Foundation had taken the risk out of vaccine research and development." Jonas E. Salk prepared "a fifty-page protocol for the manufacture of the vaccine," but it remained incomplete, leaving those companies to fill in the gaps.[2]

On December 3, 1953, Thomas M. Rivers, chair of the NFIP's Committee on Active Immunization, officially gave O'Connor the green light for the national trial, assuring him about "safety," based on the following: First, that committee had reviewed and approved Salk's "protocols and procedures for the production of poliomyelitis virus and its inactivation for the preparation of a vaccine." Second, Salk had successfully tested it on animals. Third, Salk had "vaccinated almost 700 humans in Allegheny County" without incident. Fourth, the committee had advised Salk to conduct a round of injections on an additional 5000–10,000 "human subjects" prior to the "inauguration of large scale inoculations." Fifth, following these deliberations, the committee members, at their November 13th meeting in New York City, had unanimously recommended the National Foundation "proceed with the test of this vaccine in a sufficient number of children to insure that the trials would result in the determination of the ability of the vaccine to produce sufficient immunity to protect against paralytic poliomyelitis as a result of natural infection."[3] Therefore, based on injections of only several hundred individuals, even before Salk had completed his expanded Pittsburgh trial, which would operate "as a precautionary test" of the commercially produced vaccine, committee members and Foundation leaders planned to forge ahead with the national experiment.[4]

The National Foundation scheduled it to commence on February 8, 1954. However, the pharmaceutical companies could not complete the first batch of vaccine until mid-March, forcing a timetable revision. Another delay occurred when Lilly and Parke-Davis experienced manufacturing problems. This grew even more serious when several of Lilly's test monkeys seemingly developed paralysis from that experimental mixture.[5]

The National Institutes of Health (NIH), one of three safety testing laboratories—with Salk and vaccine manufacturers serving as the other two—raised concerns about the safety of the commercially prepared antigen after detecting lesions on dissected monkeys. O'Connor sent Salk and David Bodian, of Johns Hopkins University, to examine the monkeys in question. Bodian declared a false positive; while lesions certainly existed, they were not due to poliomyelitis. Intense discussions ensued. William G. Workman, director of the Laboratory of Biologics Control, remained skeptical about pharmaceutical production and "wanted each batch ... tested on 350 monkeys, which would have slowed the trial by months; the National Foundation agreed instead that 11 batches in a row had to test free of live virus before any of the batches would be released." In addition, James Shannon, NIH's chief deputy, "succeeded in getting thimerosal added to the vaccine as an antiseptic."[6] Finally, on March 28, all parties decided to wait four more weeks for the results of Salk's field trial in the Pittsburgh area. This additional holdup eliminated Arizona, the District of Columbia, Georgia, and Maryland from the sample population since their polio seasons began early. O'Connor quickly substituted "46 health districts in Canada ... and two small areas in Finland."[7]

A University of Pittsburgh press release announced the expansion of "vaccine testing" in the Pittsburgh Public Schools. Salk followed up with two appeals for "volunteers from school children in Pittsburgh and Allegheny County." He provided no explanation for the selection of school buildings and students in this expanded sample. In his February 10th letter to parents, he outlined their choices: "If you prefer not having your child inoculated, please indicate so at the bottom of this letter and return it so that we may know you have seen this communication. If you feel you would like to have your child participate in the extension of our studies, please fill out, sign, and return the enclosed Consent Form." He further assured them the "vaccine [had been] made and safety-tested in our laboratory...," noting that "700 persons have [already] been inoculated with a variety of preparations...." Salk only wanted to know "'how much,' 'how often,' and 'how far apart'" to give the injections. This conveyed the sense of a fine-tuning exercise, operating merely, he stated, as an "essential prerequisite for providing much of the information that will be necessary before [national] tests can be undertaken." Throughout both of his communiques, he conflated the terms inoculations, testing, and vaccinations.[8]

Max Elder, Director of the University of Pittsburgh's News Service, sent a "note" to newspaper, radio, and television editors and producers announcing that on Monday morning, February 14th, Salk would brief "a group of public and parochial principals" about the forthcoming "inoculations." While Elder granted permission for photographs he forbade any "recordings" of this presentation. This press release in no way conveyed a sense that this would be an experiment. Again, the terminology not only implied unequivocal safety but certain protection.[9]

A local newspaper article publicly announced the forthcoming event. Nineteen Pittsburgh public schools, along with five schools in four districts outside the city, would participate in "an extended laboratory testing program of the polio vaccine developed at the University of Pittsburgh Virus Research Laboratory," involving three injections and four blood tests. Only those children who had their parents' consent could take part in the Pittsburgh trial, and the laboratory's staff had received 3656 consent forms that indicated "yes" out of a total of 5148 students, or 71%, enrolled in grades one through three. In response, Salk's team sent an information card home with second graders to gather information regarding family members' ages and children's allergies.[10]

Salk's sample included 2500 students enrolled in seven Roman Catholic parochial schools. Colette Freiwald attended St. Stephens and recalls the anxiety she felt one day, in first grade, as Sister Christina, "holding a small stack of index cards in her right hand," began to call "names of students to come forward and form a single line." A puzzled Colette looked "around at the children whose names were not" called, and "wondered what they did to escape the horror." Had they said more rosaries? "Or could it have been that they contributed more pennies to the 'Pagan Babies Fund?'" Their teacher then led them to the cafeteria. "We were told not to talk and to do whatever the doctors and nurses asked us to do…. There was no explanation what was to come." Colette sobbed as she sat in a chair while an "unsmiling nurse" held her arms. A cold, clinical attitude prevailed. "For those of us who got faint … there was the wonderful, warm gesture of laying us on the lunch tables with no blanket until we could sit up with clear heads." Afterward, she continues, "we were marched back to our rooms where we were expected to behave as if nothing happened and continue on with our daily activities," just a routine day at school! Consent forms had been distributed, but Freiwald only realized this two years later: "I figured out that the white paper I was carrying home from school to my parents was a permission slip. And I was angry when I figured out that my parents were actually volunteering me for this treatment!" Her mother tried to reason with her, revealing that polio had caused her father's limp. "I was a test subject and they said this was their way of helping to save other children. To this I responded that, if they felt that way, they should sign up for the march to the cafeteria to receive the shots themselves." Reflecting on this experience over fifty years later, Freiwald resented not being informed that she and other children were "used as human pin cushions and guinea pigs without their knowledge or permission." By the end of March, Salk had completed his jabs of 5320 children residing in Pittsburgh and surrounding Allegheny County communities using his own immunogen and an additional 5000 with commercially made vaccine. No problems had arisen.[11]

The Pittsburgh school superintendent's 1954 annual report downplayed the scope of this trial, nonchalantly mentioning it in a subsection titled "Poliomyelitis Immunization Project." It consisted of two short paragraphs that broadly outlined what had transpired that school year. First, he boasted, the Pittsburgh public schools had been "chosen as the starting point for poliomyelitis (infantile paralysis) immunization," a clear mark of honor. Second, this initiative had been reviewed by him and the district's medical staff and approved by the school board. Third, Salk's staff, with the "excellent administrative cooperation of Pittsburgh public school personnel," facilitated the injections of children in the first three grades of nineteen schools. "The response of the parents in requesting the vaccination," the superintendent added, "was gratifying."[12] The significance here is his reference to the forthcoming national experiment as an immunization endeavor.

The central role of education personnel became clear during this initial foray into the public schools. Max Elder shed light on this in his February 16, 1954, letter to Mr. Herman Ziel, Principal, Arsenal Elementary School, expressing appreciation and gratitude to Ziel for allowing the university's research team to use the school building. In the face of this apparent positive, yet ongoing, local trial, NIH officials met again with Basil O'Connor on April 25, 1954, and gave their final approval. The American public had high expectations, as "*Life* [magazine] flagged the story on a cover in February 1954: 'How Polio Vaccine Was Found.'"[13]

A Most Important Project

Thomas Rivers formally announced the national field trial in a letter to the editor of *Journal of the American Medical Association*, emphasizing that "safety is the utmost concern." The National Foundation's

Vaccination Advisory Committee after review and consultation, Rivers continued, established a "series of minimum standards." That Committee wanted to assure the American Medical Association "every reasonable safeguard possible has been incorporated in these standards."[14] Nonetheless, as John R. Paul observed later: "It was risky doing so many vaccinations with an unknown product that might be potentially dangerous…."[15]

Thomas Francis, Jr., overseeing the evaluation process, designated April 26, 1954, as "V-Day."[16] Such a moniker had to resonate with the American public, evoking memories of D-Day, June 6, 1944, when the Allies landed in Normandy to liberate of Europe. The analogy of fighting an evil menace would not be lost on this generation of parents.

To smooth the way for such a grand experiment, Foundation leaders executed the national trial through the public education system where they could count on highly organized record keeping and predictable attendance, all of which made for "'maximum ease and convenience.'" Moreover, the rhythm of this massive trial blended well with the academic year. Polio usually struck during the summer months; administering this immunogen during the spring could—possibly—protect many children. The public schools, in short, would operate as an extension of the medical laboratory.

Foundation planners utilized an extensive sample to ensure statistical reliability, employing Gabriel Stickle, a statistician, and Thomas Dublin, an epidemiologist, to process data of outbreaks from the previous thirty years to locate high-incidence areas. They matched these with local health departments that could handle a placebo-controlled study. They further narrowed this list to counties with populations ranging from 50,000 to 200,000 with high epidemic frequencies between 1946 and 1950. The lone exception was New York City because of "strong political and popular pressure to take part in the field trial as well as a superbly well-organized health department, capable of handling a sophisticated study on any scale."[17] This encompassed 700,000 first, second, and third graders in eighty-four areas in eleven states. Researchers injected one-half with the vaccine and the other half with a placebo. The observed-control component involved 1.1 million children in 127 areas of thirty-three states. Second graders received the vaccine while their first- and third-grade classmates served as the observed group. By any measure, "[t]his was virgin territory, the biggest medical gamble in history."[18]

Some uncertainty belied the confident, public face of this daring endeavor. A series of liability issues, according to a flurry of internal memoranda, began to emerge following the formal announcement of the national trial. Key members decided, on November 18, 1953, to secure liability insurance to protect the National Foundation and, by the end of December, had a policy amounting to $300,000 per claim; however, it did not cover public health officials, pharmaceutical companies, and physicians. Other insurance questions plagued the Foundation at the eleventh hour. Serious concerns arose at a meeting, in April 1954, between NFIP representatives and the city of Rochester and Monroe County school administrators. Feeling vulnerable, they wanted to be absolved of any responsibility. Following "confidential" consultations with legal counsel, Hart van Riper, that organization's medical director, sent a memorandum, dated April 28, 1954, to state health officials informing them that Foundation carried "comprehensive general liability insurance to indemnify it against all claims, whether groundless or otherwise, with the policy limits, which it may become legally obliged to pay for bodily injury arising out of the distribution or use of the vaccine and the placebo fluid." However, the National Foundation could not cover all the physicians involved in the national trial but did secure malpractice insurance to supplement their existing policies. Van Riper cautioned these health authorities that this coverage information had not, and would not, be made public in order to avoid incentives for unwarranted law suits.[19]

The federal government, in spite of the scope and potential dangers of this undertaking, remained on the sidelines. Following William Workman's concern about the safety of commercial production, the Vaccination Advisory Committee and US Public Health Service released a joint announcement on April 24, 1954, endorsing the experimental solution as "sound."[20] NIH, in an official and formal capacity, neither approved nor disapproved of the Salk vaccine. This decision would be forthcoming, only after the national trial and pharmaceutical companies sought permission to license it. Hence, any clinical trials were the "responsibility of the responsible investigator." The only exception occurred when NIH insisted that immunogen vials bore the label: "Caution: New Drug–Limited by Federal Law to Investigational Use." In sum, NIH only responded to a possible liability concern; the ethics of using human beings in experiments did not enter the picture.[21]

The National Foundation rallied public support and participation, publishing a pamphlet, "What You Should Know About the Polio Vaccine Tests," for national distribution. The Foundation also sent packets, containing a cover letter and a parent request form, to the participating school buildings for teachers to pin on children's coats, jackets, and sweaters so they could take them home. Basil O'Connor's note, consisting of a 160-word paragraph, described the blood test and injection processes, but remained mum about potential side effects or dangers. His wording lent a sense of gravitas: "This is one of the most important projects in medical history. Its success depends on the cooperation of parents. We feel sure you will want your child to take part." The "Parental Request Form" avoided the term *consent* placing the onus on parents: They were not granting authorization but asking to participate. These documents circumvented the terms "subjects" and "volunteers," instead substituting the phrase "polio pioneers." Moreover, it treated the words: field trial, inoculation, field study, and vaccination as synonyms. The NFIP public communications apparatus avoided, as much as possible, using the term *experiment*, replacing it with the somewhat ambiguous *trial*. Finally, the "National Foundation skillfully integrated metaphors of American initiative and heroism into popular discourse on its polio-vaccine trials."[22]

School officials also sent announcements home. Chester W. Holmes, superintendent of the Malden, Massachusetts, district, directed his to the parents of first-, second-, and third-grade students describing the three-injection process and assuring them that it would be voluntary. He affirmed the antigen's safety, stressing the Pittsburgh-area experiments: "Tests already made of this vaccine on some ten thousand young school children in Allegheny County, Pennsylvania, lead the National Infantile Paralysis Foundation [sic] to believe that country-wide inoculations are the only reliable way of testing the effectiveness of this new vaccine."[23]

New York City represented a special case. As the largest school district in the nation, hosting a superb health department and the NFIP's corporate offices, the Foundation granted it considerable leeway. While it printed a pamphlet on blue paper for national use with the observed-control sites, it designed a yellow, customized version, "What You Should Know About the Polio Vaccine Tests in New York City," distributed through its health department. This four-page, pocket-sized document assured safety based on injections of "over 7,500 volunteers," including Jonas Salk, his wife, and "three young sons." Each family's packet also contained an introductory note from the health commissioner, Leona Baumgartner, consisting of twenty-five frequently asked questions, ranging from the trial's logistics to "vaccine" development.[24] Although all other placebo-controlled areas received yellow pamphlets, few saw "purple instructions.... Those were the instructions that would have been passed out if something had gone terribly wrong."[25]

Some parents struggled with their decisions. Louise DeMartino, in retrospect, questions that consent process. The "wording of the permission form was chosen to make parents feel that it was an honor for our children to be part of the 1954 trial.... The form letter, composed by Mr. O'Connor," she continues, "indicated that participation in the trials was a moral act benefitting not only our children and their peers but future generations of children as well." She ultimately volunteered her children. Seven-year-old Brenda Serotte's mother spent a week agonizing over granting consent; although ultimately relenting, she continued to harbor serious misgivings. Serotte, herself, recalls the ambiguity of the entire process. "Along with the rest of the children who made history as polio pioneers, my parents *requested* that I be inoculated with the experimental serum. That was how it worked; it was a vaccine trial that may or may not save your child from contracting polio–who wouldn't want to try that? Although the field trial was strictly a voluntary participation, and it was never referred to as an 'experiment,' we truly were the subjects of one." In a similar vein, Dee Van Balen's oldest brother, another Polio Pioneer, provides unique insight into how it felt being volunteered. On the one hand, he believed his mother made the right decision because the vaccine worked but, on the other hand, he would not submit his own children to participate in a similar experiment.[26]

Securing parent permissions raised concerns from the beginning. The National Foundation's Advisory Group on Evaluation met, January 11, 1954, at New York City's Commodore Hotel. A key item involved the acquisition of signed consent forms. Francis wanted the "parents of second grade children come to school and fill out the request forms in the presence of the classroom instructor....." He felt strongly about this because the "teacher would act as an unofficial witness to the parents [sic] signature and obtain information from the mother on past history of poliomyelitis." The majority of that committee's members opposed this idea, "feeling it would cut down on the number of parents who would volunteer if they had to come to the school...." Hart Van Riper reinforced this point by citing educators who felt that sending forms home had always functioned well in the past and, furthermore, "they would have problems if the parents came to the schools."[27] Francis's concern, in one instance, appeared justified. In Lexington, Kentucky, as the trial began, mothers discovered that 100 out of 949 first, second, and third graders at that city's largest public elementary school were missing signed request forms. Four of those parents, in spite of inclement weather, visited each of the families and acquired those signatures.[28] In the end, not all parents allowed their children to participate. The majority, 455,474 or 61%, granted consent to participate in the placebo study. The remainder, either refused, 280,868 or 37%, or remained "unrecorded," 12,894 or 2%.[29]

To coordinate this enormous undertaking, the National Foundation published a manual, *Operational Memoranda: Vaccine Field Trial, 1954*, that contained detailed instructions for state and local health leaders, medical personnel, and fieldworkers, a multilayered set of responsibilities. Six of the seventeen roles delineated in this spiral booklet focused on school staff members and volunteers. All participants thus depended on access to school facilities and required the assistance of educators. Little doubt existed that the public school operated as an indispensable site for this unprecedented medical experiment.

"Memorandum No. 6" detailed medical arrangements. The public education system easily responded with its medicalized infrastructure and culture. Building principals arranged their clinic's physical layout in an assembly-line fashion and compiled "vaccination time tables." School nurses assisted the physicians throughout the entire injection routine. Teachers alphabetically arranged their students for efficient processing and, just prior to injection, verbally identified each child for the data recorder. Each vaccination team consisted of five individuals: the "physician-vaccinator," a nurse, clinic recorder, and two "clinic aides."

The National Foundation, according to the *Operational Memoranda* manual, placed the responsibility for enlisting volunteers on the local "Public Education Committee." It began community recruitment campaigns, sponsoring information meetings with scheduled speakers and distributing NFIP pamphlets, like "What You Should Know About the Polio Vaccine Test," and showing filmstrips, such as the "Polio Vaccine Trial." It also appealed to labor unions and businesses. With a ready pool of workers, the "School Volunteers Chairman" consulted with individual school building administrators and Parent–Teacher Associations (PTA) to appoint a "School Unit Chairman" who appointed clinic recorders, organized two to three classroom mothers per teacher, and designated clinic aides, who oversaw the acquisition and disposal of medical supplies, and ensured a smooth flowing traffic of test subjects through the clinic. Finally, the NFIP developed elaborate flowcharts, stipulating responsibilities, for the school setting.

An article in the December 1954 issue of *Ladies Home Journal* described, in vivid detail, behind-the-scene efforts of parents, in Lexington, Kentucky. In February, a dozen mothers, members of the PTA, and a local National Foundation representative assembled at one of the homes, examined printed materials about the trial, and compiled a list of clinic volunteers. The following month, they each met with groups of twenty-eight volunteers to review various documents (e.g., health records, permission slips, and request forms), ultimately holding a total of twenty-eight evening meetings. While some parents expressed how unfair it was that every child did not receive the new antigen but had to donate blood, they nevertheless agreed to comply, reflecting an overwhelming consensus. Parent-volunteers took home forms to complete and addressed envelopes. On the first day of injections, 500 mothers, fathers, teachers, and doctors participated. Mothers had, beforehand, strung sheets over clotheslines to create makeshift partitions. Finally, older students wearing "Polio Volunteer" arm bands provided assistance that day.[30]

Nevertheless, school personnel played central roles. "Memorandum No. 3" directed school superintendents to assume "responsibility for the administration of the tests within [the] school system." They operated as site managers, serving as liaisons with health officers and volunteer leaders, training building leaders and teachers, ensuring that clear lines of communications existed with building principals, volunteers, and parents, overseeing the testing timetable, preparing individual buildings to receive and store all of the "record forms, educational materials and clinic supplies," and arranging for storage of all consent documents for a period of three years. "Memorandum No. 4" focused on school principals whose duties fell into two main areas: personnel and the physical plant. The former involved volunteers such as classroom mothers as well as first-, second-, and third-grade teachers. They also organized all of the "forms and educational materials," permission slips, medical forms, and filmstrips. Moreover, they initiated a training program and assembled parents for informational and educational meetings. Finally, principals had to ensure the safekeeping of medical records and reconfigure physical space within buildings to create a clinic-like setting to accommodate a series of three, mass injections.

Teachers assumed demanding roles. They prepared students by scheduling lessons to build "enthusiasm" for their participation in this medical trial, showing them, for example, a filmstrip titled "Bob and Barbara," produced by the National Foundation. They also displayed a poster in their classrooms, depicting a Polio Pioneer button, dominated by a researcher peering into a microscope next to a rack test tubes, a clear laboratory setting, inspiring their young charges to see themselves as little scientists. "Memorandum No. 5" specified teachers as official record keepers. They had to prepare the packets students carried home, collect all of the completed paperwork, compile medical profiles of each child, a daunting responsibility, and check, maintain, and transfer all of these documents. They had to further supervise classroom mothers, distribute "educational materials and Parental Request Forms … follow up on unreturned forms, assist in setting up Parents [sic] meetings, [and] act as monitors for clinics." On V-Day, classroom instructors alphabetically organized all of the children in a line, escorted them to the designated injection areas, and walked each one into the "vaccination room." The same procedures applied to the two subsequent injections. As recognition awards, they distributed Polio Pioneer Cards and buttons to each child. Finally, and most seriously, classroom instructors had to meticulously track all absences, note any signs of illnesses, and immediately report them, a potentially life-and-death situation.[31]

In New York, where more than 20,000 school children participated, *New Yorker* magazine sent reporters to Public School No. 61, located at 610 East Twelfth Street. They observed many mothers fulfilling a variety of roles. Medical personnel drew blood just before giving children their injections. These writers also interviewed six of the participants, who likened themselves to pioneers, with one qualification, as one precocious child remarked: "Like the old pioneers, only not on land but in knowledge."[32]

What unfolded was anything but the movements of a fine Swiss timepiece, as several flaws emerged. First, stealth injections unfolded across the country, involving hospital personnel, medical researchers, and public health officials, as well as their families. Some North Carolina doctors, for example, surreptitiously removed vials to administer the experimental immunogen to their own children and those of close friends, totaling in the hundreds. Those physicians returned the unused portions only after the National Foundation threatened them with prosecution. In other cases, security surrounding vaccine storage appeared informal, if not outright lax. In Lexington, Kentucky, some vials had to be stored in household refrigerators because schools did not have enough space, leaving them available for use by anyone. Moreover, even before the completion of the national trial, Thomas Francis received countless requests for additional vaccine from medical centers, research laboratories, and state public health departments to inoculate them and their children. Francis, acting on behalf of the NFIP, acceded to all of these pleas. Second, sloppiness jeopardized safety. In "Schenectady, New York, nurses carelessly reused syringes still wet with liquid, giving a 'significant dose of immunizing vaccine to children supposed to receive the placebo' (and vice versa)." Third, data security appeared lax. For instance, someone stole the vaccination records from the unlocked principal's office of Davenport, Iowa, school building. Fourth, unanticipated shorter school years in Florida, Iowa, New Hampshire, Oklahoma, South Carolina, and Washington generated many requests to shorten the intervals for the third injection, creating adaptations by the Vaccine Evaluation Center.[33] This adjustment remained unmentioned in the final report, leaving questions about the appropriate timing of the third injection. Fifth, because of oversights, blood samples were not always collected.[34]

Thomas Francis nevertheless persevered, struggling to maintain a systematic approach that included four sequential phases. Mass injections consisting of three shots occurred between April and May of 1954. Laboratory analyses of the subjects' blood samples followed between June and December of 1954. Francis and his assistants compiled and evaluated the statistical data. Finally, Francis could have chosen an understated, traditional scholarly approach by publishing the trial's results through a medical journal; instead, "he allowed the drug company Eli Lilly, one the main manufacturers of the vaccine, to broadcast a special meeting held in the University of Michigan to 54,000 doctors in sixty-one cities by closed-circuit television at a cost of a quarter of a million dollars."[35]

Notes and Sources

1. William J. Curran, "Governmental Regulation of the Use of Human Subjects in Medical Research: The Approach of Two Federal Agencies," *Daedalus* 98 (Spring 1969): 542–43, 546, 548. Moreover, see James Colgrove, *State of Immunity: The Politics of Vaccination in Twentieth-Century America* (Berkeley: University of California Press, 2006): 119, 129.
2. Paul A. Offit, *Cutter Incident: How America's First Polio Vaccine Led to the Growing Vaccine Crisis* (New Haven: Yale University Press, 2005): 45 and 47, respectively. Colgrove, *State of Immunity*, refers to six companies (117).
3. MOD; Medical Programs Records; Series 5: Committees; Rivers to O'Connor, December 3, 1953, Vaccine Advisory Committee, 1951–1960.
4. Thomas M. Rivers, *Reflections of a Life in Science and Medicine* (Cambridge: Massachusetts Institute of Technology, 1967): 510. Also, refer to MOD, Medical Programs Records; Series 5: Committees; Rivers to O'Connor, December 3, 1953, Vaccine Advisory Committee, 1951–1960.
5. Victor Cohn, *Four Billion Dimes* (Minneapolis: 1955): 106; Jane S. Smith, *Patenting the Sun: Polio and the Salk Vaccine* (New York: William Morrow and Company, 1990): 253.
6. Arthur Allen, *Vaccine: The Controversial Story of Medicine's Greatest Lifesaver* (New York: W. W. Norton Company, 2007): 192.
7. Smith, *Patenting the Sun*: 254. Refer to pages 251, 253, as well as Jonas E. Salk, et al., "Studies in Human Subjects on Active Immunization Against Poliomyelitis, Part II. *American Journal of Public Health* 44 (August 1954): 1005–6; Cohn, *Four Billion Dimes*: 105–6, 112–14, 117; and Offit, *Cutter Incident*: 48–49.
8. Press Release from Max Elder, Director of the University of Pittsburgh's News Service [undated], Press Release by Jonas Salk through the University of Pittsburgh, January 16, 1954, Salk's draft illustrates his revisions to this final statement. He inserted obvious punctuation changes, but also included a few short phrases that infused a sense of certitude, reducing any comments that may have conveyed a sense of uncertainty. Jonas Salk to "Parents or Guardian of the Bearer," February 10, 1954, Jonas Salk Papers, Box 7, File Folder, Salk Polio Vaccine Testing (Collection #90/36/7), UPITT.
9. Press release from Max Elder, Director, Pitt's News Service, to Newspaper, Radio and Television Editors, February 1954, Jonas Salk Papers, Box 7, File Folder, Salk Polio Vaccine Testing (Collection #90/36/7), UPITT.

10. Pat McCormack, "Grades 1 to 3 Will Receive Dr. Salk's Vaccine Shots," (1954): 1 and 4, Jonas Salk Papers, Box 8, File Folder, Newspaper Clippings (Collection #90/36/7), UPITT; MOD, Chapter Administration, Pennsylvania, Allegheny Chapter, May 1953, "Poliomyelitis Immunization Project" (February 26, 1954); "News Release, Allegheny County Medical Society" (January 15, 1954). Elementary buildings included Arsenal, Belmar, Bennett, Burgwin, Colfax, Conroy, Connelly, Crescent, Clayton, Davis, Dilworth, Forbes, Frick, Fulton, Linden, Madison, Rogers, Sunnyside, and Whitman. The four suburban districts included Edgeworth, Fair Oaks, Leetsdale, and Sewickley. Further consult Press Release from the University of Pittsburgh News Service, March 4, 1954, Jonas Salk Papers, Box 7, File Folder, Salk Polio Vaccine Testing (Collection #90/36/7), UPITT.
11. Colette Freiwald, "Dr. Salk's Guinea Pig," *Pittsburgh Post-Gazette*, April 9, 2011: Section A-15. Refer as well to a Press Release from the University of Pittsburgh News Service (1954), Jonas Salk Papers, Box 7, File Folder, Salk Polio Vaccine Testing (Collection #90/36/7). Blurred copy, UPITT. Van Riper to Osborne, April 7, 1954, Box no. 14, Vaccine Safety, BHL. These parochial schools were Holy Rosary, Resurrection, Sacred Heart, St. Basil, St. George, St. Lawrence; and St. Stephens. Finally see Cohn, *Four Billion Dimes*: 113; Charlotte DeCroes Jacobs, *Jonas Salk: A Life* (New York: Oxford University Press, 2015): 130, 137; John R. Paul, *A History of Poliomyelitis* (New Haven: Yale University Press, 1971): 420; and John Troan, *Passport to Adventure: Or, How a Typewriter from Santa Led to an Exciting Lifetime Journey* (Pittsburgh, PA: Geyer Printing, 2000): 180.
12. *Annual Report of the Superintendent of the Pittsburgh Public Schools, 1953–54* (Pittsburgh: Board of Public Education, 1954): 8, Pennsylvania Room, CLP.
13. Bert Hansen, *Picturing Medical Progress from Pasteur to Polio: A History of Mass Media Images and Popular Attitudes in America* (New Brunswick, NJ: Rutgers University Press, 2009): 247. Also see Elder to Ziel, February 26, 1954, Jonas Salk Papers, Box 7, File Folder, Salk Polio Vaccine Testing (Collection #90/36/7). Finally, refer to Cohn's reconstruction of these events in *Four Billion Dimes*: 123.
14. Thomas Rivers, on behalf of the NFIP's Vaccination Advisory Committee, which he chaired, to the editor of *Journal of the American Medical Association*, February 4, 1954, Jonas Salk Papers, Box 7, File Folder, Salk Polio Vaccine Testing (Collection #90/36/7), UPITT.
15. John R. Paul, "Thomas Francis, Jr., July 15, 1900–October 1, 1969," *Biographical Memoirs* 44 (1974): 78–79.
16. *Operational Memoranda: Vaccine Field Trial, 1954*. Box no. 14, BHL. Further refer to Allan M. Brandt, "Polio, Politics, Publicity, and

Duplicity: Ethical Aspects in the Development of the Salk Vaccine," *International Journal of Health Services* 8 (1978): 260–61; David M. Oshinsky, *Polio: An American Story* (New York: Oxford University Press, 2005): 197.

17. Smith, *Patenting the Sun*: 231–32. Especially consult Thomas Francis, Jr., Robert F. Korns, Robert B. Voight, Morton Boisen, Fay M. Hemphill, John A. Napier, and Eva Tolchinsky, "An Evaluation of the 1954 Poliomyelitis Vaccine Trials: Summary Report." Vaccine Evaluation Center, April 12, 1955. Box 8, File Folder, University of Michigan (Collection #90/36/7), UPITT.
18. Oshinsky, *Polio*: 186, 187, 189, respectively. Refer, moreover, to Francis, et al., Evaluation of 1954 Field Trial of Poliomyelitis Vaccine": 1; Sarah Marie Lambert and Howard Markel. "Making History: Thomas Francis, Jr., M.D., and the 1954 Salk Poliomyelitis Vaccine Field Trial," *Archives of Pediatrics and Adolescent Medicine* 154 (May 2000): 513–15; Arnold S. Monto, "Francis Field Trial of Inactivated Poliomyelitis Vaccine: Background and Lessons for Today," *Epidemiological Reviews* 21 (1999): 13–14; Paul, "Thomas Francis": 78; and Smith, *Patenting the Sun*: 229.
19. MOD; Medical Programs Records; Series 3: Salk Vaccine; Van Riper to State Health Officers, April 28, 1954, "Additional Insurance Coverage Poliomyelitis Vaccine Field Trial"; Holloway to Field Trial Staff, November 18, 1953, "Insurance for Field Trial"; Cummins to Mrs. White, Ryan, Glasser, McGinnes, Culp, December 31, 1953, "Liability Insurance"; Potthoff to Culp, January 20, 1954, "Insurance Protection"; Wagner to Engelbach, April 6, 1954, "Rochester, NY"; Ryan and Cummins to Glasser, April 7, 1954; Engelbach to Regional Medical Consultants, April 8, 1954, "Statement on Liability"; and Public School Vaccination Programs; Salk and Sabin Vaccine Records.
20. MOD; Medical Programs Records; Series 2: Committees; Conferences and Meeting Records; Vaccine Advisory Committee; "Poliomyelitis Vaccine Types 1, 2, and 3: Recommendations of Vaccine Advisory Committee," April 25, 1954.
21. Sebrell to Norton, December 22, 1953, Vaccine Safety, Box no. 14, BHL.
22. Sidney A. Halpern, *Lesser Harms: The Morality of Risk in Medical Research* (Chicago: University of Chicago Press, 2004): 113. Refer also to Brandt, "Polio, Politics, Publicity, and Duplicity": 264; Paul Meier, "Polio Trial: An Early Efficient Clinical Trial," *Statistics in Medicine* 9 (January/February 1990): 16. The Foundation materials are: Van Riper to Hutcheson, December 29, 1953, Forms, general, Box no. 14, Thomas Francis Papers, BH, and MOD; Medical Programs Records; Salk and Sabin Vaccine Records; Series 3: Salk Vaccine; "What You Should Know About the Polio Vaccine Tests," "A Message to Parents," and "Parental

Request for Participation of Child in Poliomyelitis Vaccination Trial," Field Trials: Forms, Records, Notices (blanks), 1954. The italics are mine.

23. MOD; Medical Programs Records; Salk and Sabin Vaccine Records; Series 3: Salk Vaccine; Public School Vaccination Programs; "Office of the Superintendent of Schools, Malden, Massachusetts," May 6, 1954.
24. MOD; Medical Programs Records; Salk and Sabin Vaccine Records; Series 3: Salk Vaccine; New York City Vaccination Program, 1954; "What You Should Know About the Polio Vaccine Tests in New York City." "Facts for You about the Polio Vaccine Trials in New York City"; "News Release: From Leona Baumgartner, Commissioner, New York City Department of Health," March 22, 1954. Also see Colgrove, *State of Immunity*: 118.
25. Smith, *Patenting the Sun*: 238.
26. DeMartino in Janice Flood Nichols, *Twin Voices: A Memoir of Polio, the Forgotten Killer* (Bloomington, IN: iUniverse, Inc., 2007): 111–12; Brenda Serotte, *The Fortune Teller's Kiss* (Lincoln: University of Nebraska Press, 2006): 78–79. The italics are in the original text. Further consult Dee Van Balen, *The Closing Door* (Pittsburgh, PA: CreateSpace, 2010): 17.
27. MOD; Medical Programs Records; Conferences and Meetings Records, 1938–2002; "Meeting of Advisory Group on Evaluation, January 11, 1954," Committee on Virus Research and Epidemiology, April 14, 1953. Series 2: Committees.
28. Margaret Hickey, "Have We Won the Fight Against Polio?" *Ladies Home Journal* (December 1954): 150. Oshinsky, *Polio*, also describes this episode (197).
29. Francis, et al., "Evaluation of 1954 Field Trial of Poliomyelitis Vaccine": 2.
30. Hickey, "Have We Won the Fight Against Polio?" 150. Further consult page 25.
31. Specifically refer to "Memorandum No. 3," "Memorandum No. 4," "Memorandum No. 5," "Memorandum No. 6," "Memorandum No. 8," "Memorandum No. 11," "Memorandum No. 12," "Memorandum No. 13," and "Memorandum No. 15," Operational Memoranda: Vaccine Field Trial, 1954. Box no. 14, BHL, No pagination. Dr. Albrecht, New York State Department of Health, June 8, 1954, Forms, general. Box no. 14, BHL. "Polio Vaccine Bulletin No. 1.2," March 25, 1954, Field Trial: Operational Memoranda, Box no. 14, BHL. Francis, et al., "An Evaluation of the 1954 Poliomyelitis Vaccine Trials": 5; Refer also to Liza Dawson, "The Salk Polio Vaccine Trial of 1954: Risks, Randomization and Public Involvement," *Clinical Trials* 1 (2004): 128; Hickey, "Have We Won the Fight Against Polio?" 150.

32. "O Pioneers!" *New Yorker* (May 1954): 24–26.
33. Oshinsky, *Polio*: 198. Also see Press to Engelbach, May 27, 1954, Voight to Francis, June 3, 1954, Field Trial: Irregularities, Box no. 14 and Vaccine: Requests for Immunization, Box no. 14, and Francis to Dr. David Hudson, Tulsa City Health Department, May 26, 1954, Dublin to Francis, May 20, 1954, Field Trial: Irregularities, Box no. 14, BHL. Finally, refer to Hickey, "Have We Won the Fight Against Polio?" 150.
34. Albrecht to Francis, May 27, 1954, Field Trial: Irregularities, Box no. 14, Thomas Francis Papers, BHL. See also Francis, et al., "Evaluation of 1954 Field Trial of Poliomyelitis Vaccine."
35. Tony Gould, *A Summer Plague: Polio and Its Survivors* (New Haven: Yale University Press, 1995): 149–50. Further consult Offit, *The Cutter Incident*: 54; David W. Rose, *Images of America: March of Dimes* (Charleston, SC: Arcadia Publishing, 2003): 66–67.

PART IV

Line Up and Roll Up Your Sleeves

CHAPTER 13

"Operation Needle"

On April 12, 1955, the tenth anniversary of President Franklin D. Roosevelt's death, some 500 people, including newspaper, radio, and television reporters, with their recording equipment, crowded into the auditorium of the University of Michigan's Rackham Hall to hear Thomas Francis, Jr., deliver his report on the national experiment. Based on the data, he revealed the polio vaccine was 60–70% effective for Type I and 90% for Types II and III. Church bells pealed and sirens blared celebrating this historical, medical breakthrough for children. The largest medical trial in history had "transformed the [National Foundation for Infantile Paralysis] from a fund-granting agency to an active implementor of health strategies."[1]

Francis's declaration produced a media sensation and pubic euphoria. As Pittsburgh school's Superintendent proudly wrote in his 1954–1955 annual report, the "poliomyelitis vaccination program started last year in the Pittsburgh schools as a controlled study has been continued. Following the release of the poliomyelitis vaccine for widespread use in

According to Edith Powell and John F. Hume, *A Black Oasis: Tuskegee Institute's Fight Against Infantile Paralysis, 1941–1975* (Tuskegee, AL: Tuskegee University, 2008), medical authorities in Montgomery, Alabama, designated the immunization campaign as "Operation Needle" (131).

R. J. Altenbaugh, *Vaccination in America*,
Palgrave Studies in the History of Science and Technology,
https://doi.org/10.1007/978-3-319-96349-5_13

the first and second grade school children in our nation, all children in the first and second grades in this city whose parents so requested were inoculated during May 1955." The stage had been set for an unprecedented effort to immunize all American children.[2]

Mass Immunization

The federal government had to certify the polio vaccine before releasing it to the public. President Dwight D. Eisenhower in 1953 had expanded the New Deal's Federal Security Agency and renamed it the Department of Health, Education, and Welfare (HEW), raising it to a cabinet-level post. It absorbed the Social Security Administration, Office of Education, US Public Health Service, and National Institutes of Health (NIH). "One of those institutes was the National Institute of Microbiology, which had jurisdiction over what was then called the Laboratory of Biologics Control." Under its 1902 charter, it held the power for the final approval—and licensing—of that vaccine.[3]

Surgeon General Leonard Scheele had dispatched William G. Workman, director of the Laboratory of Biologics Control, and a fifteen member licensing advisory committee to Ann Arbor. Following Francis's presentation, they met at a nearby hotel. "Each member ... was given a copy of the Francis report ... (113 pages long) and protocols that detailed how each of the forty lots of vaccine had been manufactured (each protocol was about 50 pages long). Most ... were looking, for the first time, at approximately 2,000 pages of information." They felt pressured because they faced a tight schedule: HEW Secretary Oveta Culp Hobby had scheduled a press conference that day to make the vaccine authorization announcement in Washington, D.C., and eagerly waited by the phone for their decision. "After two and one-half hours of discussion, it unanimously recommended licensure" for Cutter, Eli Lilly, Parke-Davis, Pitman-Moore, and Wyeth pharmaceutical companies. This superficial decision-making process augured problems. First, only "Parke-Davis and Eli Lilly had made vaccine for the field trial...." They therefore had the most experience with the complex production process. Second, Workman failed to inform committee members that his laboratory had found live virus in the Cutter samples.[4]

As polio immunization moved forward, the medicalized public schools, once again, played a key role, as Ardean Marting, General Chair of the Mothers March on Polio for San Diego County recalls, "We set up stations in different areas and the school nurses cooperated with us under the supervision of a doctor in each area. We had certain ages, like five to nine, that received the shots first, and then the older ones." *Life* magazine's May 1955 issue projected a spectacular visual of this dramatic event, with children formed in block-long queues outside of their schools awaiting their vaccinations.[5]

Suddenly, on April 14, 1955, a *New York Times* front-page headline screamed, "Vaccine for All in Peril of Polio Likely this Year." With the summer polio outbreaks looming, desperate parents wanted protection for their children, creating an insatiable appetite. Moreover, parents of children who had participated in the nationwide test besieged Thomas Francis, begging to know if their children had been immunized. If not, that meant they had received the placebo and were now vulnerable to infection, paralysis, and even death. Black market operations also fed public hysteria over possible vaccine shortages. New Jersey had to delay the inoculation of 300,000 children because of a failed delivery. The entire matter spiraled into a political football. New York City Commissioner of Health, Leona Baumgartner, and Mayor Robert F. Wagner, Jr., urged President Eisenhower to oversee an equitable distribution program. Wagner asked that priority be given to first and second graders followed by fourth and fifth graders. Local and state control proved inadequate as meetings began in Washington about federal oversight. Eisenhower hesitated, though; his free-market outlook abhorred government interference. Furthermore, the American Medical Association (AMA) lobbied against such a measure since it smacked of socialized medicine. Confusion reigned. This frenzy continued for ten days before everything screeched to a halt.[6]

On April 25, 1955, William Workman received information about a one-year-old Chicago child developing paralysis. The following day, he heard similar news about two seven-year-old boys in San Diego. Three additional cases popped up in other parts of California. Since the polio season had not yet unfolded, something was amiss. As researcher Paul A. Offit describes it in his comprehensive analysis of this tragedy: "Although five companies had made and distributed polio vaccine, all six paralyzed children had received vaccine made by only one company–Cutter Laboratories." A pioneer in veterinary medicine, Cutter "was among the first companies to make vaccines against diphtheria, tetanus, and pertussis (whooping cough), and it was the first company in the world to combine these three vaccines into a single shot." Its stock of vaccine appeared to contain live virus, thus causing infections. What would infamously become known as the Cutter Incident had begun.

Institutional responses proved inadequate. Leonard Scheele, Offit continues, reviewed Workman's unsettling report, but decided to delay any action until he received the recommendations of an emergency meeting of representatives from NIH and the Centers for Disease Control (CDC). They met for seven hours and concluded Cutter was the common denominator. Not wanting to raise a "false alarm," on April 26 they informed Scheele they were deadlocked: They could not decide to recall just the Cutter product or all of it. Scheele faced an additional obstacle. Federal agencies did not have the "authority to stop a company from selling a vaccine that was already licensed.... [T]he company would have to withdraw it voluntarily." He sent a telegram to Cutter executives the next day making that request; they complied. That same day two vaccinated children in Idaho became paralyzed. By April 29, eleven cases had been verified. The following day, this reached the grim total of twenty-five children who had been paralyzed or died, all in the Midwest and West.[7] Scheele dispatched the Epidemic Intelligence Service (EIS) "to review Cutter's protocols with company officials in Berkeley."[8]

EIS emerged as a public health agency within the US Public Health Service in July 1950, under CDC authority, serving as an "early-warning system" against contagions. Its experts investigated all outbreaks anywhere in the nation as part of Cold War fears of the Soviet Union launching a biological attack. Polio, in particular, "was often mentioned as a possible biological-warfare agent...." The Service's first assignment came in September of 1950 with reports of infantile paralysis in Paulding County, Ohio. The EIS fielded a twenty-member team, consisting of clerks, engineers, entomologists, nurses, statisticians, surgeons, veterinarians, and virologists, to investigate food, milk, sewerage, and water supplies as well as the insect and rodent populations. "It was the world's first epidemiological survey of polio." Their thorough study revealed that "epidemiologists did not have any clearer notion of why Paulding County had been singled out for a polio episode then they had before they went...." *Life* magazine dedicated an article, "On the Trail of an Epidemic," to cover the story.[9]

Although the federal government had recalled the Cutter vaccine, albeit indirectly and informally, "within forty-eight hours of the first reports of suspected cases," Offit states, "it was too late; 380,000 children had already been inoculated" in over twenty-six states, with

300,000 concentrated in Arizona, California, Hawai'i, Idaho, Nevada, and New Mexico. After an investigation that included abortive paralysis, "at least 220,000 people were infected with live polio virus contained in Cutter's vaccine: 70,000 developed muscle weakness, 164 were severely paralyzed, and 10 were killed. Seventy-five percent of Cutter's victims were paralyzed the rest of their lives." In sum, Cutter's product was at the center of "a man-made polio epidemic.... [O]ne of the worst biological disasters in American history."

But it did not end there. Further inquires revealed the "Wyeth problem." One lot of its vaccine, used in Delaware, Maryland, Ohio, and Pennsylvania, had caused over twenty infections. NIH and CDC directors as well as Scheele knew about this danger but quashed any announcements, quietly removing it from use because they "wanted to maintain the public's trust in the polio vaccine program."[10]

On April 29, FDA investigators reviewing inactivation protocols "discovered that [Cutter] had changed to a cheaper filter, one that incidently allowed an occasional line particle of poliovirus to leak into the vaccine."[11] Even more alarming, they ascertained that all five pharmaceutical companies had difficulty killing this virus. Careful development of small batches in Jonas E. Salk's laboratory was one thing, but mass production at pharmaceutical plants introduced a complex calculus that no one had anticipated. The "vaccine that companies deemed safe–vaccine that was approved by the federal government–might still contain live polio virus." On May 6, 1955, Scheele temporarily suspended all polio immunizations.[12]

Following a thorough review, the federal government, on June 14, released one million doses judged safe. Nevertheless, the American public appeared distrustful. As in the aftermath of the failed Park-Brodie and Kolmer experiments, it seemed many parents feared the vaccine as much as the disease. Statistics bore this out. The number of children receiving injections plummeted. Further, a "number of state and local health departments declined to use the Salk vaccine" because of safety concerns.[13] As a result, morbidity rates wildly fluctuated during the 1950s. Between 1950 and 1954, the annual number of cases averaged 39,000. This figure dropped to 29,000 in 1955 with the successful tests. "By the end of 1957, the annual total was below 6,000. In 1958, however, the downward trend reversed itself, with the number of cases climbing back above the 6,000 mark. Even more significant, the number of paralytic cases was up 44 percent over 1957...." The Detroit area alone reported

876 cases of infection with twenty-three deaths.[14] The real tragedy was all of this could have been avoided. Salk's antigen worked, as Offit points out. Outbreaks during the summer of 1955 revealed that "[p]eople who were not vaccinated were four times more likely to get polio than those who were vaccinated."

In the immediate aftermath, everyone pointed to each other as culprits. "Cutter blamed Salk for devising a process that was inconsistent, and ... the federal government for setting up standards of manufacture and testing that were inadequate." Meanwhile, during a Congressional hearing, held June 22–23, 1955, "... many physicians, scientists, professional organizations, public health officials, and politicians felt that the National Foundation had rushed the research, staged a clinical trial of a vaccine that wasn't ready, and forced the government to license a product that lacked adequate safety tests." Basil O'Connor fixed responsibility on the federal government. Researchers John F. Enders, William McDowell Hammon, and Albert B. Sabin faulted Salk for relying on a flawed inactivation process. In the end, the federal government bore the brunt of the guilt. Hobby, Workman, and Scheele, among others in the chain of command, lost their jobs because they had failed to provide sufficient oversight of the production process.[15]

In 1957, the National Foundation called on media outlets and health departments to mobilize another vaccination initiative, pointing to Allegheny County's Polio Vaccination Program, immunizing all infants and adults, as an "example or guide for other cities and areas throughout the entire country." That county's medical society and health department, financed by the NFIP, private donations, and public funds, assisted by 3400 volunteer nurses, physicians, and volunteers, conducted 638,295 inoculations by June, using schools, colleges, and "mass clinics" throughout the city of Pittsburgh and county.[16]

Nevertheless, polio vaccination failed to become universal. By 1959, 98 million Americans had not received even one of the three-shot series. This had less to do with fallout from the Cutter Incident and more to do with the inability of poor Americans to pay a private physician. The National Foundation enlisted the American Legion, AMA, Junior Women's Clubs, Parent Teacher Association (PTA), and US Public Health Service in a "joint effort" to promote the need for immunization. Public school buildings, once again, functioned as the epicenter for this campaign—supplemented by community centers, "mobile trailers," and religious institutions.[17] The whole effort fell apart. Cajoling only went so far. Socioeconomic background remained an obstacle. Because compulsion remained off the table, "most cities and states … continued to rely on a voluntaristic approach."[18]

Albert Sabin's new, live-virus option presented yet another element. Born in 1906 in Bialystok, Russia, his family emigrated to the USA in 1921. Strongly influenced by Paul de Kruif's *Microbe Hunters*, Sabin's career trajectory led directly to research. He completed his studies at New York University's medical school in 1931, having worked with William H. Park. Sabin followed this with a "fellowship in virology at London's Lister Institute." Upon returning to New York City, he assumed a position at the Rockefeller Institute for Medical Research, subsequently joining the University of Cincinnati's faculty in 1939.[19]

Sabin tested his initial batch on 10,000 monkeys and 160 chimpanzees for the three polio types before proceeding with human trials in 1956. He had planned an initial experiment at Willowbrook State School, but the National Foundation's Virus Research Committee nixed it. He turned to another captive audience, 200 prisoners at the Federal Reformatory at Chillicothe, Ohio. Following a successful outcome there, he had to resort to using Soviet children because too many American youngsters had been immunized with the Salk vaccine. The CDC sent an observer. An estimated 10 million children received the solution, either through a medicine dropper or saturated candy. Sabin also turned to children in Mexico and Singapore and in fact received credit for ending a polio epidemic in the latter.[20]

Although Surgeon General Leroy Burney approved Sabin's vaccine on August 24, 1960, NIH officials expressed reluctance about using it, fearing it would cause secondary infections. Moreover, it remained unavailable. Manufacturing moved at a snail's pace because processing this live-virus version proved difficult. "Rigid government safety standards, purposefully set very high in an effort to prevent another Cutter Incident, also slowed production." In the interim, CDC authorities "launched an intensive campaign to boost use of the Salk vaccine, which they called 'Babies and Breadwinners,' hoping to reach the two elements of the population least likely to have been vaccinated." They used Columbus, Georgia, as a model for the nation, blitzing it with information in local newspapers and churches as well as placing inoculation booths on sidewalks. "The idea did not catch on."[21]

The Salk option had lost its public appeal. Multiple injections, its disappointing 80% rate of effectiveness, and the Cutter Incident collectively had taken their toll. "Of those who needed vaccination, only 40 percent had received the full course of [now] four injections–and another 40 percent had not been vaccinated at all." Sabin's version became the preferred choice, with 100% effectiveness. Further, his live-virus vaccine quickly induced immunity, could be easily taken, and proved cheaper to produce. Finally, in "June 1961, the American Medical Association approved the recommendation of its Vaccine Committee that Sabin's vaccine should replace Salk." Emergency Intelligence Service investigators found that Sabin's oral antigen caused polio in rare cases, but buried any evidence because, like the Cutter Incident, it would seriously damage that immunization program. In August of 1962, the Food and Drug Administration formally withdrew Salk's vaccine from use.[22] Once again, the public schools became the main site to dispense the attenuated vaccine. Promotion followed the usual pattern. "Especially popular were community-wide 'Sabin on Sunday' events, which were highly successful at reaching large percentages of the population."[23] Sabin's sugar cubes completely replaced Salk's hypodermic needles.

The polio immunization story continued to unfold over the next four decades. In 1962, one year after it had been licensed, a committee convened by the surgeon general found that eighteen vaccinated individuals had been infected with paralytic polio. "The risk was about one per million overall, but it was somewhat greater than those over thirty." By June 1964, the CDC's polio surveillance team had recorded 123 infections occurring within thirty days after receiving the oral polio vaccine (OPV). "The risk was greatest for Type III polio, about two and half times greater than for Type I, and negligible for Type II." As a result, the CDC "recommended that the order of distribution of OPV be changed so that Type II was given first and that teenagers and adults be given vaccine only when there was special risk of exposure." Albert Sabin strongly objected "to changing the order which the vaccines were given" and "excluding individuals over fifteen." He especially "regretted that so much emphasis was put on immunizing infants under one year without specifying the oral vaccine as opposed to the killed-virus vaccine."[24] History ultimately exonerated Salk. His killed-virus approach, when properly manufactured, worked. "Between 1956 and 1961, 400 million doses of Salk's polio vaccine were administered in the United States without causing a single case of paralysis." On October 20, 1998, the federal government withdrew its support of OPV, resorting exclusively to the dead-virus version because the attenuated-virus antigen proved to be too dangerous, increasing the chances of secondary infections. "Sabin's vaccine is no longer available in the United States."[25]

National Foundation officials and researchers, operating in a largely self-regulatory environment, were fortunate. International guidelines covering the use of human subjects certainly existed, but they did not scrupulously follow them. Yet, even in the lax ethical context of the USA, they still managed to cobble together and implement a hybrid, research design, resulting in a safe and effective vaccine.

Scientific Progress

Researchers worked on other immunogen and continued to test them on vulnerable populations. In 1954, Harvard University's John Enders began development of a measles vaccine. Born in 1897 to a wealthy West Hartford, Connecticut, family, Enders graduated from Yale University and completed his graduate work at Harvard. With National Foundation funding, Enders, Frederick C. Robbins, and Thomas H. Weller, in 1949, discovered how to cultivate poliovirus in a test tube, marking a major advance in virology. Researchers, until then, produced viruses from animal tissues, a contaminant, or grew them in eggs, a potential source of an allergic reaction. This new procedure provided unlimited production of safer viruses, paving the way for vaccine development.

Enders and his assistants injected the experimental measles antigen into monkeys and themselves before using it on children. In 1958, they turned to a local institution, the Walter E. Fernald State School for the Feebleminded, where measles outbreaks occurred every few years resulting in "serious morbidity and a number of deaths." Its director granted the team permission "to meet with the parents of several dozen children who had not yet suffered measles." They consented, and investigators injected half with the vaccine and half with a placebo. "Buoyed by these initially successful studies," writes Samuel L. Katz, a member of that research team, "we enlisted colleagues in Denver, New Haven, Cleveland, New York, and Boston to conduct similar studies among home-dwelling children under their care."[26] In New York, Saul Krugman experimented on forty-six children at Willowbrook State School. He injected half with the experimental vaccine while using the other half as observed controls. Six weeks later, when an outbreak occurred "infecting hundreds of children and killing four," none of the inoculated children fell ill.[27] By the end of the trials, experimenters had administered it to

303 children: 31 had it squirted into their mouths, or nasal passages, or inserted it as eye drops (conjunctival) while 272 received injections. Of these, 101 had preexisting antibodies and exhibited no visible response. The remaining 171 experienced a fever (a mean of 102.8 Fahrenheit) within a seven- to eight-day period, or a slight rash within eleven days, but 97% of them developed antibodies.[28]

Maurice Hilleman tackled the fever and rash side effects. Born in 1919 in Montana, he graduated from Montana State University in 1941 and attended the University of Chicago's medical school. He worked at E. R. Squibb pharmaceutical company for four years before moving to Walter Reed Army Medical Institute. In 1957, Hilleman began his career at Merck Research Laboratories, located in New Jersey. There, he and Joseph Stokes, Jr., refined Enders's attenuated vaccine and supplemented it with gamma globulin. They initiated their first human tests of this combination at the Clinton Farms for Women, a prison in central New Jersey. Since many inmates gave birth after being incarcerated, it had a nursery. "Hilleman and Stokes injected six infants with … vaccine in one arm and gamma globulin in the other." None of them displayed adverse symptoms. Following successful trials on hundreds of children at other sites, this measles immunogen received its license on March 21, 1963.[29]

Research on a mumps vaccine took the better part of two decades. Work began during World War II, financed by the federal Office of Scientific Research and Development. John Enders worked with Joseph Stokes to develop a short-acting antigen. In 1946, Enders's Harvard team pursued a permanent version at Children's Hospital in Boston. "In a subsequent set of experiments, conducted by both the Harvard group and by [Karl] Habel at the National Institutes of Health, vaccines containing weakened mumps virus were produced and tested in institutionalized children and [West Indian] plantation laborers in Florida."[30] It appeared safe and somewhat effective but needed refinement. Meanwhile, in the 1960s, "mumps virus infected a million people in the United States every year." Permanent protection, once again, fell to Maurice Hilleman. When Hilleman's older daughter contracted it in 1963, he swabbed her throat and, along with Robert Weibel and Stokes, prepared a live-virus vaccine. Hilleman inoculated his younger daughter with it, and she developed antibodies. In June 1965, Hilleman and his colleagues tested it on sixteen children at the Trendler School, in Bristol, Pennsylvania, an institution that housed "thirty severely retarded children," administering additional injections on thirty children at the

Merna Owens and St. Joseph's homes, located in northeast Pennsylvania. All of them acquired mumps antibodies. Next, they experimented with 400 children in Havertown, Pennsylvania, a Philadelphia suburb. After parents signed vague consent cards, Hilleman's team gave it to half of them, leaving the other half uninoculated. When a mumps outbreak occurred a few months later, 63 of these children contracted disease; 61 had not been vaccinated. It had worked and received its license in 1967.[31]

However, the medical community met it with disinterest. "Mumps was not a top public priority in 1967–in fact, it was not even a reportable disease–but the licensure of [Merck's] Mumpsvax would change the disease's standing over the course of the next decade." The CDC in 1968 reinstated "mumps surveillance, which had been implemented following World War I but suspended after World War II." Although mumps most often struck young boys, registering few annual deaths, it did pose a sterility threat for adolescents and adults. Health officers reasoned that children should be inoculated to protect everyone else. This decision seemed routine at the time since the "polio immunization drives ... had helped forge the impression that vaccines were 'for children' as opposed to adults." Still, it did not attain universal adoption until it was combined with measles and rubella.[32]

The 1963–1964 rubella epidemic spurred research since pregnant women infected by it gave birth to 20,000 children with hearing disabilities. Maurice Hilleman developed one in 1969, and Merck distributed 100 million doses. Stanley Plotkin's work improved it. "[I]nspired to a life of science by Sinclair Lewis's novel *Arrowsmith*....," he graduated from New York University and Brooklyn's Downstate Medical Center, he joined the CDC's EIS before going to Wistar Institute of Anatomy and Biology, at the University of Pennsylvania, to work with Hilary Koprowski. After investigating polio and anthrax, Plotkin began research on a rubella immunogen. Successfully tested, it was licensed in 1978, replacing existing supplies.[33]

Merck combined measles, mumps, and rubella into a single immunization in 1971—otherwise known as MMR. This proved more efficient and a cheaper approach, with only one doctor's visit. "By 1974, 40 percent of U.S. children had been vaccinated against [mumps]." In 1977, the CDC, recognizing mumps as a low-priority disease, felt the convenience of a single injection overrode that status, and endorsed it as part of "'routine immunization activities'.... The combined vaccine enabled mumps to piggyback on acceptance of the vaccines against measles and rubella and overrode for good any questions about the necessity of universal protection against mumps."[34]

Clearly, children needed to be protected from dangerous diseases. However, while these vaccines significantly reduced morbidity and mortality rates, these gains came at a steep price. Scientists routinely risked children's health and lives to test the safety and efficacy of immunogen before they became licensed. This contradiction became evident when abuses were made public. Vulnerable populations suddenly became visible.

Human Guinea Pigs

Medical experiments at two institutions, when made public, deeply and permanently damaged researchers' credibility. The first revealed the dire conditions of institutionalized children tapped as human subjects, specifically Krugman's hepatitis observations and measles trials at Willowbrook and Hilary Koprowski's polio tests at Letchworth Village. A second revelation exposed the decades-long neglect of scientists who prolonged the suffering of ill African Americans at Tuskegee Institute in the face of an effective cure.

The public's first glimpse inside Willowbrook occurred in the fall of 1965 when Robert F. Kennedy, a US Senator representing New York, toured it. He declared it a "snakepit," with conditions for children worse than those for zoo animals. Six years later, Geraldo Rivera, a "young and obscure reporter," received a tip from a concerned Willowbrook staff member. Rivera and a film crew, using that employee's key, slipped into one of its buildings and quickly recorded what they saw.[35] This "unauthorized" visit depicted that institution under "uncontrolled conditions." The subsequent television documentary, "Welcome to Willowbrook," introduced a national audience to Dachau-like scenes of skinny, naked, and filthy children, sitting in their own feces, moaning persistently, and drinking water out of toilets; it profoundly moved viewers and provoked organizations and politicians to act. The Richmond County Society for the Prevention of Cruelty to Children sent an investigative team. They found a "dimly lighted, foul-smelling" facility, earning the pejorative moniker as New York City's "leper colony."[36] That same year, Rivera made a documentary about a similar setting, Letchworth Village, in Westchester County. Rivera and his crew "filmed conditions every bit as awful as those at Willowbrook," including a graveyard with numbered tombstones. These combined revelations profoundly challenged the tradition of using institutionalized children with disabilities for medical experiments.[37] The final blow soon followed.

The research community knew about the Tuskegee experiment from the beginning. The first report surfaced as a presentation at the 1936 AMA annual meeting, "with subsequent papers issued every four to six years, through the 1960s."[38] In 1965, a public health advisor caught wind of the experiment and raised ethical and racial concerns with CDC authorities. They convened a panel of experts to review this project. It completed its deliberations in early 1969, concluding "that the knowledge gained by the Tuskegee Study was great enough to warrant its continuation. In late July 1972, Jean Heller of the Associated Press broke the story of the forty-year-old study of untreated syphilis in black males in Macon County, Alabama."[39] HEW halted that research and formed the Tuskegee Syphilis Study Ad Hoc Advisory Panel on August 28, consisting of nine members, five of whom were African Americans. It released its Final Report a year later, finding "the study to have been 'ethically unjustified,' and argued that penicillin should have been provided to the men."[40]

Safeguards

Although ethical questions about the use of human subjects arose as early as the 1960s, no fundamental changes took place until the 1970s, unfolding as an international movement. According to philosopher Robert B. Baker, it began with "Post-Hiroshima skepticism about science and technology…." Further, some new medications caused serious scandals. Finally, England's Maurice Pappworth and America's Henry Beecher, who Baker labels "[w]histleblowing reformers," had carried on trans-Atlantic correspondence about human-subject abuses in each of their countries, setting the stage for their revealing and pathbreaking articles. The research community soon found it difficult to ignore its own transgressions.

Scattered and uncoordinated developments at different levels of society during the 1960s, Baker argues, built the "scaffold for bioethics." First, the Federal Drug Administration in 1962 barred the "introduction of [the] fetus-crippling drug, thalidomide, into America." This led to the Kefauver-Harris Act that "reformed the process of researching and marketing new drugs." Second, the New York State Division of Professional Conduct in 1965 "suspended the medical licenses of two physicians who had injected live cancer cells into incapacitated terminal patients without informing them or their families that the injection contained cancer cells." Third, Beecher's article, "deemed 'the most influential single paper ever written about experiments involving human subjects,'" appeared the following year. Fourth, during that same year, because of "stories in the popular press about a supply chain for animal experimentation that resorted to dognapping and by a graphic *Life* magazine article on doggy concentration camps, Congress enacted the Laboratory Animal Welfare Act." Finally, the AMA, ever so gradually, began to reassess its position.[41]

The Association prohibited the use of convicts in 1952 but never enforced that rule. As a result, prisoners became unwilling or subtly coerced research subjects, as consent was neither needed nor given. Furthermore, no real opposition arose in using them as research subjects, as Harriet Washington points out: "... prisoners were vulnerable, stigmatized, and expendable; they tended to be poor and uneducated; they were likely to belong to despised and powerless minority groups; they had already lost most important civil rights; and their crimes or alleged crimes made them feared and hated." Many examples of abuse existed. Chester M. Southam of the Sloan-Kettering Institute injected Ohio State Prison inmates with live cancer cells in 1952. In another case, Albert M. Kligman, a University of Pennsylvania dermatologist, went to nearby Holmesburg Prison in 1951 "to treat an outbreak of athlete's foot." In the process, he realized he could freely conduct experiments on prisoners. He induced foot infections, applied stringent chemicals to their skin, performed partial dissections, implanted subdural plant and human tissue, and infected them with various skin diseases, including herpes. Between 1962 and 1966, while working for thirty-three cosmetic and pharmaceutical corporations, Kligman tested 153 products on 75% of that prison's population. Many subjects suffered permanent scarring and hair loss. By the late sixties, under contract for the Central Intelligence Agency, Kligman began to give them mind-altering drugs. Meanwhile, around the same time, another researcher submitted Alabama prisoners to experimental blood transfusions.[42]

The World Medical Association's 1964 Declaration of Helsinki advanced the Nuremberg Code in two ways. First, it "distinguished clinical research combined with patient care and nontherapeutic human experimentation." The former absented consent while the latter required it. Second, and unlike the Nuremberg Code, Helsinki "permitted experimentation on individuals unable to exercise informed consent, including children, whose parents or legal guardians agreed to allow their participation in an experiment." However, American representatives opposed restrictions on institutionalized children and prison populations, producing an ambiguous position. International delegates resented this "American influence."[43] Moreover, the "AMA did not endorese the concept of prior peer review of research proposals; it endorsed research ethics in principle but, true to laissez-faire tradition, without practical enforcement mechanisms."[44]

The turning point came in 1972 when Senator Edward Kennedy held public hearings to investigate the forty-year-old Tuskegee Syphilis Study. Two actions resulted. By December 1974, the federal government paid the survivors $10 million in reparations. More important, these "hearings dramatically underscored the need for federal guidelines on human experimentation," culminating in the National Research Act.[45] This created the National Commission for the Protection of Human Subjects of Biomedical and Behavioral Research, assigned to identify basic ethical principles and develop guidelines that conformed to these principles. The eleven-member board initially assembled as a group in February 1976 for four days at the Smithsonian Institute's Belmont Center; thereafter, it met on a monthly basis, releasing a formal document on April 18, 1979.

The Belmont Report identifies three key principles. First, it distinguishes research from therapy, although it acknowledges that they could overlap given special circumstances. Second, it enumerates basic ethical principles. Researchers must respect basic human rights: Only volunteers can participate, and vulnerable populations have to be protected. It also incorporates the concept of beneficence and embraces justice, with all participants treated equally. Third, it delineates how these ethical guidelines should be implemented. This document ends by citing the Nuremberg Code of 1947, Helsinki Declaration of 1964 (revised in 1975), and the 1971 Guidelines (codified into Federal Regulations in 1974) issued by HEW.[46] While the Nuremberg Code and the Helsinki Declaration saw consent as fundamental, the Belmont Principles linked it to "risk of physical harm or death," closing a loophole uncovered by investigations of the Tuskegee Study. It thereby provided a template for institutional review boards to evaluate research proposals involving human subjects. Committee review and oversight represented a profound break from past practices, when "… a researcher's character and conscience were traditionally regarded as the primary safeguards against abuse of research subjects…."[47] Such an approach too often failed, as we have seen. While the protection of human subjects was settled, the question of vaccinating children remained open.

Notes and Sources

1. Naomi Rogers, *Dirt and Disease: Polio Before FDR* (New Brunswick, NJ: Rutgers University Press, 1996): 179. Moreover, check Arthur Allen, *Vaccine: The Controversial Story of Medicine's Greatest Lifesaver* (New York: W. W. Norton Company, 2007): 195; Stuart Blume, *Immunization: How Vaccines Became Controversial* (London: Reaktion Books, 2017): 79; Tony Gould, *A Summer Plague: Polio and Its Survivors* (New Haven: Yale University Press, 1995): 150; John R. Paul, "Thomas Francis, Jr., July 15, 1900–October 1, 1969," *Biographical Memoirs* 44 (1974): 80–81; and John R. Wilson, *Margin of Safety: A Doctor Tells the Controversial Inside Story of the Development of the Polio Vaccine* (Garden City, NY: Doubleday and Co., 1963): 92–93.
2. *Annual Report of the Superintendent of the Pittsburgh Public Schools, 1954–55* (Pittsburgh: Board of Public Education, 1955): 35, CLP. Further consult Wilson, *Margin of Safety*: 98.
3. David M. Oshinsky, *Polio: An American Story* (New York: Oxford University Press, 2005): 217. Refer to pages 218–19, 248, as well as Edward D. Berkowitz, "Rehabilitation: The Federal Government's Response to Disability, 1935–1954" (1976: Rpt. New York: Arno Press, 1980): xii, 2.
4. Paul A. Offit, *The Cutter Incident: How America's First Polio Vaccine Led to the Growing Vaccine Crisis* (New Haven: Yale University Press, 2005): 61–62.
5. Marting in Nina Gilden Seavey, Jane S. Smith, and Paul Wagner, *A Paralyzing Fear: The Triumph Over Polio in America* (New York: TV Books, 1998): 81, 83. See pages 325, 354–55, as well as Allen, *Vaccine*: 196; Allan M. Brandt, "Polio, Politics, Publicity, and Duplicity: Ethical Aspects in the Development of the Salk Vaccine," *International Journal of Health Services* 8 (1978): 265, 268; James Colgrove, *State of Immunity: The Politics of Vaccination in Twentieth-Century America* (Berkeley: University of California Press, 2006): 117, 122; Bert Hansen, *Picturing Medical Progress from Pasteur to Polio: A History of Mass Media Images and Popular Attitudes in America* (New Brunswick, NJ:

Rutgers University Press, 2009): 250–51; Offit, *Cutter Incident*: 65; and Oshinsky, *Polio*: 220–21.

6. Robert K. Plumb, "Vaccine for All in Peril of Polio Likely This Year," *New York Times*, April 14, 1955: 1; "Inoculation Start Delayed in Jersey," *New York Times*, April 19, 1955: 16; Bess Furman, "U.S. to Set Up Unite to Allot Vaccine," *New York Times*, April 23, 1955: 1. Vaccine: Inoculations, Box no. 14, Thomas Francis Papers, BHL. Colgrove, *State of Immunity*: 118, 121–22.
7. Offit, *Cutter Incident*: 66, 73, 68, and 71–72, respectively. Moreover, see Aaron E. Klein, *Trial by Fury: The Polio Vaccine Controversy* (New York: Charles Scribner's Sons, 1972): 117, 119, 121; Arnold S. Monto, "Francis Field Trial of Inactivated Poliomyelitis Vaccine: Background and Lessons for Today," *Epidemiological Reviews* 21 (1999): 20; and Oshinsky, *Polio*: 221–23, 225, 238.
8. Oshinsky, *Polio*: 225.
9. Elizabeth W. Etheridge, *Sentinel for Health: A History of the Centers for Disease Control* (Berkeley: University of California Press, 1992): xvi and 42–43, respectively. Further check Amy L. Fairchild, Ronald Bayer, and James Colgrove, *Searching Eyes: The State, and Disease Surveillance in America* (Berkeley: University of California Press, 2007): 17, 247.
10. Offit, *Cutter Incident*: 82, 89, 87, and 102–3, respectively. Consult pages xii, 3, 86, as well as Brandt, "Polio, Politics, Publicity, and Duplicity": 265; Klein, *Trial by Fury*: 119, 122. Also see Seth Mnookin, *The Panic Virus: The True Story Behind the Vaccine-Autism Controversy* (New York: Simon and Schuster, 2011): 54; and Paul A. Offit, *Deadly Choices: How the Anti-Vaccine Movement Threatens Us All* (New York: Basic Books, 2011): 55.
11. Philip J. Hilts, *Protecting America's Health: The FDA, Business, and One-Hundred Years of Regulation* (New York: Alfred A. Knopf, 2003): viii.
12. Offit, *Cutter Incident*: 97. See pages 94–96, 98–99, as well as Monto, "Francis Field Trial of Inactivated Poliomyelitis Vaccine": 20.
13. Oshinsky, *Polio*: 237. Further refer to Monto, "Francis Field Trial of Inactivated Poliomyelitis Vaccine": 20; Offit, *Cutter Incident*: 120.
14. "1959 Polio Vaccination Facts: How YOU Can Provide the Protection Your Community Needs (23 March 1959)," Jonas Salk Papers, Box 7, File Folder, Salk Polio Vaccine Facts (Collection #90/36/7), UPITT. According to Colgrove, *State of Immunity*, "The Cutter incident shook, but did not destroy, public confidence in the vaccine" (120).
15. Offit, *Cutter Incident*: 124, 114 and 116, respectively. Consult pages xii, 3, 117–19, as well as Monto, "Francis Field Trial of Inactivated Poliomyelitis Vaccine": 7; Oshinsky, *Polio*: 238, 241.

16. Allegheny County Medical Society, "Report of the Allegheny County Polio Vaccination Program" (typescript), File Folder, Chancellor Edward H. Litchfield Papers (1956–65), Box 106, UPITT. Colgrove, *State of Immunity*: 120, 127.
17. "1959 Polio Vaccination Facts," Salk Papers. Further consult Colgrove, *State of Immunity*: 113–14, 119, 129–30, 132–34.
18. Colgrove, *State of Immunity*: 134.
19. Charlotte DeCroes Jacobs, *Jonas Salk: A Life* (New York: Oxford University Press, 2015): 77. Also check Klein, *Trial by Fury*: 137–38; and Rogers, *Dirt and Disease*: 180–81.
20. Etheridge, *Sentinel for Health*: 92–93. Further refer to Colgrove, *State of Immunity*: 140; Klein, *Trial by Fury*: 139–40, 142; Oshinsky, *Polio*: 245–46, 252; Gareth Williams, *Paralyzed with Fear: The Story of Polio* (New York: Palgrave Macmillan, 2013): 237–38. Dorothy M. Horstmann directed the WHO review of field trials in Soviet Union of Sabin's live-virus vaccine, finding it safe and effective. "Obituary: Polio Pioneers Joseph Melnick, Dorothy M. Horstmann Die," *Washington Post*, January 22, 2001: B-6, http://pitt.idm.oclc.org/login?url=http://search.proquest.com/docview/409060435?accountid=14709 (Accessed November 7, 2015).
21. Etheridge, *Sentinel for Health*: 93, 141, and 94, respectively. Also consult page 147, as well as Allen, *Vaccine*: 203; Colgrove, *State of Immunity*: 140, 188; Elena Conis, *Vaccination Nation: America's Changing Relationship with Immunization* (Chicago: University of Chicago Press, 2015): 187; and Klein, *Trial by Fury*: 144.
22. Williams, *Paralyzed with Fear*: 243, and, 245, respectively. Refer to pages 244, 246. Further consult Colgrove, *State of Immunity*: 139–41, 188; Klein, *Trial by Fury*: 148, 153; and Mnookin, *Panic Virus*: 54.
23. Colgrove, *State of Immunity*: 143. Further resort to pages 131–42; Conis, *Vaccination Nation*: 28.
24. Etheridge, *Sentinel for Health*: 145. Consult page 144, as well as Colgrove, *State of Immunity*: 173.
25. Offit, *Cutter Incident*: 120. Also check page 127.
26. Samuel L. Katz, "John F. Enders and Measles Virus Vaccine—A Reminiscence," in Diane E. Griffin and Michael B. A. Oldstone, eds., *Measles: History and Basic Biology* (Berlin: Springer, 2009): 6. Further consult page 6, as well as Colgrove, *State of Immunity*: 151; Jacob, *Jonas Salk*: 88; Klein, *Trial by Fury*: 55–56; "Obituary: Frank C. Robbins," *Pittsburgh Post-Gazette*, August 6, 2003: B-5; Paul A. Offit, *Vaccinated: One Man's Quest to Defeat the World's Deadliest Diseases* (New York: Smithsonian Books, 2007): 40, 47; John F. Enders, Thomas H. Weller, and Frederick C. Robbins, "Cultivation of the Lansing Strain of

Poliomyelitis Virus in Cultures of Various Human Embryonic Tissues," *Science* 109 (January 28, 1949), refer to it as "extra-noural multiplication" (85–87).

27. Offit, *Vaccinated*: 47–48. See also Susan E. Lederer and Michael A. Grodin, "Historical Overview: Pediatric Experimentation," in Michael A. Grodin and Leonard H. Glantz, eds., *Children as Research Subjects: Science, Ethics, and Law* (New York: Oxford University Press, 1999): 17–18.
28. Samuel L. Katz, et al., "Studies on an Attenuated Measles-Virus Vaccine," *New England Journal of Medicine* 263 (1960): 180–82. In addition, see Etheridge, *Sentinel for Health*: 170; Andrew Cliff, Peter Haggett, and Matthew Smallman-Raynor, *Measles: An Historical Geography of a Major Human Viral Disease, from Global Expansion to Local Retreat, 1840–1990* (Cambridge, MA: Blackwell Publishers, 1993): 218.
29. Offit, *Vaccinated*: 49. See pages 3, 10–11, 22, 48, as well as Colgrove, *State of Immunity*: 151; Katz, "John F. Enders and Measles Virus Vaccine": 8.
30. Conis, *Vaccination Nation*: 66–67. Further check page 63.
31. Offit, *Vaccinated*: 21 and 28, respectively. See pages 22–24, 29–30, as well as College of Physicians, *History of Vaccines* (Philadelphia: College of Physicians, 2013): 33.
32. Conis, *Vaccination Nation*: 72–73, 70, and 75, respectively. Moreover, consult pages 65, 69, 74–75; Colgrove, *State of Immunity*: 180; Alan R. Hinman and Walter A. Orenstein, "A Shot at Protection: Immunizations Against Infectious Disease," in John W. Ward and Christian Warren, eds., *Silent Victories: The History and Practice of Public Health in Twentieth-Century America* (New York: Oxford University Press, 2007): 75; Klein, *Trial by Fury*: 56.
33. Offit, *Vaccinated*: 78. Also see pages 79–80, 81–82, 88.
34. Conis, *Vaccination Nation*: 80–81. See page 88, as well as Colgrove, *State of Immunity*: 180.
35. David J. Rothman and Sheila M. Rothman, *The Willowbrook Wars: Bringing the Mentally Disabled into the Community* (1984; Rpt. New Brunswick, NJ: Transaction Press, 2005): 23 and 335, respectively. Refer to page 16, as well as Colgrove, *State of Immunity*: 191; John J. O'Connor, "TV: Willowbrook State School, 'The Big Town's Leper Colony,'" *New York Times*, February 2, 1972: 78.
36. The first quote is from O'Connor, "TV: Willowbrook State School": 78, and the second is from "Volunteers Investigate Willowbrook School," *New York Times*, January 13, 1972: 45. Also consult Geraldo Rivera, "Welcome to Willowbrook," in *Childhood in America*, Paula S. Fass and Mary Ann Mason, eds. (New York: New York University Press, 2000): 436–37.

37. Rothman and Rothman, *Willowbrook Wars*: 335–36. Further consult Jonathan D. Moreno, *Undue Risk: Secret State Experiments on Humans* (New York: Routledge, 2001): 249–50; Offit, *Vaccinated*: 27.
38. Brandt, "Polio, Politics, Publicity, and Duplicity": 392. Check also page 398.
39. Etheridge, *Sentinel for Health*: 236–37. Moreover, see to Brandt, "Polio, Politics, Publicity, and Duplicity": 392.
40. Brandt, "Polio, Politics, Publicity, and Duplicity": 392. Further consult page 399. Harriet Washington, *Medical Apartheid: The Dark History of Medical Experimentation on Black Americans from Colonial Times to the Present* (New York: Doubleday Books, 2006), characterizes the Tuskegee this way: the "most lengthy instance of experimental abuse in Western history…." (169–70; also see 181).
41. Robert B. Baker, *Before Bioethics: A History of American Medical Ethics from the Colonial Period to the Bioethics Revolution* (New York: Oxford University Press, 2013): 275–76, 279, 280, and 278, respectively. Emphasis is in original. Baker offers a comprehensive historical account, noting five interpretive approaches to explain these changes (274–75). Refer also to Herman L. Blumgart, "The Medical Framework for Viewing the Problem of Human Experimentation," *Daedalus*, 98 (Spring 1969): 256; Leonard H. Glantz, "The Influence of the Nuremberg Code on U.S. Statutes and Regulations," in George J. Annas and Michael A. Grodin, eds., *The Nazi Doctors and the Nuremberg Code: Human Rights in Human Experimentation* (New York: Oxford University Press, 1992): 185–86; Jenny Hazelgrove, "The Old Faith and the New Science: The Nuremberg Code and Human Experimentation Ethics in Britain, 1946–73," *Social History of Medicine* 15 (April 2002): 126–27; Susan E. Lederer, "Medical Ethics and the Media: Oaths, Codes, and Popular Culture," in Robert B. Baker, et al., eds., *The American Ethics Revolution: How the AMA's Code of Ethics Has Transformed Physicians' Relationships to Patients, Professionals, and Society* (Baltimore: Johns Hopkins University Press, 1999): 100–101; Lederer and Grodin, "Historical Overview: Pediatric Experimentation": 16.
42. Washington, *Medical Apartheid*: 246 and 249, respectively. Consult pages 244–45, 247, 250–53, 259–60, as well as Smith, *Patenting the Sun*: 136–37.
43. Susan E. Lederer, "Children as Guinea Pigs: Historical Perspective," *Accountability in Research* 10 (2003): 11–13. Moreover, check Sharon Perley, Sev S. Fluss, Zbigniew Bankowski, and Françoise Simon, "The Nuremberg Code: An International Overview," in Annas and Grodin, eds., *The Nazi Doctors and the Nuremberg Code*: 157–60.
44. Baker, *Before Bioethics*: 284–85.

45. Etheridge, *Sentinel for Health*: 238–39. Also refer to Brandt, "Polio, Politics, Publicity, and Duplicity": 399.
46. US Department of Health and Human Services, *The Belmont Report (1979)*, http://ori.hhs.gov (Accessed May 22, 2015). Consult as well Baker, *Before Bioethics*: 254, 287; Lainie Friedman Ross, *Children in Medical Research: Access Versus Protection* (New York: Oxford University Press, 2006): 23, 63–66.
47. Baker, *Before Bioethics*: 289. Also see pages 280–81, 288, 295, and Lederer and Grodin, "Historical Overview: Pediatric Experimentation": 18–19. For a full treatment of "proxy" consent, refer to Leonard H. Glantz, "The Law of Human Experimentation with Children," in Michael A. Grodin and Leonard H. Glantz, eds., *Children as Research Subjects: Science, Ethics, and Law* (New York: Oxford University Press, 1994): 103–24.

CHAPTER 14

The Complexities of Mass Immunization Culture

The headline, "The Last Push for a Polio-Free World," graced the front-page of April 11, 2016, issue of the *Pittsburgh Post-Gazette*. This represents a point of pride for that city, where Jonas E. Salk developed and first tested the polio vaccine. Maintaining a highly optimistic tone, the article announced that the Global Polio Eradication Initiative would employ a new inactivated Salk immunogen, now given orally, in Afghanistan and Pakistan, hosting the last reported cases of this disease. Hamid Jafari, the organization's former director, declared it as the "biggest globally coordinated project of its kind in the history of vaccine."[1] Both safety and efficacy were presumed.

Science ultimately conquered poliomyelitis. Numerous and large field trials followed by mass immunization campaigns marked a dramatic intersection of the histories of childhood, education, and medicine. First, with less than absolute certainty, investigators gambled with the well-being of America's children on a scale never seen before or since. Second, the public education system operated as a key medical instrument; without it, none of this could have occurred. Third, the national field test represented the last such effort by a private philanthropy in the USA. Fourth, subsequent mass inoculations brutally exposed the flaws of unregulated vaccine research and production. This marked the climax of an era, as medical historian Bert Hansen observes:

R. J. Altenbaugh, *Vaccination in America*,
Palgrave Studies in the History of Science and Technology,
https://doi.org/10.1007/978-3-319-96349-5_14

> The year 1955 was a great moment for medical research and for the American public.
>
> Millions of cheering ... families warmly welcomed the triumph of the Salk polio vaccine, and the media celebrated it grandly. Enthusiasm for medical progress had never been higher. It seemed that more breakthroughs were already at hand and that medical progress had no limit. Over the prior seventy years, medicine had improved immeasurably, breakthroughs had become bigger and bigger, setbacks and failures had gone unnoticed, and criticism of medical research had been largely unimaginable.[2]

It also inaugurated a shift in public health policy: "in the wake of licensing of polio vaccine, the federal government took the first tentative steps toward a substantive role in vaccination; this involvement expanded in the 1960s, as an immunization-focused bureaucracy within the U.S. Public Health Service (USPHS) became a strong force in programming around the county." Therefore, the development and distribution of the polio antigen marked a "pivotal event in the history of vaccination policy in America."[3]

Vaccination Policy

The National Foundation for Infantile Paralysis, and its immensely successful fund-raising apparatus, the March of Dimes, imprinted the need for childhood inoculations on the American public's psyche. The "ubiquitous images of poster children, the recruitment of over a million young Polio Pioneers, and the [F]oundation's plan to ration limited vaccine to five-to-nine-year-olds," historian Elena Conis writes, "placed children at the center of the polio vaccination crusade...." This led President Dwight D. Eisenhower to sign the Poliomyelitis Vaccination Assistance Act of 1955, allocating "$30 million to states to vaccinate children under twenty and pregnant women...." The National Foundation thus "move[d] vaccination from local health departments and medical societies to the national stage." From 1955 to 1957, USPHS provided immunogen free of charge to the states and, since they did not receive additional funds to distribute it, they resorted to the cheapest and most efficient means of doing this, that is, through the public schools. As a result, infants and preschool children failed to become immunized. Nevertheless, this act marked "a watershed movement in U.S. vaccination history, as it carved out, for the first time, an active role for the federal government in the funding and disseminations of a vaccine to everyday Americans. The act also established a foothold for federal health officials, who would later use it to argue for further federal involvement in vaccination promotion."[4]

Such was the case when President John F. Kennedy, with support from the American Medical Association, submitted the Vaccination Assistance Act (VAA) to Congress in 1962. Two significant changes in the federal program followed. First, children younger than five would be protected from four diseases: diphtheria, polio, tetanus, and whooping cough. It especially targeted poor urban and rural children, whose parents could not afford immunization. Second, the Centers for Disease Control (CDC) would provide both immunogen and field personnel. It funneled money through that agency to states which would carry out the inoculation process, covering vaccine and personnel costs. "Nothing on the scale of this immunization program had been attempted before, and no one knew how or whether parents would respond." Still, the VAA had created a de facto federal presence and leadership in vaccination campaigns.[5] "The act," Conis points out, "specified that states were not required to force vaccines on people who didn't want them for themselves or their children." This left the CDC staff with little choice but to continue relying on the decades-old approach of persuading parents to voluntarily submit their children. "By nationally coordinating immunization efforts, the CDC in effect began to do for all vaccines what the March of Dimes had once done for the polio vaccine." It supplied "educational materials, courses, seminars, and even government-trained personnel to states and metropolitan areas," emphasizing that inoculations led to fewer illnesses and deaths. It ordained "Wellbee, a smiling, round-faced cartoon bumblebee designed by a Hollywood artist," as its good-health mascot. "Wellbee urged children in Atlanta and Tampa to 'drink the free [Sabin] polio vaccine' and appeared on billboards and pin-on buttons in Chicago." He visited schools in Honolulu and posed with the Boston Red Sox baseball team for a publicity shot.

The Vaccination Assistance Act proved successful. "Between 1962 and 1964," Conis writes, "50 million children and adults were vaccinated against polio, and 7 million children were immunized against diphtheria, pertussis, and tetanus. As a result, annual cases of polio dropped from 910 to 121, and … diphtheria fell from 404 to 306." Congress extended the VAA in 1965 for an additional three years and expanded it to include measles.[6] "Grants were made to forty states and fifty-four cities and counties for immunization of year-old children, and to twenty-nine state and city-county programs for immunization of children in Kindergarten [sic] and the first and second grades." CDC officials also sent an Epidemic Intelligence Service team into the field, in August 1966, to monitor the program in case the measles vaccine induced an outbreak. While the Polio Vaccine Assistance Act served as a one-time federal initiative, the VAA, in its comprehensiveness and longevity was different.[7]

The Vaccination Assistance Act rested on the notion of "eradicationism," as medical historian James Colgrove labels it. This marked a shift, in the minds of health professionals, from the control of infectious diseases to their complete elimination. The sharp decline in polio infections during the 1960s fed this confidence, leading to a similar attitude toward other contagion.[8] Reaching this goal required herd immunity, with a vaccination rate as high as a 96 percent for measles and pertussis and as low as 85 percent for mumps and rubella. This meant giving children the full battery of antigens. According to Conis, they were seen "… as reservoirs of infection in their communities…. Children were vaccinated to protect them from disease, but also to protect their communities from disease, and the state and nation from the medical and economic burdens of disease."

This agenda followed a trajectory similar to the polio effort, especially with the education blitz that ensued. Merck pharmaceutical company, hoping to exploit its measles vaccine, revived the iconic poster child—a young girl who had suffered catastrophic mental and physical disabilities—and used magazines, like *Good Housekeeping* and *McCall's*, to promote immunization. The CDC "produced public service announcements, billboard ads, films, comic strips, and coloring books…."[9] A *Peanuts* comic strip, ubiquitous in newspapers across the country, portrayed Charlie Brown receiving his measles inoculation. Television spots also promoted protection. "The remarkably successful campaign made measles more prominent than any disease since polio." With 7 million children immunized, the incidence rate fell 70 percent in one year. "A measles epidemic was halted in Mason County, Kentucky, with the vaccination of children in just the first and second grades, proving for the first time that the epidemic spread of measles could be stopped by vaccinating only a limited group." Some 260,000 cases were reported in 1965 but only 22,000 in 1968. In that "three-year period, the massive vaccination program prevented an estimated 8.5 million cases of measles,

850 deaths from such complications as encephalitis, 2,800 cases of mental retardation, 500,000 days of hospitalization, and 17 million days of absence from school." However, unlike the polio inoculation effort, this vaccine and its persuasive techniques failed to wipe out this disease. While 90 percent of all Americans had been immunized, 40 percent of poor children remained untouched.[10] "Measles, which once struck all children, became a disease of the disadvantaged."

President Richard M. Nixon's administration refused to renew funding for the Vaccination Assistance Act and, as a result, "immunization stagnated or fell in the seventies." Without federal assistance, Conis points out, many parents could not afford to have their children inoculated. Expenditures had risen from four vaccines in the 1950s that cost about two dollars to seven in the 1970s for about fifty dollars. Measles infection rates wildly fluctuated, jumping to 75,290 in 1971, falling to 22,094 three years later, and rising again to 57,345 cases in 1977.[11]

President Jimmy Carter's administration attempted to rectify this situation by passing the Childhood Immunization Initiative (CII), "providing increased government support for immunization and undertaking major efforts to identify and immunize schoolchildren who had not been vaccinated."[12] Once again, states instituted a new, federally funded program. Joseph Califano, Secretary of Housing, Education and Welfare, spearheaded the effort by invoking, Conis stresses, the "polio immunization campaigns of the 1950s and 1960s, which drew on widespread popular investment and the help of tens of thousands of volunteers." The Carter administration held parents responsible for inoculating their children and targeted them. Califano "called on communities to form Immunization Action Committees," while his office sent "letters and brochures ... to local Parent Teacher Associations, women's clubs, chapters of the Red Cross, the National League for Nursing, and other community groups...." He appealed to television executives to run public service announcements, implored popular newspaper columnists to encourage their readers to participate, reached out to doctors to advise their patients, and persuaded the National Football League to adopt it as a public relations tool, as well as published an informational article in *Parents* magazine.[13] All of these informal education efforts fizzled.

This forced health officials to resurrect an abandoned tactic: compulsion. "Between 1968 and 1981, the legal infrastructure supporting immunization underwent its most thorough transformation of the century"[14] Public education again became the linchpin for this medical initiative. Laws already required school attendance. To enforce inoculations, state governments made them mandatory for admissions. By 1981, "all 50 states had laws requiring measles immunization (or history of diseases) prior to first entry to school."[15]

As earlier in that century, such coercion bred conflict. It alienated many students and their families from school officials, triggering bureaucratic tensions. Depending on the state, building principals could expel unvaccinated students from school, or state attorneys could pursue legal action against school districts, even building principals themselves, if they admitted these violators. "The CDC coordinated a massive drive to help states audit millions of student immunization records." This heaped a plethora of problems on students and their families. "Education for some ... was temporarily disrupted if they could not provide proof of their protection; some parents opposed to vaccination were denied the right to make decisions about an aspect of their children's health care. Most health officials, politicians, and jurists considered these acceptable trade-offs in order to control infectious diseases." In the end, enforcement proved extremely uneven among the fifty states. State education agencies granted exemptions based on religion, preferences for alternative medical beliefs, or in the name of individual freedom.[16]

In spite of such disarray, compulsion worked. When Carter left office in 1980, "upwards of 96 percent of all children entering school were vaccinated against measles, rubella, polio, diphtheria, pertussis and tetanus, marking the highest rates of vaccine coverage the country had ever seen."[17] Measles, in particular, appeared to be almost eradicated, with 2600 reported cases in 1981, dropping to 1500 two years later, the lowest in history. Sharply reduced mortality rates testify to this point: "... in a single peak month in 1917 over 2,000 measles deaths were recorded...." Seventy years later measles-related deaths had virtually disappeared.[18]

Another development during the 1970s, Conis points out, involved the pharmaceutical industry's drive for profits. Throughout that period, it shaped policy homogenizing " ... all of the vaccine-preventable infections." Mumps, a "childhood nuisance," now held the same hazards as diphtheria, polio, and smallpox. This transformation grew out of two influences: the development of the MMR single vaccine and drop in immunization rates during the 1970s. The latter led public health advocates to cultivate parent anxiety, if not outright fear, by portraying it as a dangerous disease. This ploy succeeded. Finally, from a health standpoint, inoculating children against rubella protected pregnant women and their fetuses. This "marked the first time that vaccination was deployed in a manner that offered no direct benefit to the individuals vaccinated...."

Yet another shift in federal policy occurred when President Ronald Reagan's administration failed to increase funding for child immunization programs, or eliminated them altogether, Conis notes, "even as the prices of vaccines rose." This produced disastrous outcomes. "In 1983, CDC officials complained that they'd be able to vaccinate only half as many children with federal funds as they had in 1981." This made young children extremely vulnerable. It came as no surprise that,"[b]eginning in 1989, a series of measles outbreaks struck cities across the USA, causing a total of 18,000 cases that year and over 27,000 the next. By the end of 1991, more than 50,000 children had caught the disease, more than 11,000 had been hospitalized, and over 150 had died." Children of color younger than five suffered the most. This caused health experts to recommend children receive their measles injections before age two, well before school age. During these epidemics, "health officials had repeatedly pointed out that the United States' immunization rate was worse than that of nearly every Latin American country."[19]

President William J. Clinton's 1993 initiative, Vaccines for Children Program, finally made immunization a federal entitlement. Unlike its predecessors, the Vaccination Assistance Act and the Childhood Immunization Initiative, it provided direct federal funding for immunogen and the administration of injections. This helped to level the socio economic playing field. By 2000, "[d]epending on which combination vaccines were used, a typical child received a total of eleven vaccines in a possible twenty injections by two years of age."[20] As usual, the CDC provided oversight. The program proved highly successful. Inoculation "rates 'reached record levels'" in 1999 with 96 percent against "diphtheria, pertussis, and tetanus; 93 percent were vaccinated against Hib [*Haemophilus influenzae b*]; 91 percent had shots against measles, mumps, and rubella; 90 percent against polio; and 88 percent against chicken pox and hepatitis B," the "lowest vaccination rates since the Carter administration." This program also sharply reduced "disparities in vaccination rates" among children.[21] Finally, at the beginning of the twenty-first century, "measles was no longer an endemic disease in the United States and that the continuing small numbers of cases represented importation or limited spread from importations."[22] But public opinion proved fickle.

Crosscurrents

Today vaccines are under fire. Two fundamental questions drive this opposition. First, how safe are they? Commercialization shapes this concern, one that has been raised repeatedly by immunization opponents. Early patent medicine manufacturers clearly exploited illnesses for profit. Meanwhile, the concept of chemotherapy (i.e., "artificial antibodies") that emerged during the late nineteenth and early twentieth centuries led to the development of immunogen to counter many contagious diseases. In addition, drugs like aspirin, which relieved pain and reduced fevers, and antibiotics such as sulfa, that checked the spread of bacterial infections, helped launch what would become an international pharmaceutical industry. Businesses like Pfizer (1840), American Squibb (1858), Parke-Davis and company (1867), Eli Lilly (1876), and the German Merck and company (1891) introduced the mass production of new medicines.[23]

In 1938, Congress significantly tightened the 1906 Food and Drug Act through restrictive labeling criteria that required all ingredients be listed and curative claims removed. "Government regulation had laid the necessary base for weeding out fraud and false advertising," effectively killing *magic* elixirs. Manufacturers now had to submit medications to that agency to assure safety and efficacy before marketing them to the public. This prompted these companies to build, or expand existing research facilities to discover and test new products. This approach "has provided the framework for drug creation and sales in the United States since that time."[24]

The second question: Should immunization be compulsory? For Conis, it operates as a "social contract among citizens; if most everyone gets vaccinated, everyone is protected." Nevertheless, during the Progressive Era, this approach largely failed, often sparking violent confrontations. Public health officials turned to education which worked for decades, but it too fell short, forcing policymakers to once more employ coercion. This, in turn, renewed resistance.

Although anti-immunization attitudes appeared quiescent by the 1930s, they had not completely disappeared. Ann Riley Hale, in *The Medical Voodoo* (1935), "continued to attack vaccination as a form of tyranny propped up by false science and capitalism." By the 1950s, chiropractor R. G. Wilburn had founded Health Research, "a small press that began reprinting nineteenth- and early twentieth-century works on teetolism, fasting, natural hygiene, and other nature cures." He also published *The Poisoned Needle*, by Eleana McBean, a naturopath, in 1957. Her thesis focused on immunization "as a direct affront to the laws of nature." She bolstered her argument by reviving and citing the critiques of the Progressive anti-vaccination movement's leading lights. She also embraced an early, rudimentary notion of organic agriculture, seeing food additives, insecticides, and preservatives as poisons, harming humans in particular and nature in general. "To McBean, widespread pesticide application, which citizens were powerless to avoid, were, like vaccination, crimes committed by government acting in the interest of powerful corporations with no regard for human health." Her book sold 5000 copies, with another 5000 reprinted two years later, and re-released in 1974. McBean authored three more anti-vaccination tracts between 1977 and 1980. For Conis, *The Poisoned Needle* "served as a bridge between anti-vaccinationism that faded in the 1930s and the renewed vaccine skepticism that began to gain momentum in the last decades of the century." Furthermore, the original publication date did not occur serendipitously: "McBean's book was a response to the nation's massive polio vaccination efforts of the 1950s."[25]

The fight against infantile paralysis more than culminated broader cultural changes toward medicine. Indeed, a consensus and trust had been cultivated and the public mobilized. It also profoundly "changed vaccination programs" themselves. How they were conceptualized and administered underwent an unparalleled transformation. "The decline of polio helped to fuel a new ambition among public health professionals to seek out not merely the control of disease but its complete eradication."[26] The federal government, along with state and municipal authorities, directed them. And compulsion, with the introduction of new immunogen, once again, characterized public health policy, contributing to the emergence of a new and more complex anti-vaccination movement.

Notes and Sources

1. Virginia Lin, "The Last Push for a Polio-Free World: Health Officials to Launch Global Effort to Wipe Out Polio Using Jonas Salk's Original Formula," *Pittsburgh Post-Gazette*, June 11, 2016: A-1, 7. Also refer to Edward P. K. Parker, and Nicholas C. Grassley, "Polio Vaccination: Preparing for a Change of Routine," *Lancet*, 388 (July 9, 2016): 107–8.
2. Bert Hansen, *Picturing Medical Progress from Pasteur to Polio: A History of Mass Media Images and Popular Attitudes in America* (New Brunswick,

NJ: Rutgers University Press, 2009): 256. See pages 1–3, as well as Alexandra Minna Stern and Howard Markel, eds., *Formative Years: Children's Health in the United States, 1880–2000* (Ann Arbor: University of Michigan Press, 2004): 11.

3. James Colgrove, *State of Immunity: The Politics of Vaccination in Twentieth-Century America* (Berkeley, CA: University of California Press, 2006): 14–15.
4. Elena Conis, *Vaccination Nation: America's Changing Relationship with Immunization* (Chicago: University of Chicago Press, 2015): 23, 22, and 6, respectively. See page 5, as well as Elizabeth W. Etheridge, *Sentinel for Health: A History of the Centers for Disease Control* (Berkeley: University of California Press, 1992): 143.
5. Etheridge, *Sentinel for Health*: 143–44. Further consult Colgrove, *State of Immunity*: 122, 145–46; Conis, *Vaccination Nation*: 29, 30, 32.
6. Conis, *Vaccination Nation*: 26, 32, and 48–49, respectively.
7. Etheridge, *Sentinel for Health*: 147 and 173, respectively. Further consult page 171; Andrew Cliff, Peter Haggett, and Matthew Smallman-Raynor, *Measles: An Historical Geography of a Major Human Viral Disease, from Global Expansion to Local Retreat, 1840–1990* (Cambridge, MA: Blackwell Publishers, 1993): 218–19; Colgrove, *State of Immunity*: 146; Conis, *Vaccination Nation*: 35–37, 51.
8. Colgrove, *State of Immunity*: 149–50, 153–56.
9. Conis, *Vaccination Nation*: 11 and 55–56, respectively. See pages 50–51, 57–58, as well as Colgrove, *State of Immunity*: 149, 152; College of Physicians, *History of Vaccines* (Philadelphia: College of Physicians, 2013): 31; Alan R. Hinman, and Walter A. Orenstein, "A Shot at Protection: Immunizations Against Infectious Disease," in John W. Ward and Christian Warren, eds., *Silent Victories: The History and Practice of Public Health in Twentieth-Century America* (New York: Oxford University Press, 2007): 70; Paul A. Offit, *Deadly Choices: How the Anti-Vaccine Movement Threatens Us All* (New York: Basic Books, 2011): xvii. The CDC illustration Colgrove uses in *State of Immunity* appears reminiscent of the March of Dimes poster child (168).
10. Etheridge. *Sentinel for Health*: 173–74. See page 172, as well as Cliff, Haggett, and Smallman-Raynor, *Measles*: 220; Colgrove, *State of Immunity*: 166–67, 172.
11. Conis, *Vaccination Nation*: 40 and 80–81, respectively. Consult pages 57–59, 93, as well as Cliff, Haggett and Smallman-Raynor, *Measles*: 221; Colgrove, *State of Immunity*: 149, 151, 167; Haggett and Smallman-Raynor, *Measles*: 221; Hinman and Orenstein, "A Shot at Protection": 70; Paul A. Offit, *Vaccinated: One Man's Quest to Defeat the World's Deadliest Diseases* (New York: Smithsonian Books, 2007): 56, 159, 177.

12. Hinman and Orenstein, "A Shot at Protection": 70–71. Also see Conis, *Vaccination Nation*: 89, 91–92, 93–94.
13. Conis, *Vaccination Nation*: 99–100. Moreover, see pages 97–98; Cliff, Haggett, and Smallman-Raynor, *Measles*: 221–22; Colgrove, *State of Immunity*: 199–200.
14. Colgrove, *State of Immunity*: 150. Further consult Conis, *Vaccination Nation*: 60–61.
15. Hinman and Orenstein, "A Shot at Protection": 70. Further check Colgrove, *State of Immunity*: 175–77; Conis, *Vaccination Nation*: 8, 101; Etheridge, *Sentinel for Health*: 291; Offit, *Deadly Choices*: 138.
16. Colgrove, *State of Immunity*: 202 and 208, respectively. Also see pages 200, 201, 205.
17. Conis, *Vaccination Nation*: 88. Moreover, consult Etheridge, *Sentinel for Health*: 291; Samuel L. Katz, "John F. Enders and Measles Virus Vaccine—A Reminiscence," in Daine E. Griffin and Michael B. A. Oldstone, eds., *Measles: History and Basic Biology* (Berlin: Springer-Verlag, 2009): 9.
18. Cliff, Haggett, and Smallman-Raynor, *Measles*: 224. Refer to pages 221, 225, as well as Conis, *Vaccination Nation*: 163; Hinman and Orenstein, "A Shot at Protection": 70–71.
19. Conis, *Vaccination Nation*: 81–82, 76, 164–65, 163, and 171, respectively. Consult pages 77, 80, 83 as well as Colgrove, *State of Immunity*: 172, 178, 219, 224; College of Physicians, *History of Vaccines*: 8, 32; Cliff, Haggett, and Smallman-Raynor, *Measles*: 220; Hinman and Orenstein, "A Shot at Protection": 75. Colgrove, *State of Immunity*, makes a similar observation: "But the most controversial aspect of universal rubella immunization was the fact that children who received the shot did not themselves derive a significant benefit from it" (179).
20. Colgrove, *State of Immunity*: 227. Further check page 223, and Conis, *Vaccination Nation*: 161–62, 175–76.
21. Conis, *Vaccination Nation*: 179, and 177, respectively. Also check Colgrove, *State of Immunity*, for similar statistics (248).
22. Hinman and Orenstein, "A Shot at Protection": 71–72.
23. Roy Porter, *The Greatest Benefit to Mankind: A Medical History of Humanity* (New York: Norton, 1997): 448, 449, 454.
24. Philip J. Hilts, *Protecting America's Health: The FDA, Business, and One-Hundred Years of Regulation* (New York: Alfred A. Knopf, 2003): 100 and 93, respectively. Further consult pages 78–79, 95. Emphasis is mine.
25. Conis, *Vaccination Nation*: 11 and 134–38, respectively.
26. Colgrove, *State of Immunity*: 114 and 147–48, respectively.

PART V

Intellectual Authority?

CHAPTER 15

A Little Knowledge Is a Dangerous Thing

Robert De Niro, an Academy Award-winning actor, rocked both the medical world and entertainment industries during the last week of March 2016 when he withdrew *Vaxxed: From Cover-Up to Catastrophe* from, his fifteenth annual Tribeca Film Festival, reversing his earlier decision to include it. That purported documentary claimed to have evidence of CDC authorities covering up a MMR-autism link. News of its premier at such a high-profile event generated an immediate outcry, the *New York Times* reported, from "unnerved and angered doctors, infectious disease experts and even other film makers."[1] Social media intensified these feelings. In the face of this uproar, De Niro canceled it. According to the *Los Angeles Times*, he stated: "'after reviewing it … with the Tribeca Film Festival team and others from the scientific community we do not believe it contributes to or further the discussion I had hoped for.'" Opponents saw *Vaxxed* challenging vaccinations and attributed recent increases in measles outbreaks to this mistrust. The American Academy of Pediatricians applauded De Niro's action.[2]

However, De Niro's sudden flip-flop sparked a sharp response from Philippe Diaz, chair of Cinema Libre Studio, that movie's distributor. He denied it promoted anti-vaccination; rather, it focused on the "'idea of a government cover-up.'" He went on to explain it "'seemed a convincingly meaty topic for a film….'" As Diaz labeled Tribeca's decision "censorship," he announced it would premier at the Angelika Film Center

R. J. Altenbaugh, *Vaccination in America*,
Palgrave Studies in the History of Science and Technology,
https://doi.org/10.1007/978-3-319-96349-5_15

on April 1. Thus, as the *New York Times* pointed out, *Vaxxed* was never denied a platform, it would be shown seven times a day at that festival.[3]

Pointed commentary continued for several days, following additional twists and turns. De Niro and his wife, Grace Hightower, who is African American, have an eighteen-year-old son, Elliot, who is autistic. De Niro's initial support of that film may have reflected, according to one *Guardian* reporter, an effort to expose yet another example of "systemic racism" regarding the "welfare of black bodies," since it alleges the federal government "intentionally concealed evidence from a 2004 study indicating that African-American boys are more likely to be diagnosed with autism after receiving the MMR." At first blush, this seemed to echo the infamous Tuskegee syphilis study. However, this reporter stressed, the autism claim had been "debunked." More importantly, Andrew Wakefield had co-written and directed this production and been scheduled to speak at the Tribeca Film Festival. His involvement undermined any legitimacy it may have had. The *Guardian* article characterized him as a "discredited, self-serving doctor … stripped of his medical license" who "then joined other anti-vaccination campaigners to make a documentary under the guise of fostering a 'debate.'" It simply represented propaganda, the *Guardian* added, analogous to "'Leni Riefenstahl making a movie about the Third Reich….'"[4] What sparked such rash reactions and generated equally strong language?

The Usual Suspect

Vaxxed, a highly polished production, focuses solely on alleged, secret information about the relationship between MMR and autism. Del M. Bigtree, a former Emmy Award-Winning producer of *The Doctors* television show, co-wrote and produced it. He is identified as a medical journalist, provides some of the narration, and appears in many clips. This film's premise rests heavily on the recorded phone calls of William W. Thompson, a scientist at the Centers for Disease Control (CDC), portrayed as a whistle-blower, providing insider revelations of corruption and tainted research. In 2013, Thompson contacted Brian Hooker, identified as an environmental biologist, and who serves as the main narrator. Hooker, who has an autistic son, taped their phone conversations for several months. Based on Thompson's claims, the pharmaceutical industry wields a great deal of influence over CDC officialdom and their decisions. This production also relies heavily on parent testimonials, cementing the central thesis as frontline experts, bearing the emotional brunt of caring for children and adults with autism.

Following its release in New York City, *Vaxxed* toured numerous American cities during that summer, generating controversy wherever it appeared. An independent, Pittsburgh movie theater owner, lobbied by a local group of anti-vaccine advocates, decided to show it over a seven-day period, arranging a live question-and-answer period via Skype with Andrew Wakefield. No stranger to controversy, he figures prominently in *Vaxxed*. That city's medical community objected to the film, as one physician stated: "'There aren't two sides to this issue, and we need to be careful about putting movies out there and portraying them as truthful when they aren't.'" Another simply declared "'there's just no science backing it up.'"[5]

Andrew Wakefield thinks differently. In 1998, associated with the Royal Free Hospital and University College Medical School in London as a gastroenterologist, he and twelve colleagues coauthored an article, "Lleal-Lymphoid-Nodular Herplasia, Non-Specific Colitis, and Pervasive Developmental Disorder in Children." Published in *Lancet*, a highly respected British medical journal, it described a sample of twelve children, between the ages of three and six (eleven of them boys), who received a battery of tests, including blood, stool, and urine samples, colonoscopies, EEGs, endoscopies, lumbar punctures, and magnetic resonance imaging. Using tables of test results, endoscopic photographs, and slides of tissue samples, the study hinted at a correlation between digestive disorders and certain neurological conditions, but asserted this does "not prove an association between measles, mumps, and rubella vaccine and the [autistic] syndrome...."[6] Nevertheless, "[m]any people still view this article as establishing the scientific grounds for such a conclusion, and, as a result, the Wakefield, et al., article is commonly identified as the starting point of the autism vaccine controversy."[7] How did this happen?

To boost the study, the hospital's public relations team arranged a press conference on February 4, 1998, inviting journalists from London's daily newspapers. It also prepared "... a twenty-minute promotional video for the occasion and assembled a panel of five of the hospital's researchers to address the report's implications." Wakefield starred in that video and, as team leader, fielded reporters' questions.[8] Asked if his findings revealed a link between MMR and autism, Wakefield responded, "the work certainly raises a question mark over

MMR vaccine.... It is our suspicion that there may well be but that is far from being a causal association that is proven beyond a doubt." A follow-up question pushed for clarification. Wakefield replied less opaque: "... I have to say that there is sufficient anxiety in my own mind of the safety ... that I think it should be suspended in favour of the single vaccines, that is continued use of the individual measles, mumps and rubella components." This interviewer pressed Wakefield, who then responded with certitude, "the risk of this particular syndrome developing is related to the combined vaccine, the MMR, rather than the single vaccines." Wakefield had unleashed a firestorm.[9]

The research community responded. Between 2002 and 2004, the *British Medical Journal*, *Lancet*, *New England Journal of Medicine*, and other journals published numerous studies finding "no link between MMR and autism."[10] Furthermore, Brian Deer, a British investigative reporter, began to raise other "doubts ... by uncovering a conflict of interest on Wakefield's part."[11] Continued accusations and controversy led ten of Wakefield's co-authors, in a statement published in the March 6, 2004, issue of *Lancet*, to withdraw their support from the study. In 2007, moreover, the UK's General Medical Council (GMC) "proved that Wakefield authored the paper alone" and filed charges of professional misconduct against him.[12] First, he had conducted research "without proper ethical approval and failed to carry out the research as described in the application to the [Royal Free Hospital] ethics committee." Second, Wakefield had pursued "potentially harmful tests on children that were not clinically indicated, including colonoscopies and lumbar punctures."[13] Third, he received payment from a solicitor to "investigate patients involved in litigation over alleged reactions to the [MMR] vaccine," a serious conflict of interest. His population sample thus proved problematic: several of the children used in the published study were litigants in legal action against a manufacturer, and at least one child allegedly had symptoms of autism before receiving that inoculation.[14] Fourth, to further compound matters, after undermining the existing MMR immunogen, Wakefield had "filed a patent for a new vaccine." On January 28, 2010, the GMC found Wakefield guilty because of "dishonesty and flouting ethics protocols." On February 2, *Lancet* formally retracted Wakefield's manuscript, noting its false findings. Professionally denounced and his medical credentials revoked, Wakefield resigned from the Royal Free Hospital and moved to Austin, Texas, to became executive director at the Thoughtful House Center for Children with Autism. He has since then continued his campaign of a vaccine-autism link.[15]

In *Vaxxed*, Wakefield refers to the discredited 1998 study and implies it suffered from misinterpretations and exaggerated conclusions. That article, Wakefield calmly explains, never claimed a causal relationship between MMR and autism. He further mentions the press conference that introduced it, and a brief clip is shown from it, further reinforcing what he had said. In sum, this film implicitly portrays Wakefield as a martyr because William Thompson's telephone disclosures about the connection between MMR and autism appear to vindicate him.[16]

Vaxxed, Wakefield's alleged documentary, dramatically returned him to the national stage, according to one observer, "'kind of opening the scab on an old wound.'"[17] That ninety-minute film, which reiterated all of the information that had preceded it, linking MMR with autism, had revitalized the anti-vaccination movement.

Modern Roots of Opposition

This crusade appears more nuanced than its Progressive Era predecessor. It consists of individuals completely opposed to immunization, typically labeled anti-vacs. It also includes those who support inoculations but raise serious concerns about dangerous ingredients in vaccines: They are known as anti-toxs. Finally, delayers, while supporting the immunization menu, argue for spreading out the schedule. The general perception, however, oversimplifies these differences and lumps all three groups into *the Anti-Vaccination Movement.*[18]

The origins of this phenomenon, regardless of its emphasis, appear diffuse, unfurling over decades, characterized by profound social, demographic, and educational changes. First, as historian Elena Conis notes, the feminist movement not only raised concerns about women's health but also directed more attention to children's well-being. Based on a socially constructed notion of domesticity that stretched back to the mid-nineteenth century, "[v]accination campaigns from the fifties through the seventies ... routinely emphasized maternal responsibility for obtaining needed vaccines for children." Jimmy Carter's Childhood Immunization Initiative, built on this notion, holding mothers accountable in its drive to reach a 90% inoculation rate. But this was already an outdated presumption. "Second-wave feminism" refashioned the notion of gender, one that "gave vaccination skeptics a framework for criticizing vaccines and the ways they were used." This represented a movement that "encouraged women to take control of their own health." The Boston Women's Health Board Collective published *Our Bodies, Ourselves: A Book for and by Women* in 1973, reflecting growing disillusionment with a profit-driven pharmaceutical industry and male-dominated medical profession. It expanded this agenda with a sequel, *Ourselves and Our Children: A Book by and for Parents.* Mothers knew what was best for their children, giving domesticity a new twist. "Such doubts lay at the heart of the organized vaccine-safety movement that emerged in the 1980s." From this point on, opposition became largely genderized. Middle- and upper-class, largely European American, often liberal, mothers became the "nation's premier vaccine refusals, putting the rest of society at risk of infectious disease epidemics in their narrow-minded quest to protect their own children from vaccine injury." Although mumps outbreaks occurred in 2006 and 2009, pertussis in 2010, and measles in 2011, they remained steadfast, trusting "'alternative medicines, organic food, and yoga'" while distrusting "'Big Pharma and their lackeys in the media.'"

Environmentalism, for Conis, represents a second way to frame criticism of immunogen safety and policy: "... vaccines were akin to environmental risks inasmuch as they were products of industry with uncertain and potentially harmful long-term consequences." This was nothing new. The American public could just look at the health impact of cigarettes, exposure to asbestos and lead, as well as the use of DDT. Industry in general and science in particular could not always be trusted to protect humans. The seeds of doubt had been sown. Antigens, containing chemicals that could inflict serious harm, started to become unknown quantities in the minds of many parents. As humans polluted the environment, so they contaminated their bodies through artificial immunization; thus, the toxicity of chemical agents used in vaccines raised concerns.[19] Thimerosal became the most visible example. Frequent punctures in the rubber seals of vaccine vials stored in doctor's offices resulted in the growth of bacteria. In response, beginning in the 1930s, pharmaceutical companies added ethyl mercury, otherwise known as thimerosal, a safe preservative the human body quickly purges. Highly vocal critics did not see it that way and, in 1999, Public Health Service officials ordered this substance, while harmless, removed from immunogen. Pharmaceutical companies complied by switching to single-dose ampules.[20]

Third, perceptions of childhood underwent another transition during the last quarter of the twentieth century. The "parent-child relationship" replaced "marriage as [the] primary social and emotional connection." The value of children soared as divorces and increasing numbers of unmarried couples redefined the notion of family. Spousal relationships appeared temporary while offspring bonds remained permanent, demonstrating the ever "pressing centrality of children in the lives of adults."[21] A child's welfare now loomed ever larger. This resulted in parents giving closer scrutiny to previous, unquestioned medical treatments, like immunization, reasserting control over their children's lives. Deeply rooted, inalienable rights give them, "not the state, ... responsibility for and authority over decisions concerning the raising of their children—including vaccination choices."[22] Based on national and international laws, like the Declaration of Independence and the United Nations Charter, this position is broader and more sophisticated than its Progressive counterpart. Legal precedents regarding compulsion, such as the 1905 *Jacobson v. Massachusetts* Supreme Court decision, no longer apply because, parents reason, "no infectious disease epidemics exist in the United States...."[23] Further, since children remain a vulnerable group, since vaccination equates with experimentation, "[f]ree and informed consent" protections offered to human subjects and shaped by the Nuremberg Code and Tuskegee experience, apply to immunization. The "distinction between research and therapy" no longer exists, requiring the "need for individuals to have adequate information about all medical interventions."[24] Finally, mandatory inoculations deny due process guaranteed by the US Constitution. Individuals convicted of a crime and assigned the death penalty have more rights "than any child receiving mandatory vaccines...."[25]

Fourth, and like the Progressive anti-vaccination movement, religion and belief systems continue to trump medical "reason." God represents the final arbiter of health and illness, life and death. Only prayers and anointed individuals have the power to heal and, ultimately, save people. Medical researcher Paul A. Offit describes what has unfolded within this context. In 1990, Philadelphia confronted the "worst measles epidemic in U.S. history." Members of the Faith Tabernacle Congregation, who "did not believe in medical care" of any kind, chose not to immunize their children. Once they fell ill, parents' refused to render any medical remedies, including Tylenol to lower their fevers and obviate dehydration. The results proved tragic. While mortality rates for this disease hovered around one in a thousand during that outbreak, "4 of 150 children with measles" died, for a death rate of "one in thirty-five." That congregation remained intransigent in spite of public health appeals, forcing a local court to order medical care. Hundreds of children received attention in emergency rooms while dozens had to be hospitalized. On February 27, 1991, that city's mayor obtained a "court order to forcibly vaccinate children." Church members turned to the ACLU to intervene on their behalf, invoking the First Amendment. Its lawyers refused, seeing children's health and lives as paramount. Failing that, they hired their own attorney to oppose that court order, but Pennsylvania's Superior Court overruled their appeal. With immunization underway, the measles epidemic, after infecting 1400 individuals, subsided by June. The damage had been done, however. "Among church members, 486 people were infected and six were killed by measles," all of them children. This exception for belief systems continues. "In 2013, the CDC identified 30,000 children whose parents had chosen not to vaccinate them for religious reasons."[26] This has led, Offit adds, to exemptions for "philosophical" reasons: Vaccines represented unnatural substances introduced into children's bodies. "By 2012, twenty-one states allowed philosophical exemptions to vaccination."[27]

Avenues of Persuasion

Education, which had helped build public trust in medicine, now undermines it. Modern technology has expanded the scope of information and increased the speed of its distribution, inaugurating the age of Internet medicine. Although the public feels empowered with access to more information, it may be misleading, wrong, or biased; put simply, the democratization of knowledge has devalued expertise. Fear grows out of the merest suspicion of harm; it is assumed dangerous until proven *absolutely* safe. Possibility supersedes probability; any uncertainly discredits the product. Much of this is based on anecdotal evidence, emotional testimonials, and overgeneralizations. Sinister theories also fit into this web of anxiety, linking the CDC, Federal Drug Administration (FDA), National Institutes of Health, and physicians with pharmaceutical corporations. This collusion occurs, plotters insist, through private research funding that obfuscates vaccine threats. Mere association connotes guilt, resulting in public uncertainty and suspicion.

In place of mainstream doctors and researchers, according to cultural studies scholar Graeme Turner, highly visible individuals have become authoritative substitutes, instant experts, posting information and advice online. Although famous entertainers and professional athletes have always existed, this most recent example of hyper-celebrity is "unprecedented." Fame now is not necessarily dependent on accomplishments. What separates them from the masses, but holds their attention and loyalty, is their conspicuousness across media; that is, today's "highly convergent media environment, where cross-media and cross-platform content and promotion has become the norm…." The celebrity news industry constitutes a $3 billion enterprise. Celebrity lives, actions, and perceptions often overshadow television news, with Hollywood gossip intruding on regular coverage, gracing grocery-store tabloids, dominating Internet blogs, and flooding social media. Facebook in 2015 had 1.5 billion monthly users; Twitter, 304 million; and Instagram, 300 million. This exposure lends celebrities a sense of authority; these public platforms give them power. This explains much of the controversy surrounding Robert De Niro's on-again, off-again relationship with *Vaxxed*. "In effect, we are using celebrity as a means of constructing a new dimension of community through the media."[28]

Traditional media only sent information in one direction with the audience as the recipient. The Internet "permits movement of information in both directions and in many of its forms can be defined as a many-to-many form of communication…." This process cements connections, creating communities of like-minded individuals sharing ideas, information, and experiences; that is, celebrities and audiences become linked on an emotional level.[29] This shapes, as always, clothing and hairstyles, music trends, and other consumer appetites, but also extends to health information dispensed by syndicated and local daytime television shows and digital forms.

A host of these individuals opposes immunization: Jessica Alba, Jim Carrey, Cindy Crawford, Barbara Loe Fisher, Bill Maher, and Matthew McConaughey. Jenny McCarthy, a model, television host, and actress, represents the quintessential example of a "celebrity role model" who dispenses health advice. With ready access to mass audiences, she considerably raises the profile of the vaccine issue, giving her the aura of certainty on television shows like *20/20*, *Ellen*, *Good Morning America*, *Larry King Live*, *Oprah*, and *The View*, as well as appearances on CNN, reaching an estimated 15–20 million viewers. She promotes her books, as a *New York*

Times best-selling author, linking autism to immunogen, advancing treatments that include alternative diets, vitamin supplements, and hyperbaric oxygen treatments, and declaring her opposition to vaccinations. This is a personal crusade for McCarthy since her son, Evan, is autistic.[30]

Her books, television appearances, and Internet presence tap into the "celebrity mom profile," long portrayed in traditional women's magazines, like *Good Housekeeping* and *Ladies Home Journal*. This has morphed, according to communications expert Elizabeth Hatfield, into a profound force to influence consumption patterns, ones that define them as "a 'good mother.'" Her image and opinions "negotiate with lived experiences to shape readers' [and viewers'] understandings of their own experience as mothers."[31] This applies, of course, to the immunization of young children. Given McCarthy's example, what should mothers do?

Warrior Mothers: A Nation of Parents Healing Autism Against All Odds recounts McCarthy's personal journey, frustrations, and actions as a parent of a child with autism and highlights testimonials from numerous "warrior mothers" (and a warrior father) and their experiences with autism. McCarthy describes them as a network of parents whose children have been diagnosed with autism. They have built a support system through social media as well as organizations, like Defeat Autism Now and Talk About Curing Autism. McCarthy sees herself as one of these warrior mothers; together they seek "change" and the "truth." As this book unfolds, she reveals her twofold agenda: "Change this insane vaccine schedule" and "... GREEN our vaccines. Take the crap out! Enough is enough."[32]

Celebrification especially sharpens the debate over thimerosal. Robert F. Kennedy, Jr., an attorney and well-known environmental advocate from a high-profile political family, has been at the forefront. Paul Offit claims Kennedy has not only given moral support to McCarthy but they seem to embrace the same goal: "McCarthy, too, stresses that she wasn't anti-vaccine but 'anti-toxin,' and she lent her support to a rising popular movement that marched on Washington in 2008 to demand that government and industry 'Green our Vaccines.'"[33]

Kennedy drew attention when, in 2005, he published "Deadly Immunity" in *Rolling Stone* magazine, seemingly revealing a cover-up between federal regulators and pharmaceutical companies, one that assumes the mantle of a "moral crisis." Tom Verstraeten, a CDC epidemiologist, had conducted a statistical study using that agency's database, containing the medical records of 100,000 children, and discovered a significant correlation between vaccines containing thimerosal and autism, rising steadily through the 1990s. This claim sparked an immediate reaction, with CDC leaders in June 2000 convening a closely guarded meeting at the Simpsonwood Conference Center, inviting representatives from the FDA, World Health Organization, and vaccine manufacturers. If these proceedings became public, they feared, pharmaceutical companies would be hit with numerous lawsuits. According to Kennedy, CDC officials buried the Verstraeten study, the database he used, and the Simpsonwood transcript, as well as engaged the Institute of Medicine, at the National Academy of Sciences, to conduct a new study in the hopes of finding a different outcome. Kennedy marshals additional evidence of thimerosal's dangers, citing an Amish investigation conducted by a journalist in Lancaster County, Pennsylvania. Since the Amish do not believe in immunization, since that writer found no evidence of autism, he had apparently confirmed the vaccine-autism connection. Kennedy further points to Maurice Hilleman, a vaccine researcher at Merck, who warned that company about the risks of using this preservative. Kennedy's article shocked the public, to be sure.[34]

Kennedy followed this exposé in 2015 with an online film, *Trace Amounts*. In the style of a documentary, it opens with a disclaimer that it is not about vaccines, but rather focuses on thimerosal. This well-produced video, approximately ninety minutes long, centers on the personal journey of thirty-three-year-old Eric Gladen. In 2004, he received a tetanus shot. The side effects, extreme anxiety, attention deficit disorder, hearing and light sensitivity, muscle coordination and weakness, nervousness, numbness, rash, and tremors, prevented him from working. Doctors had no answers for him, but his own research pointed to thimerosal contained in that injection. In 2012, he began a nationwide tour, in a recreational vehicle, to raise awareness and seek answers to questions. The film shows Gladen phoning Eli Lilly's corporate office to speak to someone about thimerosal; Lilly declined to respond. Later, he is depicted calling CDC officials to ask questions about thimerosal; they refuse an interview. This leaves viewers with an unspoken suspicion: What do they have to hide?

Gladen ties thimerosal to autism by referencing Andrew Wakefield's 1998 article. As Gladen narrates, "hinting at a link between measles-mumps-rubella vaccines and autism" earned Wakefield enmity in the medical field, disgracing him and ending his career. Wakefield is projected as a victim for revealing the truth. Gladen reinforces this point by tracing the explosion of autism, stable from the 1930s to the 1990s at 1:10,000 but by 2014, 1:68. By the end of the film, Gladen announces he has begun to recover from toxic buildup, utilizing the same diet and regimen as the alleged autism cure.

Trace Amounts largely replicates the argument and evidence Kennedy used in his *Rolling Stone* article in addition to William Thompson's recorded claims about a CDC cover-up. *Trace Amounts* portrays thimerosal's whitewash by the FDA, CDC, the American Academy of Pediatrics, physicians, and the pharmaceutical industry as a scandal. It illustrates how this issue eerily parallels the decades-long cover-up by tobacco companies, comparing excerpts of Congressional testimony by each to reveal similarities; in both instances, corporate representatives claimed that science was on their side. Finally, short clips show both McCarthy and Kennedy at the Green Our Vaccines demonstration in Washington, DC.[35]

Celebrity status gives Kennedy access to popular television shows. On April 24, 2015, he appeared on *Real Time with Bill Maher* to discuss the thimerosal issue. Kennedy attributed his interest to concerned mothers—what McCarthy labels warrior mothers—who had lobbied him. During his interview with Maher, Kennedy highlighted points in *Trace Amounts.* As that segment closed, Maher complemented Kennedy on his film.[36]

In many ways, *Trace Amounts* and *Vaxxed* appear highly similar. They utilize much of the same information and sources. Although William Thompson's recorded, phone revelations appear quite similar, *Vaxxed* states that Thompson wants to speak out publicly but, as a federal employee, he would face criminal charges. Furthermore, *Trace Amounts* and *Vaxxed* include clips from Representative Dan Burton's 2002 Congressional hearings as he grills CDC officials. And parent observations and experiences echo each other. Finally, both hint at toxic overload yet never provide hard evidence. They are, nevertheless, quite persuasive.

The difference between these two films rests on the alarmist tone of *Vaxxed.* The growing frequency of autism foreshadows a medical catastrophe. A Massachusetts Institute of Technology statistician lends credibility to this claim by predicting that by 2032 50% of all children and 80% of boys will become autistic because of the MMR vaccination. *Vaxxed* also discloses the "African American Effect"; that is, Black children, according to the CDC's supposedly suppressed data, appear more susceptible to autism because of inoculation, especially males. The emotional toll will be immeasurable and cost American society billions of dollars in treatment and care. To resolve any doubt, a narrator calls for a research study comparing vaccinated children with unvaccinated children. Clearly, this represents a totally unrealistic option since it would allow children in the control group (i.e., not immunized) to be vulnerable to measles, mumps, and rubella. The Amish experience mentioned in *Trace Amounts* comes closest to this kind of protocol. However, a serious omission exists: some Amish communities accept immunizations, especially against polio.[37]

Finally, unlike anti-vacs and anti-toxs, vaccine delayers promote an alternative injection schedule. Southern California pediatrician Robert W. Sears has attracted a large audience, publishing a series of parenting books and maintaining a website, "Ask Doctor Sears," and Facebook page. He has appeared on *20/20*, *Good Morning America*, *Oprah*, and the *Today Show*. In *The Vaccine Book: Making the Right Decisions for Your Child*, Sears presents both his rationale and own versions of inoculation plans. Parting with the American Academy of Pediatricians and the CDC; Sears recommends spreading out shots based on the following criteria: avoid vaccines that contain mercury, begin with those which protect infants from the most threatening diseases while suspending jabs for the least dangerous ones, give no more than two different antigens at a time to reduce the intake of potentially harmful chemicals and side effects, administer live-virus immunogen one at a time, and postpone meningococcal inoculation until the age of sixteen. Within these qualifications, Sears appears tepid about MMR, since, as he sees it, so much controversy surrounds it. He strays even further from orthodoxy by presenting another "delayed selective vaccine schedule," one that begins at age five and continues for each year thereafter. MMR immunization would occur at the onset of puberty while others would be obtained during adolescence, or even adulthood. Sears favors this latter, rather casual, approach since many diseases, like polio, no longer pose a threat. He, of course, ignores the fact that this situation exists because of herd immunity, a delicate balance to be sure. Further, he staunchly believes that such "alternative vaccine schedules actually *increase* vaccination rates in our country," because they represent a compromise that will induce hesitant parents. Sears clearly sides with them: "… from a freedom-of-choice point of view, we can't really fault parents who think that vaccines are too risky and decide to put their kids first." He ignores the fate of other children in this scenario, rhetorically asking, "Is this selfish?" He answers,

"Perhaps.... Are you supposed to make decisions that are good for the country as a whole?.... Can we fault parents for putting their own child's health ahead of that of the kids around him?"[38]

Notes and Sources

1. Pam Belluck and Melena Rizik, "Tribeca Film Festival to Screen Anti-Vaccination Movie," *New York Times*, March 26, 2016: A-11.
2. Steven Zeitchik, "Tribeca Drops Anti-Vaccination Film; Robert De Niro Says His Festival Team Has Changed Plans About Showing *Vaxxed* After Review with Scientists," *Los Angeles Times*, March 27, 2016: A-8. Also check Steven Zeitchik, "*Vaxxed* Pulled from Tribeca Film Festival," *Pittsburgh Post-Gazette*, March 29, 2016: C-1, 3; and Editorial, "Nixing *Vaxxed*," *Pittsburgh Post-Gazette*, March 30, 2016: A-8. View De Niro's Televised Interviews and His Ambiguous Position on This Film. *Today Show*, April 13, 2016 (NBC News, 2016), http://www.today.com/... robert-de-niro-on-anti-vaccine-film-controversy-let-s-find-out-the-truth-665031235642 (Accessed July 4, 2016); *Today Show*, April 18, 2016 (NBC News, 2016), https://www.youtube.com/watch?v=W-8c327UcjLQ (Accessed July 4, 2016).
3. Melana Rizak, "Cut from Festival, Vaccination Film Opens Early," *New York Times*, March 31, 2016: C-3.
4. Rebecca Carroll, "Why Did De Niro Promote an Anti-Vax Film?" *The Guardian*, March 28, 2016, http://gu.com/p/4hqn6/sbl (Accessed March 31, 2016). Refer also to Belluck and Rizik, "Tribeca Film Festival to Screen Anti-Vaccination Movie": A-11; Zeitchik, "Tribeca Drops Anti-Vaccination Film": A-8.
5. Anya Sostek, "Controversial Film Linking Vaccine and Autism Coming to Pittsburgh," *Pittsburgh Post-Gazette*: B-1, 6. Further check the advertisement for this film, "*Vaxxed*." *Pittsburgh Post-Gazette*, June 10, 2016: D-3. Moreover, check Community Reel Arts Center, www.reelartscenter.org (Accessed June 15, 2016). Finally, Seth Mnookin, *Panic Virus: The True Story Behind the Vaccine-Autism Controversy* (New York: Simon and Schuster, 2011), in Chapter 8, focuses on the process of how traditional and reputable television and newspaper outlets rely on false or inaccurate information from sources on the Internet for news. They take them at face value, lending them legitimacy.
6. Andrew Wakefield, et. al., "Lleal-Lymphoid-Nodular Hyperplasia, Non-specific Colitis, and Pervasive Developmental Disorder in Children," *Lancet* 351 (February 28, 1998): 637–41. Refer specifically to page 641, as well as James Colgrove, *State of Immunity: The Politics of Vaccination*

in Twentieth-Century America (Berkeley, CA: University of California Press, 2006): 229; Elena Conis, *Vaccination Nation: America's Changing Relationship with Immunization* (Chicago: University of Chicago Press, 2015): 209–10; and Paul A. Offit, *Deadly Choices: How the Anti-Vaccine Movement Threatens Us All* (New York: Basic Books, 2011): 92. Mnookin, *Panic Virus*, demonstrates how Wakefield's supporters view him as a victim (299–302).

7. Lauren R. Kolodziejski, "Harms of Hedging in Scientific Discourse: Andrew Wakefield and the Origins of the Autism Vaccine Controversy," *Technical Communication Quarterly* 23 (2014): 165.
8. Mnookin, *Panic Virus*: 106. Further consult Colgrove, *State of Immunity*: 229; Owen Dyer, "GMC Hearing Against Andrew Wakefield Opens," *British Medical Journal* 335 (July 14, 2007): 63; and Kolodziejski, "Harms of Hedging in Scientific Discourse": 166.
9. Brian Deer, Transcript: "Royal Free Facilities Attack on MMR in Medical School Single Shots Videotape" (February 4, 1998), http://briandeer.com/wakefield/royal-video.htm (Accessed June 27, 2016). Kolodziejski, "Harms of Hedging in Scientific Discourse," cites this transcript as well, even some of the same sentences. See, for example, pp. 172–73. Refer also to Mnookin, *Panic Virus*: 108; Paul A. Offit, *Vaccinated: One Man's Quest to Defeat the World's Deadliest Diseases* (New York: Smithsonian Books, 2007): 160–62.
10. "A Timeline of the Wakefield Retraction," *Nature Medicine* 16 (2010): 248.
11. Kolodziejski, "Harms of Hedging in Scientific Discourse": 177. Refer moreover to Mnookin, *Panic Virus*: 103–5, 236.
12. Kolodziejski, "Hrams of Hedging in Scientific Discourse": 165. Also refer to Colgrove, *State of Immunity*: 249; College of Physicians, *History of Vaccines* (Philadelphia: College of Physicians, 2013): 40; Clare Dyer, "High Court Judge Crticises Andrew Wakefield for Trying to Silence His Critics," *British Medical Journal* 331 (November 12, 2005): 1104; Annabel Ferriman, "MP Raises New Allegations Against Andrew Wakefield," *British Medical Journal* 328 (March 27, 2004): 726; "Retraction of an Interpretation," *Lancet* 363 (March 6, 2004): 750; "A Statement by Dr. Andrew Wakefield," *Lancet* 363 (March 6, 2004): 823–24; Mnookin, *Panic Virus*: 301; and Offit, *Vaccinated*: 169, and *Deadly Choices*: 93–94.
13. Dyer, "GMC Hearing Against Andrew Wakefield Opens": 62–63.
14. Owen Dyer, "Andrew Wakefield Is Accused of Paying Children for Blood." *British Medical Journal* 335 (July 21, 2007): 118. Also see "Andrew Wakefield," *Nature Science* 17 (2011): 148; Clare Dyer,

"*Lancet* Retracts MMR Paper After GMC Finds Andrew Wakefield Guilty of Dishonesty," *British Medical Journal* 340 (February 6, 2010): 281.

15. Dyer, "*Lancet* Retracts MMR Paper After GMC Finds Andrew Wakefield Guilty of Dishonesty": 281. The word "retracted" appears in bold, red caps across each page of the ill-fated *Lancet* article. Further check "A Timeline of the Wakefield Retraction": 248; Conis, *Vaccination Nation*: 213; and Offit, *Deadly Choices*: 94, 96. Kolodziejski, "Harms of Hedging in Scientific Discourse," provides a clear sequence of events (178).
16. *Vaxxed: From Cover-Up to Catastrophe*. Cinema Libre Studio, 2016 (Viewed at the Parkway Theater, McKees Rocks, PA, June 16, 2016).
17. Zolan Kanno-Youngs, "Feelings Mixed as Festival Boots Film on Vaccines," *Wall Street Journal*, March 28, 2016: A-19.
18. The modern anti-vaccination movement in many ways diverges from its predecessor. In other ways, the former continues to resonate, revealing common concerns and organizational development. Compulsory immunization certainly aroused both movements. However, the literature describing the origins of the recent campaign varies. In *Vaccination Nation*, Conis generally argues that it developed as a reaction, fed by a more complex context, to federal immunization policies. Meanwhile, Colgrove's *State of Immunity* stresses abstractions, that is, individual rights versus public welfare. Finally, Offit, in *Deadly Choices*, largely sees individuals, namely Barbara Loe Fisher, Jenny McCarthy, and Andrew Wakefield, fueling this phenomenon, for whatever reasons. In Chapter 7, "Past Is Prologue," Offit summarizes the similarities and differences of the old and new anti-vac movements.
19. Conis, *Vaccination Nation*: 105, 108, 106–07, 219, and 139, respectively. Refer to pages 114, 129, 140, 143–44, 150–53, 211–12, 215–16, as well as Stuart Blume, *Immunization: How Vaccines Became Controversial* (London: Reaktion Books, 2017): 216, 232; Mnookin, *The Panic Virus*: 18–19, 217. Colgrove, *State of Immunity*, also sees a confluence of similar social changes driving vaccine concerns, namely feminism and environmentalism (236). For a description and analysis of the role of mothers as health care-givers, see Richard J. Altenbaugh, *The Last Children's Plague: Poliomyelitis, Disability, and Twentieth-Century American Culture* (New York: Palgrave Macmillan, 2015). Finally, Barbara Welter's pioneer article, "The Cult of True Womanhood: 1820–1860," *American Quarterly* 18 (1966): 151–74, insightfully outlines the origins of this domestic role.
20. Offit, *Vaccinated*: 171–72. Also see College of Physicians, *History of Vaccines*: 40–41; Conis, *Vaccination Nation*: 103; and Mnookin, *Panic Virus*: 243.
21. Paula S. Fass and Mary Ann Mason eds., *Childhood in America* (New York: New York University Press, 2000): 1. Conis, *Vaccination Nation*, makes a similar observation (198–99).

22. William Wagner, "God, Government, and Parental Rights," in Louise Kuo Habakus and Mary Holland, eds., *Vaccine Epidemic: How Corporate Greed, Biased Science, and Coercive Government Threaten Our Human Rights, Our Health, and Our Children* (New York: Skyhorse Publishing, 2012): 53.
23. Louise Kuo Habakus, "A Human Rights Assessment," in Habakus and Holland, *Vaccine Epidemic*: 23.
24. Sookyung Song, "The International Human Rights Standard," in Habakus and Holland, *Vaccine Epidemic*: 15 and 18, respectively.
25. James Turner, "Due Process and the American Constitution," in Habakus and Holland, *Vaccine Epidemic*: 36. Refer to pages 34–25, as well as Sookyung Song, "The International Human Rights Standard," in Habakus and Holland, *Vaccine Epidemic*: 15, 18.
26. Paul A. Offit, *Bad Faith: When Religious Belief Undermines Modern Medicine* (New York: Basic Books, 2015): 72, 95, 101, 107, and 112, respectively. Further check pages 94, 97, 102–3, 108–9, 110–11, as well as Offit, *Deadly Choices*: 139–42.
27. Offit, *Deadly Choices*: 142 and 143, respectively. Likewise, refer to Colgrove, *State of Immunity*: 181–82.
28. Graeme Turner, *Understanding Celebrity* (Los Angeles: Sage, 2014): 4, 9, and 6, respectively. Check pages 3, 26–27, and Elizabeth Fish Hatfield, "Celebrity Influence on Audiences' Consumption Practices as Parents," in Jackie Raphael, Basuli Deb, and Nidhi Shrivastava, eds., *Building Bridges in Celebrity Studies* (Toronto: Waterhill Publishers, 2016): 16; Raphael, Deb and Shrivastava, eds., *Building Bridges in Celebrity Studies*: 1, 3; Judith Roberts, "Commodifying Celebrity: Social Media, Sensationalism, and How the Media Plays a Role in Creating Celebrities." in Raphael, Deb, and Shrivastava, *Building Bridges in Celebrity Studies*: 47. Robert A. Goldberg thoroughly critiques this celebrification phenomenon in *Tabloid Medicine: How the Internet Is Being Used to Hijack Medical Science for Fear and Profit* (New York: Kaplan Publishing, 2010). Emphasis is mine. Especially refer to Chapters 1, "Prospect Theory: The Risks We Choose to Live With and Why;" 2, "Precautionary Principle: The Politics of Pseudocertainty;" and 3 "Insta-Americans: The Rise of Online Self-Diagnosis." Goldberg's *Talcoid Medicine* nicely explicates the interaction between new and orthodox media in Chapter 1. Refer also to Mnookin, *Panic Virus*: 8.
29. P. David Marshall, "New Media—New Self: The Changing Power of Celebrity," in *The Celebrity Culture Reader*, ed. P. David Marshall (New York: Routledge, 2006): 637. Investigate pages 634–35, as well as Goldberg, *Tabloid Medicine*, especially Chapter 1.
30. Hatfield, "Celebrity Influence on Audiences' Consumption Practices as Parents": 8. Also check College of Physicians, *History of Vaccines*: 41;

Conis, *Vaccination Nation*: 214–17; Jenny McCarthy, *Mother Warriors: A Nation of Parents Healing Autism Against All Odds* (New York: Plume Books, 2008): 4–5, 7, 8–9, 18, 23, 26, 32, 43, 44, 69–70, 82, 98; and Offit, *Deadly Choices*: ix, 149–50, 151, 154, 162–63, 165–68. Joe Moran, commenting on celebrity culture, explains how media conglomerates smoothly connect their publishing houses with their television operations arranging for their authors to promote their books on popular shows. Moran refers to this phenomenon as the "literary celebrity" (340) and "literary star system" (341). Refer to Chapter 17, "The Reign of Hype: The Contemporary Star System," in P. David Marshall, ed., *The Celebrity Culture Reader* (New York: Routledge, 2006). Goldberg, *Tabloid Medicine*, also provides a long list of individuals who oppose vaccines and profit from their apparent expertise and activities in Chapter 5, "Web of Fear: Vaccines, Autism, and the Emergence of 'Instant Experts.'"

31. Hafield, "Celebrity Influence on Audiences' Consumption Practices as Parents": 10, 17, respectively. Further see page 9. Mnookin, *Panic Virus*, reviews her career (249–50).
32. McCarthy, *Mother Warriors*: 217 and 215, respectively. Emphasis is in original. Refer, moreover, to pages 29, 86, 91. McCarthy maintains Facebook, Instagram, and Twitter accounts as well as a blog. Her official website is www.jennymccarthy.com. Finally, consult Mnookin, *Panic Virus*: 256–57.
33. Offit, *Deadly Choices*: 152. Also see Conis, *Vaccination Nation*: 220; Goldberg, *Tabloid Medicine*, especially Chapter 5.
34. Robert F. Kennedy, Jr., "Deadly Immunity," *Rolling Stone*, no. 977/978 (June 30, 2005): 57–65. *Academic Search Premier*, EBSCO*host* (Accessed May 26, 2016). Further consult Offit, *Vaccinated*: 173 and Chapter 18 in Mnookin, *Panic Virus*.
35. Gathr Films, *Trace Amounts: Autism, Mercury, and the Hidden Truth* (Trace Amounts, LLC, 2015), https://traceamounts.vhx.tv/packages/trace.amounts (Accessed May 27, 2016). According to Goldberg, Offit has become a special target of anti-vacs. They portray him as a tool of pharma because of his rotavirus vaccine licensed by Merck. However, he no longer receives financial benefits from it. Still, he and his children have received threats. Refer also to Chapter 2 in *Tabloid Medicine*. To grasp a clear understanding of the differences between ethylmercury and methylmercury, see Chapter 10 in Mnookin, *Panic Virus*.
36. *Real Time with Bill Maher* (April 24, 2015) HBO Productions. Roberts, in "Commodifying Celebrity," asserts: "Celebrities, of course, can also be politicians" (49).
37. Tom Ferrell, Virginia Adams, and Michael Wright, "Amish Polio Cases Cause New Alarm," *New York Times*, June 3, 1979: E9,

http://pitt.oclc.org/login?url=http://search.proquest.com/docview120827897?accountid=14709 (Accessed June 25, 2016); "Disease Center Reports 10 Amish Polio Cases." *New York Times*, June 9, 1979: 8, http://pitt.idm.oclc.org/login?url://search.proquest.com/docview/1284141329?=14709 (Accessed June 25, 2016); and Vaishali Honawar, "Four Amish Children Contract Polio Virus." *Education Week* 25 (November 30, 2005): 15, http://go.galegroup.com/ps/i.do?id=GALE%7CA215123766&v=21&u=upitt_main&it=r&p=AON-E&sw=w&asid=56abda1befbce83479e5a37ebeffaa62 (Accessed June 25, 2016).

38. Robert W. Sears, *The Vaccine Book: Making the Right Decisions for Your Child* (New York: Little, Brown and Company, 2011): 272, 278, and 279, respectively. Emphasis is in original. Further refer to pages 251–52, 262–63, 274 and consult his website www.askdrsears.com. Finally, check Mnookin, *Panic Virus*: 266.

CHAPTER 16

What Is Science?

In September of 1920, Porter F. Cope, Secretary of the Anti-Vaccination League, proclaimed that vaccinations caused infantile paralysis. Public health ideas changed significantly during the next several decades. Nothing illustrated this more than a 1999 retrospective published by the US Department of Health and Human Services, proclaiming that century's greatest public health achievements, among them, fluoridated water, prenatal and infant care, and vaccinations. The development of immunogen and sera and their expanded use had eliminated chronic epidemics of contagious childhood diseases and related deaths.

What this report overlooked was the long process of persuasion that had overcome the public's doubts and opposition. Its impact could be seen as early as 1930 when "opinion polls showed that Americans had more respect for doctors than other professionals…." William H. Park and Hermann Biggs, who had eliminated cholera in New York City in 1892 and appeared well on their way to controlling diphtheria, tuberculosis, and typhoid, became public health champions. Others had become martyrs, like Walter Reed whose death advanced research in the battle against malaria. Seventeen additional "health soldiers" would join him by 1941. These pioneers "were able to claim a new measure of cultural authority over a public impressed by the achievements of the laboratory…."[1] A heavily vaunted and largely trusted medical culture existed by the mid-twentieth century. As medical historian Naomi Rogers points out, the "polio vaccines [both Salk and Sabin] appeared at a high water

R. J. Altenbaugh, *Vaccination in America*,
Palgrave Studies in the History of Science and Technology,
https://doi.org/10.1007/978-3-319-96349-5_16

mark in the history of American biomedical medicines, and they became one of the great symbols of the impressive potential of modern medical research."[2] As its knowledge and tools grew more refined with each passing year, the ability to combat diseases appeared unfettered.

However, their conquest has bequeathed an ironic legacy. Mandatory immunization accounts for much of this progress, but it has led to resentment and resistance. Moreover, the very absence of diseases has extinguished any need for urgency. Parents believe they have the luxury of choice; that is, if the disease does not appear to exist, why become inoculated against it? Or, absent any threats from contagion, they can pick and chose among antigen and decide when to have them administered. Immunization opponents and proponents point to a host of studies to support their respective claims. Each blithely throws around the phrase, *good science versus bad science*. Both, of course, embrace the former and accuse their adversaries of accepting the latter. But what is so-called good science?

A Penny for Your Thoughts

According to science historian John C. Burnham, too many Americans lack any concept of the intellectual rigors of this field. He points to the shifting nature of the curriculum in the public schools as one reason. On the one hand, between 1900 and 1959, this content area significantly improved with specialization and the implementation of a distinct sequence. General science classes began to appear at the seventh and eighth grades while biology, chemistry, and physics courses cemented their places in the secondary curriculum. The only debate arose over methodology; that is, "'laboratory' versus 'demonstration.'" The Soviet Union's 1957 launch of the Sputnik satellite prompted a shift: scientists rather than educators took the lead. They used this Cold War setback to push for "massive curricular changes and [introduced] new ways of thinking that they believed were appropriate for students in the atomic age." Federal funding, vis-a-vis the National Science Foundation, facilitated these profound alterations. However, two movements modified all of this. First, in the late 1960s, student development began to supplant academic content. Second, during the early 1980s, "environmental concerns gave science negative connotations"; scientific and engineering discoveries not only fueled modern economies but also imperiled the environment. The impact of these two phenomena increased through that decade, by which time “scientists and science teachers” began to lament “their unpopularity” as students avoided its content and research methods.

Further, the pressures of a consumer-oriented society have commodified technological breakthroughs. The product alone is important, not the process that developed it. Acquiring the latest electronic game systems and forming long queues to purchase the newest smart phone define technical savvy. Formerly, Burnham adds, “[e]vangelists of science had … used the heroes and history of scientific discoveries to dramatize the battle against superstition and intolerance as well as to suggest that progress made in the past could be projected into the future.” The new outlook sees science as “part of civilization and that every cultured person, benefitting from the ‘general education’ movement of the midcentury period, ought to be able to understand science without having to learn vast amounts of learning.”[3] This debasing process leads us to the crux of the current anti-vaccination movement.

Jenny McCarthy represents the most egregious example. In *Warrior Mothers*, she expresses outright disdain for science and the scientific method. This becomes clear when she writes, in response to the CDC’s findings contradicting her claims, “Who needs science…? At home, Evan [her son] is my science.” McCarthy explains how diet was leading to her child’s recovery, likening it to “chemotherapy.” The “gluten-free, casein-free diet was helping kids with autism, a theory that had always been controversial.” She presents anecdotes as proof: “Mom after mom reported similar improvements after changing their child’s diet and trying other biomedical treatments like oxygen therapy and metals detoxification.” McCarthy is acutely aware that “old-school conservative policies,” as she terms them, rely on “double-blind research before promoting nontraditional treatments like specialized diet and supplements.” She dismisses this approach out of hand. First, she bases many of her claims on research obtained through her “favorite university: ‘University of Google.’” Second, she writes, “[p]eople can say that there is no science to support our beliefs about the causes of autism and ways to treat it, but there is plenty of evidence. Just walk into the homes of families who have children with autism. They will be happy to introduce you to their science.”

For McCarthy, federal regulators, the medical community, and pharmaceutical industry comprise the antagonists while she and determined parents, and their organizations, operate as the protagonists:

"The medical community is terrified to come within ten feet of detoxing metals out these kids because it will point the finger at what everyone is so afraid to admit. Vaccines CAN trigger autism." McCarthy sees the "medical community" profiting from both autism and "pharmaceutical politics," thus explaining its reluctance to accept her claims and those of her constituents. Because the American Academy of Pediatrics dismissed the findings of Defeat Autism Now, McCarthy declares that the "American Academy of Pediatrics Sucks." By the end of her book, McCarthy challenges the credibility of federal regulators and the medical sector: "Are we to believe that ALL thirty-six vaccinations given now are ALL safe with no side effects? Give me a break. Are we supposed to buy that these shots are one-size-fits-all? Or that every child is born with a perfect immune system? Wake the hell up, America, and think hard about the logic in this."[4]

In *Thimerosal: Let the Science Speak*, Robert F. Kennedy Jr., employs a different strategy, lending a sober and respectful tone as he pleads for the elimination of thimerosal. In the introduction to his edited book, Kennedy claims a "crack team of respected scientific researchers" confirm that "mercury in thimerosal can in fact cause brain injury in children." To develop his argument, he divides seventeen essays into three parts: the first covers thimerosal's dangers to human health, especially the brain; second challenges its use in any medications; and third refutes its exoneration by the Institute of Medicine's 2004 study. He includes a sizable number of digitized sources as evidence. However, the patina of science quickly disappears. While he twice repeats the phrase "correlation does not prove causation," he also notes the growth in the number of vaccines during the last quarter of the twentieth century "coincided closely with increased case reports of neurodevelopmental disorders…."[5] Coincidences are not proof; this closely resembles Andrew Wakefield's approach.

Lauren R. Kolodziejski's narrative analysis of Wakefield's 1998 essay explains how semantics can twist science. She points to two categories of scientific writing: the "technical sphere" and "public sphere." The former presumes that readers possess expertise and thus presents a "rigorous level of evidence" necessary to an "audience well versed in the field being discussed." The latter addresses a broader community in which "expectations for valid arguments can blur"; that is, research claims become elastic, expanding without presenting supporting evidence. These distinctions "prove useful in studying the rhetoric of science-based controversies" such as the MMR-autism relationship. "A close reading reveals how the rhetorical elements of that article, especially hedge words, enabled two different readings." Wakefield buries his MMR assertion, as Kolodziejski highlights in her excerpt: "'We have studied a chronic enterocolitis that *may* be related to neuropsychiatric dysfunction. *In most cases*, onset of symptoms was after measles[,] mumps, and rubella immunization. *Further investigations are needed* to examine this syndrome and its *possible* relation to this vaccine.'" Kolodziejski sees this "cautious hedging allowing Wakefield to discuss the possibility of a link without overstepping the data presented in the report…." Using this analysis, Wakefield, by omission, is being disingenuous. On the one hand, Kolodziejski remarks, such evasion "marks them as speculative rather than declarative, [and] fall within the norms of scientific discourse." This would explain the study's acceptance by a peer-reviewed journal, like *Lancet*. On the other hand, the "polysemous nature of the hedged statements … allowed for an alternative interpretation of the article, one that led to the [autism-vaccine controversy] and to more than a decade of debate regarding the safety of childhood immunization." Wakefield's public pronouncements follow the same pattern. Kolodziejski's deconstruction of the MMR-autism phenomenon explicates this: "… Wakefield utilized the media, through a press conference, to extend his scientific authenticity and promote his 'findings.'" And he continues to use various media outlets in the same way. However, Kolodziejski asserts, "science by press conference" threatens the basic credibility of the field. "This shift in communicating science to the public means that more science-based controversies may take hold, stirring up debate and possibly undermining public trust in science," leading inexorably to "bad science."[6]

Such high-profile individuals, along with many parents, have redefined, or outright discarded, the scientific process. They believe "anecdotal evidence" plays a pivotal role in "scientific inquiry." Personal testimonials about children's experiences appeal to the general public, who emotionally connect with them, becoming a powerful tool of persuasion. More importantly, it ignores the fundamental notion and process of gathering, analyzing, and utilizing empirical evidence, the very basis of the scientific method in search of ultimate truths.[7] As science writer Seth Mnookin neatly summarizes it: "Anecdotes and suppositions, no matter how right they feel, don't lead to universal truths; experiments that can be independently confirmed by impartial observers do. Intuition leads to the flat earth society and bloodletting; experiments lead to men on the moon and microsurgery."[8] In the case of vaccine-autism concerns, anecdotes may lead to *possible* correlations but do not demonstrate direct causation.

Universal Internet access has intensified this line of thinking, giving users immediate access to postings about the dangers of artificial immunization. Without any scientific training, they not only can digest it, and spout supposed truisms, but they can offer their own opinions and share their stories. This has contributed to laicization of knowledge, as medical historian James Colgrove terms it.[9] Web sites, online message boards, and blogs provide immense sources of information and endless perspectives. These new media, historian Elena Conis adds, educate a generation with alternative information: "as vaccine worries were being amplified in the news, the rise of the Internet created yet another forum for parents' suspicions to circulate and gain momentum." Autism sites grew, further building alarm. Instructions on them warned about the dangers of thimerosal, to "space out vaccines ... and opt for separate shots" instead of the "combined MMR vaccine." None of it was based on scientific evidence. The emergence of social media accelerated this phenomenon. This leads political scientist Robert Goldberg to ask a fundamental question: Is this the end of science? Given that science cannot produce 100% certainty about anything, then any doubt spells danger, generates distrust, and breeds opposition. This "precautionary principle," as he labels it, does not accept anything but 100% certainty.[10]

Finally, medical science's once-gleaming image has been tarnished. Just as public expectations peaked, Burnham points out, "disillusionment" began to surface. "By the 1960s ... the popular image of the physician as portrayed on television reflected a ... new type of hero, one with only ordinary endowments and who potentially could behave unheroically." Public deference previously accorded them began to evaporate: "....trust and freedom" no longer characterized medicine.[11] Today, physicians host popular daytime shows to dispense advice to their studio and viewing audiences. According to medical researcher Paul A. Offit, Mehmet Oz, host of the nationally televised *Dr. Oz Show*, has expressed "disdain for vaccines."[12] This only adds to the confusion. Furthermore, exposés revealed fallibility. "In the third quarter of the twentieth century there were no fewer than twenty investigations of the New York City health system, and in 1966 ... it was clear that the medical profession was in trouble...."[13] Much of this arose from the misuse of human subjects.

The thalidomide tragedy offers a clear example. Vicks Chemical Company, producer of cough drops and other over-the-counter aids, attempted to crack into the pharmaceutical niche without the requisite research facility by introducing thalidomide, a German-made drug to ease morning sickness during pregnancy, to the American market in 1959. According to historian Philip Hilts, initial animal tests, on rats and dogs, resulted in deaths. Nevertheless, that company pressed ahead with human experiments in order to meet its early 1961 marketing deadline. This consisted of sending 2.5 million tablets to "1,267 doctors who in turn gave them to about 20,000 patients." This supposed clinical trial lacked even the patina of medical research, with no control group, placebos, or consent forms; moreover, sales representatives, who supplied that product, collected no data. Profit clearly superseded safety and efficacy. "The company had not yet spoken to the FDA, nor did the law require it to. Under the 1938 statute, doctors could experiment on patients…, in any numbers and with any chemical, so long as they called the work an experiment. There was also no requirement that patients give their consent to be part of the experiment. They were simply given the drug." Presuming *pro forma* acceptance, Vicks sent the Federal Drug Administration (FDA) its "application to officially market thalidomide in the United States" in September of 1960. This created turmoil within that agency as some regulators questioned the data while others lobbied for approval. Company representatives added pressure to break the stalemate. Meanwhile, news from Great Britain and Germany, where the drug had been used extensively, revealed serious toxicity levels, causing severe nerve damage. Reports of babies born with birth defects followed in late 1961. Vicks continued with its "clinical trials." Finally, on July 15, 1962, newspapers broke the story of American birth defects. FDA officials attempted to halt the experiment but were too late. "It was found that several hundred of the doctors in the 'clinical trial' made no reports to the company about their use of thalidomide, and about one-third of the doctors involved made no attempt to talk to the patients who had taken the drug, or to retrieve the pills from them." A regulatory loophole resulted in forty cases of death or birth defects. "If the drug had made it to the American market by its target date, it was estimated that an additional 10,000 babies might have been born grossly deformed from the drug." The fundamental problem focused on a clear concept of *safety* as well as the procedural framework to ensure it. Because of vagueness, a dangerous medication fell through the bureaucratic cracks. Congress tightened regulations in 1962. "Companies wanting to sell new drugs would now have to demonstrate that they could be used safely, and that they worked for the stated purposes." No companies could "give some samples of the drug to patients for experimental purposes." Safety had to rely on "'substantial'" evidence based on controlled, scientific experiments overseen by "'experts.'"[14] But it was too late. Public faith in regulatory agencies began to wane.

Of course, the 1972 revelations of the hepatitis experiments on children housed at Willowbrook and syphilis study involving African Americans at Tuskegee, conducted by the US Public Health Service, further shook the very foundation of medicine. Not only did pharmaceutical companies face public accountability but the medical community did as well. Scientific experts, through their abuses of human subjects, had "violated the public trust" that had been so carefully constructed over previous decades.[15] These incidents, in turn, sowed skepticism of vaccine development and vaccines themselves. If federal agencies and scientific investigators put vulnerable populations at risk, what would stop large pharmaceutical corporations from producing and distributing unsafe vaccines? More importantly, could medical professionals be trusted with their children's well being?

Who Owns the Child?

Two parts of child welfare exist within a discussion of vaccines. First, through most of the twentieth century, children lacked protection from experimental hazards during vaccine development. Rarely did medical scientists reflect on this risk. Furthermore, most of this occurred absent public scrutiny; it represented an abstraction for most Americans, at least until the Salk polio vaccine trial. Driven by fear of poliomyelitis, the public focused on the ends rather than the means, stamping out a dreaded childhood disease. Although aware of the potential risks to their children, parents consented; after decades of persuasion, they trusted medical scientists. Because it provided efficient access to 1.8 million children, because students felt safe, because it had an educated administrative and instructional staff capable of carrying out demanding and tedious tasks, because it could facilitate parent assemblies to convey information, and because it had access to a ready supply of volunteers, and because it had been medicalized, the public school system served as an extension of the research laboratory. Nothing like it had occurred before or since. It took another two decades for a protective mechanism to be put into place, only then driven by inescapable scandals involving vulnerable populations.

Second, child welfare centers the debate over the safety and efficacy of immunization, one that has existed, in any public fashion, since the late nineteenth century. Opposition ebbed and flowed. In the beginning, public health officials employed coercion. Both sides debated safety, but opponents drew on religious and civil freedoms. Both sides used organizations to ground their campaigns and publish tracts arguing their positions. As resistance grew intractable, sometimes resulting in violence, health and civil authorities resorted to persuasion, utilizing informal and formal outlets to promote the benefits and assure the safety of vaccines. The polio immunization campaign tapped into this legacy. As the number of immunizations increased, as federal support grew, and as many common diseases began to fade, vaccinations became mandatory either because of public apathy or socioeconomic roadblocks. For the past four decades, the modern anti-vaccination campaign, fed by a variety of social movements, has displayed incredible persistence. Two variables, unique to this period, have ensured this. First, a new world of communications emerged. The Internet has provided a fount of information and built interactive ethereal communities. Celebrification enabled this campaign to cross different media platforms, combining the old with the new, profoundly raising the profile of this issue. Second, and a related phenomenon, is how opponents have oversimplified arguments. Playing on peoples' emotions and relying on anecdotes, this represents a profound epistemological shift on the part of the political right and left, implicitly and explicitly questioning the very nature of science, its processes and evidence.

We have indeed come a long way, as medical historian Naomi Rogers points out: the public's view of immunization profoundly shifted from the overwhelming consensus for the 1947 smallpox inoculations in New York City to deep skepticism during President George W. Bush's December 2002 call for mass vaccinations of Americans following the September 11 terrorist attacks. Fearing the use of weaponized smallpox, he proposed that 500,000 health workers be immediately immunized, with all Americans following by the summer of 2003. Bush's announcement elicited mixed responses. Public health authorities and the American Medical Association offered only tepid support, "hospital unions resisted," and some states temporarily suspended the program.

"By December 2003, only around 39,000 health workers had accepted the vaccine."[16]

More recently, candidate Donald J. Trump fuelled opposition to immunization. Using Twitter in 2012 and during a Republican primary debate in 2015, he linked vaccines to autism. Sharing a similar view, Robert Kennedy conferred with the President-Elect to discuss the possibility of Kennedy chairing a commission on vaccine safety. As he departed Trump Tower, following that meeting, Kennedy stated, "we ought to be debating the science." Finally, Trump and Andrew Wakefield, a high-profile opponent of MMR immunization, met at one of the inaugural balls.[17] It comes as no surprise, then, that residents of Vashon Island, Washington, who maintain one of the lowest immunization rates in the country, see Trump as a "champion" of the anti-vaccination movement. Trump's position has also energized Texans in San Antonio, who formed Texans for Vaccine Choice. Their opposition rests on personal beliefs and conspiracy theories, based on a worldwide cover-up of antigen dangers by medical scientists, pharmaceutical companies, physicians, as well as federal regulatory agencies. Largely libertarians, they lobby state legislators for a "parental choice" statute. Generally speaking, this movement has had a significant impact. Personal belief exemptions jumped statewide from 2314 during the 2003–2004 school year to 44,716 in 2015–2016, in spite of 21 measles and 4000 mumps infections in 2013. Affluent parents avoid the question altogether by enrolling their children at private schools. In Texas, one-third of those students remain unvaccinated, with one such institution reporting 40% unvaccinated. Further, Minnesota's worst measles epidemic in three decades occurred in April 2017, concentrated in its Somali-immigrant community. Twenty-five percent of infected children required hospitalization. Vaccine opponents had "targeted" that community, holding several rallies, including three guest appearances by Wakefield. They proved persuasive in their attempts to link MMR inoculation to autism since that group's MMR immunization rates fell from 92 to 42% between 2004 and 2014.[18]

The cumulative effect has been profound. A 2016 study, based on a survey of 682 pediatricians nationwide, found increasing skepticism among parents of young children. These doctors report encountering parents who, on a daily basis in their practices, choose to delay or completely skip one or more inoculations. Although less concerned about autism and thimerosal, some in the former group want to avoid discomfort while others do not want to burden their children's immune systems; meanwhile, refusals see immunization as unnecessary, a growing trend as this research concludes: The "public's collective memory of the consequences of these illnesses [have] faded, leading some parents to view vaccines as less crucial to the health of their children." This flies in the face of expected benefits: "Among U.S. children born between 1994 and 2013, vaccinations will prevent ~322 million illnesses, ~21 million hospitalizations, and ~732,000 deaths in their lifetimes."[19]

As an apt metaphor, Mark Honigsbaum, in the May 14, 2016, issue of *Lancet*, reviews an art exhibit titled, *Vaccination: Medicine and the Masses*, that opened on April 19 at the Hunterian Museum, Royal College of Surgeons, London. It featured nineteenth-century images depicting the bizarre nature of British fears, especially within the context of using cowpox immunization for smallpox. The significance of this exhibit, in his mind, refreshes the historical memory of the general public:

> ... the more that vaccines have reduced the incidence of formerly deadly childhood diseases, the harder it becomes to convince parents that it is necessary for their own child be vaccinated.... This is particularly the case with a new generation of parents too young to have experienced measles or mumps for themselves, never mind diseases such as polio.... Instead, it is the remote and unproven risks of vaccination that keep parents awake at night, not the fact that about one in every 1,000 cases of measles results in encephalitis.

It does not stretch the imagination to see how Honigsbaum links these now comical depictions to modern American, middle-class parents who "play Russian roulette with their [and other] children's health" by postponing or avoiding immunizations.[20]

Notes and Sources

1. Arthur Allen, *Vaccine: The Controversial Story of Medicine's Greatest Lifesaver* (New York: W. W. Norton Company, 2007): 130–31. See as well page 74, and "Vaccination Theory of Paralysis Denied," *New York Times*, September 29, 1916: 24. U.S. Department of Health and Human Services, "Morbidity and Mortality Weekly Report," 48 (April 2, 1999): 242–48, http://www.cdc.gov/mmwr/PDF/wk?mm4812.pdf (Accessed June 29, 2008).

2. Naomi Rogers, "Polio Can Be Conquered: Science and Health Propaganda in the United States from Polio Polly to Jonas Salk," in John W. Ward and Christian Warren, eds., *Silent Victories: The History and Practice of Public Health in Twentieth-Century America* (New York: Oxford University Press, 2007): 82–83.
3. John C. Burnham, *How Superstition Won and Science Lost: Popularizing Science and Health in the United States* (New Brunswick, NJ: Rutgers University Press, 1987): 185, 186, 187, 188, 223 and 224, respectively. Emphasis is in the text. Also see pages 183–84.
4. Jenny McCarthy, *Mother Warriors: A Nation of Parents Healing Autism Against All Odds* (New York: Plume Books, 2008): 9, 11, 17, 18, 39, 12, 214, 22, 30, 39, 57, and 214–15, respectively. Emphasis is in original. Refer also to page 10. Elena Conis, *Vaccination Nation: America's Changing Relationship with Immunization* (Chicago: University of Chicago Press, 2015), adds on page 217: "McCarthy ... asked that doctors and scientists acknowledge the expertise of parents (and mothers in particular)."
5. Robert F. Kennedy, Jr., ed., *Thimerosal: Let the Science Speak* (New York: Skyhorse Publishing, 2014): xxii, xxx, 14, xxviii, and xxi, respectively. Refer moreover to pages xxi, xxvii, xxxiii–xxxiv, 16–17, 162, 172.
6. Lauren R. Kolodziejski, "Harms of Hedging in Scientific Discourse: Andrew Wakefield and the Origins of the Autism Vaccine Controversy," *Technical Communication Quarterly* 23 (2014): 168, 167, 168, 171, 172, 180 and 181, respectively. Kolodziejski quotes these two sentences from Wakefield's study and italicizes the words and phrases to make her point; she uses additional examples to reinforce her points. Likewise, see Seth Mnookin's critique of Wakefield's article in *Panic Virus: The True Story Behind the Vaccine-Autism Controversy* (New York: Simon and Schuster, 2011): 110–11.
7. James Colgrove, *State of Immunity: The Politics of Vaccination in Twentieth-Century America* (Berkeley, CA: University of California Press, 2006): 237–38. Colgrove labels anecdotes as "individual case reports."
8. Mnookin, *Panic Virus*: 11. Check page 7 as well as see Kolodziejski, "Harms of Hedging in Scientific Discourse": 166.
9. Colgrove, *State of Immunity*: 237–38. Robert A. Goldberg, *Tabloid Medicine: How the Internet Is Being Used to Hijack Medical Science for Fear and Profit* (New York: Kaplan Publishing, 2010), likewise addresses the role of anecdotes in Chapter Seven: "Assault on Scientists: The Conflict-of-Interest Canard."
10. Conis, *Vaccination Nation*: 213 and 214, respectively. Further check page 228. Goldberg, *Tabloid Medicine*, fully develops this concept in Chapter

Two. Mnookin, *Panic Virus*, sees this process feeding "cognitive relativism" (9).

11. Burnham, *How Superstition Won and Science Lost*: 287, 286, 291, respectively.
12. Paul A. Offit, *Deadly Choices: How the Anti-Vaccine Movement Threatens Us All* (New York: Basic Books, 2011): 188–90. Also consult pages 168–69, 171–87.
13. Burnham, *How Superstition Won and Science Lost*: 286.
14. Philip J. Hilts, *Protecting America's Health: The FDA, Business, and One-Hundred Years of Regulation* (New York: Alfred A. Knopf, 2003): 151, 152, 156, 157, 158, 164, and 160, respectively. Further check pages 144, 149–50, 153–55. Emphasis is mine. Also consult Robert B. Baker, *Before Bioethics: A History of American Medical Ethics from the Colonial Period to the Bioethics Revolution* (New York: Oxford University Press, 2013): 275–76. The effects of the thalidomide scandal are still felt today. Finally, see Sarah Bosley, "Thalidomide and Journalism's Golden Era," *Lancet*, 387 (March 2016): 1151–52, www.thelancet.com (Accessed May 12, 2016).
15. Colgrove, *State of Immunity*: 191.
16. Naomi Rogers, "Polio Can Be Conquered: Science and Health Propaganda in the United States from Polio Polly to Jonas Salk," in *Silent Victories*: 81–82. Further consult Colgrove, *State of Immunity*: 245, 247; Mnookin, *Panic Virus*: 307–8.
17. Abby Phillip, Lena H. Sun, and Lenny Bernstein, "Vaccine Skeptic Robert Kennedy, Jr., Says Trump Asked Him to Lead Commission on 'Vaccine Safety.'" *Washington Post*, January 10, 2017, http://wapo.st/2jzHAs7 (Accessed June 24, 2017); Sarah Kaplan, "The Truth About Vaccines, Autism and Robert F. Kennedy, Jr.'s, Conspiracy Theory," *Washington Post*, January 10, 2017, http://wapo.st/2j5ecc0 (Accessed June 24, 2017); Daniel Summers, "Donald Trump and Robert F. Kennedy, Jr., Just Made Pediatricians' Jobs a Lot Harder," *Washington Post*, January 11, 2017, http://wapo.st/2icWI44 (Accessed June 23, 2017); Lena H. Sun, "Trump's Vaccine Views Are at Odds with Those of Most Americans, Study Says," *Washington Post*, February 6, 2017, http://www.washingtonpost.com/news/to-your-health/wp/2017/02/02/trumps-vaccine-views-at-odds-with-those-of-most-americans-study-says/?utm_term=.d2449198c49a (Accessed March 16, 2017).
18. Lena H. Sun, "Trump Energizes the Anti-Vaccine Movement in Texas," *Washington Post*, February 20, 2017, http://wapo.st/211naFp (Accessed February 20, 2017); "Vashon Island: The Place Where Parents Skip Jabs," *BBC News*, April 11, 2017; Lena H. Sun, "Anti-Vaccine Activists Spark a State's Worst Measles Outbreak in Decades," *Washington Post*,

May 5, 2017. http://wapo.st?2pLSSdd (Accessed May 5, 2017), http://www.bbc.co.uk/news/health-39558607 (Viewed April 11, 2017); "Free Speech and Facts in the Trump Era," *Lancet*, 389 (February 4, 2017): 478, http://dx.dol.org/10.1016/50140-6736(17)30265-9 (Accessed June 28, 2017).

19. Catherine Hough-Telford, et al., "Vaccine Delays, Refusals, and Patient Dismissals: A Survey of Pediatricians," *Pediatrics* 138 (2016): e20162127: 6 and 2, respectively. See also pages 3, 4, 7.
20. Mark Honigsbaum, "Vaccination: A Vexatious History," *Lancet* 387 (May 14, 2016): 1988–89, http://dx.org/10.1016?S01040-6736(16)30474-3 (Accessed May 19, 2016). That exhibit ran from April 19–September 17, 2016. Refer to www.hunterianmuseum.org (Accessed June 9, 2016). Also consult Linda C. Fentiman, "When It Comes to Vaccines, Rich Parents Get Away with Child Neglect," *Washington Post*, May 10, 2017, http://www.washingtonpost.com/opinions/rich-parents (Accessed May 11, 2017); Offit, *Deadly Choices*: x, 146–47.

Bibliography

Archives and Collections

American Philosophical Society, Philadelphia, PA (APS).
Carnegie Library of Pittsburgh, Pennsylvania Room (CLP).
Chicago Public Library, Municipal Reference Collection, Harold Washington Library Center (CPL).
Children's Institute Archives, Pittsburgh, PA (CIA).
College of Physicians, Philadelphia, PA (CP).
D. T. Watson Institute Archives (WIA).
Jonas Salk Papers, University of Pittsburgh, Health Science Library System (UPITT)
March of Dimes, White Plains, NY (MOD).
New York Academy of Medicine Archives, New York City (NYAM).
New York City Department of Records and Information Services, Municipal Archives (NYCMA).
Pennsylvania State Archives, Harrisburg, PA (PSA).
Rockefeller Archive Center, Tarrytown, NY (RAC).
Thomas Francis Papers, Bentley Historical Library, University of Michigan (BHL).
Warm Springs Archives, Georgia (WSA).

R. J. Altenbaugh, *Vaccination in America*,
Palgrave Studies in the History of Science and Technology,
https://doi.org/10.1007/978-3-319-96349-5

Court Cases

Commonwealth v. Smith, 9 Pa. D.R. 625 (1900).
Jacobson v. Massachusetts, 197 U.S. 1 (1905).
John A. Watson City v. City of Cambridge, 157 Mass. 561 (1893).
People v. McIlwain (1915).
State ex. Rel. Beattie v. Board of Education of City of Antigo, 176 Wisc. 231 (1919).

Government Documents

Advisory Committee on Human Radiation Experiments. "Research Ethics and the Medical Profession." *Journal of the American Medical Association* 276 (August 7, 1996): 403–9.

Lavinder, C. H., Allen W. Freeman, and Wade H. Frost. *Epidemiologic Studies of Poliomyelitis in New York City and the Northeastern United States During the Year 1916.* U.S. Public Health Bulletin No. 91. Washington, DC: Government Printing Office, 1918.

Solenberger, Edith Reeves. *Public School Classes for Crippled Children.* Department of the Interior, Bureau of Education, Bulletin No. 10. Washington, DC: Government Printing Office, 1918.

U.S. Department of Health, Education, and Welfare. "Gamma Globulin in the Prophylaxis of Poliomyelitis as Used in the United States, 1953." Report of the National Advisory Committee for the Evaluation of Gamma Globulin in the Prophylaxis of Poliomyelitis. Public Health Monograph No. 20. Washington, DC: Government Printing Office, 1954.

U.S. Department of Health and Human Services. *The Belmont Report (1979).* http://ori.hhs.gov (Accessed May 22, 1015).

U.S. Department of Health and Human Services. "Morbidity and Mortality Weekly Report." 48 (April 2, 1999): 242–48. http://www.cdc.gov/mmwr/PDF/wk?mm4812.pdf (Accessed June 29, 2008).

U.S. Public Health Service. Centers for Disease Control and Prevention.

Newspapers

Globe and Mail.
Los Angeles Times.
New York Times.
Pittsburgh Post-Gazette.
The Guardian.
USA Today.
Wall Street Journal.

Nonfiction Journal Articles

Altenbaugh, Richard J. "'Our Children Are Being Trained Like Dogs and Ponies': Schooling, Social Control, and the Working Class." *History of Education Quarterly*, 21 (Summer 1981): 213–22.

"Andrew Wakefield." *Nature Medicine*, 17 (2011): 148.

Anon. "Conference of Anti-Vaccinationists." *American Medicine*, 12 (October 1906): 449.

Anon. "Pennsylvania Wins Vaccination Fight." *American Medicine*, 12 (October 1906): 449–50.

Anon. "Compulsory Vaccination." *American Medicine*, 15 (June 1909): 219–20.

Anon. "The Slow Decadence of the Anti-Vaccination Propaganda." *American Medicine*, 15 (June 1909): 291.

Anon. "The Anti-Vaccinationists." *American Medicine*, 16 (November 1910): 558.

Anon. "Reports Toward Establishing Public School Dental Clinics in America." *Dental Cosmos*, 53 (March 1911): 310–11.

Anon. "Appalling Statistics Regarding Defective Children." *Dental Cosmos*, 55 (September 1913): 933.

Anon. "An Anti-Vaccination Judge." *American Medicine*, 20 (October 1914): 613–14.

Anon. "Defeat of the Anti-Vaccination Bill." *Boston Medical and Surgical Journal*, 170 (May 1914): 847.

Anon. "School Clinic at North Tonawanda." *Dental Cosmos*, 60 (September 1918): 840.

Anon. "Work of the Department of Health's Dental Clinics for School Children." *Dental Cosmos*, 61 (October 1919): 1032.

Anon. "Vaccination and Parents' Rights." *The Survey*, 47 (February 1922): 718.

Apple, Rima D. "'Advertised by Our Loving Friends:' The Infant Formula Industry and the Creation of New Pharmaceutical Markets, 1870–1910." *Journal of the History of Medicine and Allied Sciences*, 41 (January 1986): 3–23.

Apple, Rima D. "'They Need It Now:' Science, Advertising and Vitamins, 1925–1940." *Journal of Popular Culture*, 22 (Winter 1988): 65–83.

Apple, Rima D. "Educating Mothers: The Wisconsin Bureau of Maternal and Child Health." *Women's History Review*, 12 (2003): 559–76.

Barskey, Albert E., John W. Glaser, and Charles W. LeBaron. "Mumps Resurgences in the United States: A Historical Perspective on Unexpected Elements," *Vaccine*, 27 (2009): 6186–95.

Baumgartner, Leona. "Attitude of the Nation Toward Immunization Procedures: A Study Based on a Public Opinion Poll Made in 1941." *American Journal of Public Health*, 33 (March 1943): 256–60.

Beatty, Elizabeth. "Operative Work in the School Dental Clinic as Preventive Dentistry." *Dental Cosmos*, 59 (December 1917): 206–7.

Beecher, Henry K. "Ethics and Clinical Research." *New England Journal of Medicine*, 274 (June 1966): 1354–60.

Beecher, Henry K. "Scarce Resources and Medical Advancement." *Daedalus*, 98 (Spring 1969): 275–313.

Beeson, Paul B., Philip K. Bondy, Richard C. Donnelly, and John E. Smith, "Panel Discussion: Moral Issues in Clinical Research." *Yale Journal of Biology and Medicine*, 36 (June 1964): 455–76.

Blumgart, Herrman L. "The Medical Framework for Viewing the Problem of Human Experimentation." *Daedalus*, 98 (Spring 1969): 248–74.

Bodian, David. "Poliomyelitis and the Sources of Useful Knowledge." *Johns Hopkins Medical Journal*, 138 (April 1976): 130–36.

Brandt, Allan M. "Polio, Politics, Publicity, and Duplicity: Ethical Aspects in the Development of the Salk Vaccine." *International Journal of Health Services*, 8 (1978): 257–70.

Colgrove, James. "'Science in a Democracy:' The Contested Status of Vaccination in the Progressive Era and the 1920s." *Isis*, 96 (2005): 167–91.

Colgrove, James. "Foot Soldiers Against Infectious Diseases: Nurses, Families, and Immunization in the Twentieth Century." *Pediatric Nursing*, 33 (October 2007): 449–51.

Collins, Selwyn D. "Frequency of Immunizing Procedures of Various Kinds in 9,000 Families Observed for 12 Months, 1928–1931." *American Journal of Public Health and the Nation's Health*, 25 (1935): 1221–25.

Condrau, Flurin. "The Patient's View Meets the Clinical Gaze." *Social History of Medicine*, 20 (December 2007): 525–40.

Dawson, Liza. "The Salk Polio Vaccine Trial of 1954: Risks, Randomization and Public Involvement." *Clinical Trials*, 1 (2004): 122–30.

Duffy, John. "School Vaccination: The Precursor to School Medical Inspection." *Journal of the History of Medicine*, 33 (July 1978): 344–55.

Dye, Nancy Schrom, and Daniel Blake Smith. "Mother Love and Infant Death, 1750–1920." *Journal of American History*, 73 (September 1986): 329–53.

Dyer, Clare. "High Court Judge Criticises Andrew Wakefield for Trying to Silence His Critics." *British Medical Journal*, 331 (November 12, 2005): 1104.

Dyer, Clare. "*Lancet* Retracts MMR Paper After GMC Finds Andrew Wakefield Guilty of Dishonesty." *British Medical Journal*, 340 (February 6, 2010): 281.

Dyer, Owen. "Andrew Wakefield Is Accused of Paying Children for Blood." *British Medical Journal*, 335 (July 21, 2007): 118–19.

Dyer, Owen. "GMC Hearing Against Andrew Wakefield Opens." *British Medical Journal*, 335 (July 14, 2007): 62–63.

Enders, John F., Thomas H. Weller, and Frederick C. Robbins. "Cultivation of the Lansing Strain of Poliomyelitis Virus in Cultures of Various Human Embryonic Tissues." *Science*, 109 (January 28, 1949): 85–87.

Estabrooks, Carole A. "Lavinia Lloyd Dock: The Henry Street Years." *Nursing History Review*, 3 (1995): 143–72.

Ferriman, Annabel. "MP Raises New Allegations Against Andrew Wakefield." *British Medical Journal*, 328 (March 27, 2004): 726.

Flexner, Simon, and Paul A. Lewis. "Experimental Poliomyelitis in Monkeys." *Journal of the American Medical Association*, 54 (May 28, 1910): 1780–82.

"Free Speech and Facts in the Trump Era." *Lancet*, 389 (February 4, 2017): 478. http://dx.dol.org/10.1016/50140-6736(17)30265-9 (Accessed June 28, 2017).

Geison, Gerald L. "Pasteur's Work on Rabies: Reexamining the Ethical Issues." *Hastings Center Report*, 8 (April 1978): 26–33.

Ghormley, Ralph K. "History of Treatment of Poliomyelitis." *Journal of the Iowa State Medical Society*, 37 (August 1947): 343–50.

Grimshaw, Margaret L. "Scientific Specialization and the Poliovirus Controversy in the Years Before World War II." *Bulletin of the History of Medicine*, 69: 44–65.

Hansen, Bert. "The Image and Advocacy of Public Health in American Caricature and Cartoons from 1860 to 1900." *American Journal of Public Health*, 87 (November 1997): 1798–807.

Hansen, Bert. "America's First Medical Breakthrough: How Popular Excitement About a French Rabies Cure in 1885 Raised New Expectations for Medical Progress." *American Historical Review*, 103 (April 1998): 373–418.

Hansen, Bert. "New Images of a New Medicine: Visual Evidence for the Widespread Popularity of Therapeutic Discoveries in America After 1885." *Bulletin of the History of Medicine*, 73 (1999): 629–78.

Hansen, Bert. "Medical History for the Masses: How American Comic Books Celebrated Heroes of Medicine in the 1940s." *Bulletin of the History of Medicine*, 78 (2004): 148–91.

Hansen, Bert. "True-Adventure Comic Books and American Popular Culture in the 1940s: An Annotated Research Bibliography of the Medical Heroes." *International Journal of Comic Art*, 6 (Spring 2004): 117–47.

Harkness, Jon M. "Nuremberg and the Issue of Wartime Experiments on U.S. Prisoners: The Green Committee." *Journal of the American Medical Association*, 276 (November 27, 1996): 1672–75.

Hazelgrove, Jenny. "The Old Faith and the New Science: The Nuremberg Code and Human Experimentation Ethics in Britain, 1946–73." *Social History of Medicine*, 15 (April 2002): 109–35.

Hennock, E. P. "Vaccination Policy Against Smallpox, 1835–1914: A Comparison of England with Prussia and Imperial Germany." *Social History of Medicine*, 11 (April 1988): 49–71.

Hickey, Margaret. "Have We Won the Fight Against Polio?" *Ladies Home Journal* (December 1954): 25–26, 150.

Hill, William Z. "The Care of Children's Teeth." *Dental Cosmos*, 62 (May 1920): 1379–84.

Hornblum, Allen M. "They Were Cheap and Available: Prisoners as Research Subjects in Twentieth-Century America." *British Medical Journal*, 315 (November 1997): 1437–41.

Horstmann, Dorothy M., Robert W. McCollum, and Anne D. Mascola. "Viremia in Human Poliomyelitis." *Journal of Experimental Medicine*, 99 (April 1954): 355–69.

Hough-Telford, Catherine, David W. Kimberlin, Imaculada Aban, William P. Hitchock, Jon Almquist, Richard Kratz, and Karen G. O'Connor. "Vaccine Delays, Refusals, and Patient Dismissals: A Survey of Pediatricians." *Pediatrics*, 138 (2016): e20162127.

Howe, Howard A. "Antibody Response of Chimpanzees and Human Beings to Formalin-Inactivated Trivalent Poliomyelitis Vaccine." *American Journal of Hygiene*, 56 (July 25, 1952): 265–85.

Katz, Samuel L., C. H. Kempe, F. L. Black, M. L. Lepow, S. Krugman, R. J. Haggerty, J. F. Enders., et al. "Studies on an Attenuated Measles-Virus Vaccine." *New England Journal of Medicine*, 263 (1960): 180–84.

Kaufman, Martin. "The American Anti-Vaccinationists and Their Arguments." *Bulletin of the History of Medicine*, 41 (September–October 1967): 463–78.

Kaufman, Sharon. "The World War II Plutonium Experiments: Contested Stories and Their Lessons for Medical Research and Informed Consent." *Culture, Medicine and Psychiatry*, 21 (1997): 161–97.

Kennedy, Robert F., Jr. "Deadly Immunity." *Rolling Stone*, No. 977/978 (June 30, 2005): 57–65. *Academic Search Premier*, EBSCO*host* (Accessed May 26, 2016).

Kenny, Stephen C. "'I Can Do the Child No Good:' Dr. Sims and the Enslaved Infants of Montgomery, Alabama." *Social History of Medicine*, 20 (August 2007): 22–41.

Knapp, A. C., E. S. Godfrey, Jr., and W. L. Aycock. "An Outbreak of Poliomyelitis: Apparently Milk Born." *Journal of the American Medical Association*, 87 (August 1926): 635–39.

Kolmer, John A., and Anna M. Rule. "Concerning Vaccination of Monkeys Against Acute Anterior Poliomyelitis." *Journal of Immunology*, 26 (June 1934): 505–15.

Kolmer, John A. "Susceptibility and Immunity in Relation to Vaccination in Acute Anterior Poliomyelitis." *Journal of the American Medical Association*, 105 (December 1935): 1956–62.

Kolmer, John A. "Vaccination Against Acute Anterior Poliomyelitis." *American Journal of Public Health*, 26 (February 1936): 126–35.

Kolmer, John A. "Active Immunization Against Acute Anterior Poliomyelitis with Ricinoleated Vaccine." *Journal of Immunology*, 32 (May 1937): 341–56.

Kolmer, John A., George F. Klugh, and Anna M. Rule. "A Successful Method for Vaccination Against Acute Anterior Poliomyelitis." *Journal of the American Medical Association*, 104 (February 9, 1935): 456–60.

Kolodziejski, Lauren R. "Harms of Hedging in Scientific Discourse: Andrew Wakefield and the Origins of the Autism Vaccine Controversy." *Technical Communication Quarterly*, 23 (2014): 165–83.

Koprowski, Hilary, George A. Jervis, and Thomas W. Norton. "Immune Responses in Human Volunteers Upon Oral Administration of a Rodent-Adapted Strain of Poliomyelitis Virus." *American Journal of Hygiene*, 55 (1952): 109–26.

Koprowski, Hilary, Thomas W. Norton, George A. Jervis, Thomas L. Nelson, David L. Chadwick, Doris J. Nelsen, and Karl F. Meyer. "Clinical Investigation on Attenuated Strains of Poliomyelitis Virus." *Journal of the American Medical Association*, 160 (March 17, 1956): 954–66.

Kort, Michael. "The Delivery of Primary Health Care in American Public Schools, 1890–1980." *Journal of Social Health*, 54 (December 1984): 453–57.

Kramer, S. D., W. Lloyd Aycock, Charles I. Solomon, and C. L. Thenebe. "Convalescent Serum Therapy in Preparalytic Poliomyelitis." *New England Journal of Medicine*, 206 (March 1932): 432–35.

Lambert, Sarah Marie, and Howard Markel. "Making History: Thomas Francis, Jr., M.D., and the 1954 Salk Poliomyelitis Vaccine Field Trial." *Archives of Pediatrics and Adolescent Medicine*, 154 (May 2000): 512–17.

Lasagna, Louis. "Special Subjects in Human Experimentation." *Daedalus*, 98 (Spring 1969): 449–62.

Leake, James. P. "Poliomyelitis Following Vaccination Against this Disease." *Journal of American Medical Association*, 105 (December 28, 1935): 2152.

Lederer, Susan E. "Hideyo Noguchi" Luetin Experiment and the Antivivsectionists." *Isis*, 76 (March 1985): 31–48.

Lederer, Susan E. "Children as Guinea Pigs: Historical Perspective." *Accountability in Research*, 10 (2003): 1–16.

Martin, Henry A. "Anti-Vaccinism." *North American Review*, 134 (April 1882): 368–78.

Mechling, Jay. "Advice to Historians on Advice to Mothers." *Journal of Social History*, 9 (Fall 1975): 44–63.

Meckel, Richard A. "Combating Tuberculosis in Schoolchildren: Providence's Open-Air Schools." *Rhode Island History*, 53 (August 1995): 91–99.

Meier, Paul. "Polio Trial: An Early Efficient Clinical Trial." *Statistics in Medicine*, 9 (January/February 1990): 13–16.

Meldrum, Marcia. "'A Calculated Risk:' The Salk Polio Vaccine Field Trials of 1954." *British Medical Journal*, 317 (October 31, 1998): 1223–36. http://www.bmj.com/cgi/content/full/317/7167/1233?view=full&p-mid=0704869 (Accessed November 23, 2000).

Minardi, Margot. "The Boston Inoculation Controversy of 1721–1722: An Incident in the History of Race." *William and Mary Quarterly*, 61 (January 2004). http://www.historycooperative.org/journals/wm/61.1/ (Accessed April 27, 2004).

Monto, Arnold S. "Francis Field Trial of Inactivated Poliomyelitis Vaccine: Background and Lessons for Today." *Epidemiological Reviews*, 21 (1999): 7–23.

Multer, W. J. "Dental Education in the Schools of Nassau County, New York." *Dental Cosmos*, 62 (May 1920): 572–78.

Murphy, Timothy R. "The Ethics of Research with Children." *Virtual Mentor: Ethics Journal of the American Medical Association*, 5 (August 2003): 1–3. http://www.ama-assn.org/ama/pub/category/print/10817.html (Accessed April 29, 2004).

Pappworth, Maurice H. "Human Guinea Pigs: A Warning." *The Twentieth Century*, 171 (Autumn 1962): 66–75.

Parsons, Talcott. "Research with Human Subjects and the 'Professional Complex.'" *Daedalus*, 98 (Spring 1969): 325–60.

Paul, John R. "Thomas Francis, Jr., July 15, 1900–October 1, 1969." *Biographical Memoirs*, 44 (1974): 57–110.

Perillo, Jonna. "Beyond 'Progressive' Reform: Bodies, Discipline, and Construction of the 'Professional Teacher' in Interwar America." *History of Education Quarterly*, 44 (Fall 2004): 337–62.

Petrina, Stephen. "The Medicalization of Education: A Historiographical Synthesis." *History of Education Quarterly*, 46 (Winter 2006): 503–31.

"Prison Malaria: Convicts Expose Themselves to Disease So Doctors Can Study It." *Life*, 4 (June 4, 1945): 43–44, 46.

Ramsey, Paul. "The Enforcement of Morals: Nontherapeutic Research on Children." *Hastings Center Report*, 6 (August 1976): 21–30.

"Retraction of an Interpretation." *Lancet*, 363 (March 6, 2004): 750.

Reverby, Susan M. "Rethinking the Tuskegee Syphilis Study: Nurse Rivers, Silence and the Meaning of Treatment." *Nursing History Review*, 7 (1999): 3–28.

Rodríguez, Gonzalo-Herranz. "1952 Address of Pope Pius Xii to the Medical Community." *L'Osservatore Romano English Edition*, 30 (July 2003): 4. www.ewtn.com/library/ISSUES/BIORSP12.HTM (Accessed July 19, 2011).

Rosenberg, Charles E. "Martin Arrowsmith as Hero." *American Quarterly*, 14 (Fall 1963): 447–58.

Rothman, David J. "Were Tuskegee and Willowbrook 'Studies in Nature?'" *Hastings Center Report*, 12 (April 1982): 5–7.

Salk, Jonas E. "Recent Studies on Immunization Against Poliomyelitis." *Pediatrics*, 12 (November 1953): 471–78.

Salk, Jonas E. "Vaccination Against Paralytic Poliomyelitis: Performance and Prospects." *American Journal of Public Health*, 45 (May 1955): 575–96.

Salk, Jonas E., et al. "The Use of Adjutant to Facilitate Studies on the Immunologic Classification of Poliomyelitis Viruses." *American Journal of Hygiene*, 54 (September 1951): 157–73.

Salk, Jonas E., et al. "Studies in Human Subjects on Active Immunization Against Poliomyelitis. Part I. A Preliminary Report of Experiments in Progress." *Journal of the American Medical Association*, 151 (March 28, 1953): 1081–98.

Salk, Jonas E., et al. "Formaldehyde Treatment and Safety Testing of Experimental Poliomyelitis Vaccines." *American Journal of Public Health and the Nations Health*, 44 (May 1954): 563–70.

Salk, Jonas E., et al. "Studies in Human Subjects on Active Immunization Against Poliomyelitis. Part II. A Practical Means for Inducing and Maintaining Antibody Formation." *American Journal of Public Health*, 44 (August 1954): 994–1009.

Sandelowski, Margarete. "Making the Best of Things: Technology in American Nursing, 1870–1940." *Nursing History Review*, 5 (1997): 3–32.

Schlossman, Steven L. "Before Home Start: Notes Toward a History of Parent Education in America, 1897–1919." *Harvard Educational Review*, 46 (August 1976): 436–67.

Schmidt-Chanasit, Jonas, and Andreas Sauerbrei. "Evolution and World-Wide Distribution of Varicella–Zoster Virus Clades." *Infection, Genetics and Evolution*, 11 (2011): 1–10. http://www.sciencedirect.com.pitt.idm.oclc.org/science/article/pii/S1567134810002431 (Accessed March 8, 2017).

Schultz, Edwin W., and L. P. Gebhardt. "On the Problem of Immunization Against Poliomyelitis." *California and Western Medicine*, 43 (August 1935): 111–12.

Smith, J. David, and Alison L. Mitchell. "Sacrifices for the Miracle: The Polio Vaccine Research and Children with Mental Retardation." *Mental Retardation*, 39 (October 2001): 405–9.

Tomes, Nancy. "Merchants of Health: Medicine and Consumer Culture in the United States, 1900–1940." *Journal of American History*, 88 (September 2001): 519–47.

Vaughan, Henry, C. H. Halliday, John F. Kessel, Maurice Brodie, and James P. Leake. "Discussion of Poliomyelitis Papers." *American Journal of Public Health*, 26 (February 1936): 143–48.

Wakefield, Andrew, and Simon H. Murch, Andrew Anthony, David H. Casson, Mohsin Malik, Mark Berelowtz, Amar P. Dhillon, Michael A. Thomson, P. Harvey, Alan Valentine, Susan E. Davies, and John A. Walker-Smith. "Lleal-Lymphoid-Nodular Hyperplasia, Non-Specific Colitis, and Pervasive Developmental Disorder in Children." *Lancet*, 351 (February 28, 1998): 637–41.

Weller, Thomas H. "Advances in the Laboratory Diagnosis of Poliomyelitis." *Journal of the American Medical Association*, 156 (September 24, 1954): 16–18.

Welter, Barbara. "The Cult of True Womanhood: 1820–1860." *American Quarterly*, 18 (1966): 151–74.

White, Paul G. "The Practical Application of Oral Hygiene to the School Curriculum." *Dental Cosmos*, 52 (1910): 863–68.

Wilson, Daniel J. "A Crippling Fear: Experiencing Polio in the Era of FDR." *Bulletin of the History of Medicine*, 72 (1998): 464–95.

Woodhouse, Barbara Bennett. "'Who Owns the Child?:' *Meyer* and *Pierce* and the Child as Property." *William and Mary Law Review*, 33 (Summer 1992): 995.

Withers, John W. "The Dental Clinic and the Public Schools." *Dental Cosmos*, 61 (October 1919): 1015–25.

Yosifon, David, and Peter Stearns. "The Rise and Fall of American Posture." *American Historical Review*, 103 (October 1998): 1057–95.

Booklets/Pamphlets/Reports

American Humane Association. *Human Vivisection: A Statement and an Inquiry.* Chicago: American Humane Association, 1899.

American Humane Association. "AHA Report. Rpt. 1895." Memphis, TN: General Books, 2012.

Emerson, Haven. *The Epidemic of Poliomyelitis (Infantile Paralysis) in New York City in 1916: Based on the Official Reports of the Bureau of the Department of Health.* 1917; Rpt. New York: Arno Press, 1977.

Francis, Thomas, Jr., Robert F. Korns, Robert B. Voight, Morton Boisen, Fay M. Hemphill, John A. Napier, and Eva Tolchinsky. *An Evaluation of the 1954 Poliomyelitis Vaccine Trials: Summary Report.* Ann Arbor, MI: Edwards Brothers, 1955.

International Committee for the Study of Infantile Paralysis. *Poliomyelitis: A Survey.* Baltimore: Williams and Wilkins Co., 1932.

Knapp, A. C., E. S. Godfrey, Jr., and W. L. Aycock. *An Outbreak of Poliomyelitis: Apparently Milk Borne.* Chicago: American Medical Association, 1926.

Nonfiction: Books, Chapters

Allen, Arthur. *Vaccine: The Controversial Story of Medicine's Greatest Lifesaver.* New York: W. W. Norton Company, 2007.

Altenbaugh, Richard J. *The Last Children's Last Plague: Poliomyelitis, Disability, and Twentieth-Century American Culture.* New York: Palgrave Macmillan, 2015.

Altman, Lawrence K. *Who Goes First? The Story of Self-Experimentation in Medicine.* Berkeley: University of California Press, 1998.

American Medical Association. *Code of Medical Ethics.* Chicago: AMA, 1847. http://www.ama.assn.org/ama/pub/about-ama/our-history-ama (Accessed January 24, 2014).

Annas, George J. "The Nuremberg Code in U.S. Courts: Ethics Versus Expediency." In *The Nazi Doctors and the Nuremberg Code: Human Rights in Human Experimentation*, edited by George J. Annas and Michael A. Grodin, 201–22. New York: Oxford University Press, 1992.

Annas, George J., and Michael A. Grodin. Editors. *The Nazi Doctors and the Nuremberg Code: Human Rights in Human Experimentation.* New York: Oxford University Press, 1992.

Apple, Rima D. "The More Things Change: A Historical Perspective on the Debate Over Vitamin Advertising in the United States." In *Silent Victories: The History and Practice of Public Health in Twentieth-Century America*, edited by John D. Ward and Christian Warren, 193–206. New York: Oxford University Press, 2007.

Baker, Robert B. "The American Medical Ethics Revolution." In *The American Ethics Revolution: How the AMA's Code of Ethics Has Transformed Physicians' Relationships to Patients, Professionals, and Society*, edited by Robert B. Baker, et al., 17–51. Baltimore: Johns Hopkins University Press, 1999.

Bailyn, Bernard. *Education in the Forming of American Society.* Chapel Hill: University of North Carolina Press, 1960.

Baker. Robert B. *Before Bioethics: A History of American Medical Ethics from the Colonial Period to the Bioethics Revolution.* New York: Oxford University Press, 2013.

Baker, Robert B., Arthur L. Caplan, Linda L. Emanuel, and Stephen R. Latham. Editors. *The American Ethics Revolution: How the AMA's Code of Ethics Has Transformed Physicians' Relationships to Patients, Professionals, and Society.* Baltimore: Johns Hopkins University Press, 1999.

Bates, Barbara. *Bargaining for Life: A Social History of Tuberculosis, 1876–1938*. Philadelphia: University of Pennsylvania Press, 1992.

Beatty, Barbara, Emily D. Cahan, and Julia Grant. Editors. *When Science Encounters the Child: Education, Parenting, and Child Welfare in Twentieth-Century America*. New York: Teachers College Press, 2006.

Beekman, Daniel. *The Mechanical Baby: A Popular History of the Theory and Practice of Child Raising*. Westport, CT: Lawrence Hill & Co., 1977.

Berg, Roland H. *Polio and Its Problems*. Philadelphia: J. B. Lippincott, 1948.

Blume, Stuart. *Immunization: How Vaccines Became Controversial*. London: Reaktion Books, 2017.

Bookchin, Debbie, and Jim Schumacher. *The Virus and the Vaccine: The True Story of a Cancer-Causing Monkey Virus, Contaminated Polio Virus, and the Millions of Americans Exposed*. New York: St. Martin's Press, 2004.

Bowen, Catherine Drinker. *Family Portrait*. Boston: Little, Brown and Company, 1970.

Brandt, Allan M. "Racism and Research: The Case of the Tuskegee Syphilis Study." In *Sickness and Health in America: Readings in the History of Medicine and Public Health*, edited by Judith Walzer Leavitt and Ronald L. Numbers, 392–404. Madison: University of Wisconsin Press, 1997.

Brandt, Allan M., and Martha Gardner. "The Golden Age of Medicine?" In *Medicine in the Twentieth Century*, edited by Roger Cooter and John Pickstone, 21–37. Amsterdam: Harwood Academic Publishers, 2000.

Bruno, Richard L. *The Polio Paradox: Understanding and Treating "Post-Polio Syndrome" and Chronic Fatigue*. New York: Warner Books, 2002.

Burnham, John C. *How Superstition Won and Science Lost: Popularizing Science and Health in the United States*. New Brunswick, NJ: Rutgers University Press, 1987.

Burnham, John C. "American Medicine's Golden Age: What Happened to It?" In *Sickness and Health in America: Readings in the History of Medicine and Public Health*, edited by Judith Walzer Leavitt and Ronald L. Numbers, 284–94. Madison: University of Wisconsin Press, 1997.

Burns, Chester R. "Setting the Stage: Moral Philosophy, Benjamin Rush and Medical Ethics in the United States Before 1846." In *The American Ethics Revolution: How the AMA's Code of Ethics Has Transformed Physicians' Relationships to Patients, Professionals, and Society*, edited by Robert B. Baker, et al., 3–19. Baltimore: Johns Hopkins University Press, 1999.

Burt, Brian A., and Scott L. Tomar. "Changing the Face of America: Water Fluoridation and Oral Health." In *Silent Victories: The History and Practice of Public Health in Twentieth-Century America*, edited by John D. Ward and Christian Warren, 307–22. New York: Oxford University Press, 2007.

Byington, Margaret. *Homestead: The Households of a Mill Town*. 1910; Rpt. Pittsburgh: University of Pittsburgh Center for International Studies, 1974.

Cable, Mary. *The Little Darlings: A History of Child Rearing in America*. New York: Charles Scribner's Sons, 1975.

Carter, Richard. *Breakthrough: The Saga of Jonas Salk*. New York: Pocket Books, 1967.

Carter, Richard. *The Gentle Legions: National Voluntary Health Organizations in America*. New Brunswick, NJ: Transaction Publishers, 1992.

Cassell, Eric J. *The Nature of Suffering and the Goals of Medicine*. New York: Oxford University Press, 2004.

Chase, Allan. *Magic Shots: A Human and Scientific Account of the Long and Continuing Struggle to Eradicate Infectious Disease by Vaccination*. New York: William Morrow and Co., 1982.

Children's Institute. *Memories and Miracles: The Children's Institute*. Pittsburgh: Children's Institute, 2016.

Christianson, Eric H. "Medicine in New England." In *Sickness and Health in America: Readings in the History of Medicine and Public Health*, edited by Judith Walzer and Ronald L. Numbers, 47–71. Madison: University of Wisconsin Press, 1997.

Cohen, Sol. *Challenging Orthodoxies: Toward a New Cultural History of Education*. New York: Peter Lang, 1999.

Cohn, Victor. *Four Billion Dimes*. Minneapolis, 1955.

Coleman, Sidney H. *Humane Society Leaders in America: With a Sketch of the Early History of the Humane Movement in England*. Albany, NY: American Humane Association, 1924.

Colgrove, James. *State of Immunity: The Politics of Vaccination in Twentieth-Century America*. Berkeley, CA: University of California Press, 2006.

Condran, Gretchen A., Henry Williams, and Rose A. Cheney. "The Decline in Mortality in Philadelphia from 1870 to 1930: The Role of Municipal Services." In *Sickness and Health in America: Readings in the History of Medicine and Public Health*, edited by Judith Walzer and Ronald L. Numbers, 452–66. Madison: University of Wisconsin Press, 1997.

Conis, Elena. *Vaccination Nation: America's Changing Relationship with Immunization*. Chicago: University of Chicago Press, 2015.

Connolly, Cynthia A. *Saving Sickly Children: The Tuberculosis Preventorium in American Life, 1909–1970*. New Brunswick, NJ: Rutgers University Press, 2008.

Cooper, Jill E. "A Brief Romance with Magice Bullets: René Dubos at the Dawn of the Antibiotic Era." In *Silent Victories: The History and Practice of Public Health in Twentieth-Century America*, edited by John W. Ward and Christian Warren, 44–59. New York: Oxford University Press, 2007.

Coops, Helen Leslie. *Health Education in Elementary Schools: Activities, Materials, Methods.* New York: A. S. Barnes and Company, 1950.

Cooter, Roger. "The Ethical Body." In *Medicine in the Twentieth Century*, edited by Roger Cooter and John Pickstone, 451–68. Amsterdam: Harwood Academic Publishers, 2000.

Coss, Stephen. *The Fever of 1721: The Epidemic that Revolutionized Medicine and American Politics.* New York: Simon and Schuster, 2016.

Coulter, Harris L. *Vaccination, Social Violence, and Criminality.* Berkeley, CA: North Atlantic Books, 1990.

Coulter, Harris L., and Barbara Loe R. Fisher. *A Shot in the Dark: Why the P in the DPT Vaccination May Be Hazardous to Your Child's Health.* New York: Avery, 1991.

Creager, Angela N. H. *Life Atomic: A History of Radioisotopes in Science and Medicine.* Chicago: University of Chicago Press, 2013.

Cremin, Lawrence A. *The Wonderful World of Ellwood Patterson Cubberley: An Essay on the Historiography of American Education.* New York: Teachers College Press, 1965.

Cremin, Lawrence A. *Traditions of American Education.* New York: Basic Books, 1977.

Cremin, Lawrence A. *American Education: The National Experience, 1783–1876.* New York: Harper and Row, 1980.

Cutler, William W. III. *Parents and Schools: The 150-Year Struggle for Control in American Education.* Chicago: University of Chicago Press, 2000.

D'Antonio, Michael. *The State Boys Rebellion.* New York: Simon and Schuster, 2004.

Daniel, Thomas M. *Captain of Death: The Story of Tuberculosis.* Rochester, NY: University of Rochester Press, 1997.

Davidovitch, Nadav. "Negotiating Dissent: Homeopathy and Anti-Vaccinationism at the Turn of the Twentieth Century." In *The Politics of Healing: Histories of Alternative Medicine in Twentieth-Century North America*, edited by Robert D. Johnston, 11–28. New York: Routledge, 2004.

Davis, Fred. *Passage Through Crisis: Polio Victims and Their Families.* Indianapolis, IN: Bobbs-Merrill, 1963.

Draper, George. *Acute Poliomyelitis.* Philadelphia: P. Blakiston's Son & Co., 1927.

Duffy, John. *The Sanitarians: A History of American Public Health.* Urbana: University of Illinois Press, 1992.

Duffy, John. "Social Impact of Diseases in the Late 19th Century." In *Sickness and Health in America: Readings in the History of Medicine and Public Health*, edited by Judith Walzer Leavitt and Ronald L. Numbers, 418–25. Madison: University of Wisconsin Press, 1997.

Edward, P. K. Parker, and Nicholas C. Grassly. "Polio vaccination: preparing for a change of routine." *The Lancet* 388 (10040): 107–108, 2016.

Etheridge, Elizabeth W. *Sentinel for Health: A History of the Centers for Disease Control.* Berkeley: University of California Press, 1992.

Faden, Ruth R., and Tom L. Beauchamp. *A History and Theory of Informed Consent.* New York: Oxford University Press, 1986.

Fairchild, Amy L., Ronald Bayer, and James Colgrove. *Searching Eyes: The State, and Disease Surveillance in America.* Berkeley: University of California Press, 2007.

Fass, Paula S., and Mary Ann Mason. Editors. *Childhood in America.* New York: New York University Press, 2000.

Fenn, Elizabeth A. *Pox Americana: The Great Smallpox Epidemic of 1775–82.* New York: Hill and Wang, 2001.

Finkelstein, Barbara. "Uncle Sam and the Children: A History of Government Involvement in Child Rearing." In *Growing Up in America: Children in Historical Perspective*, edited by N. Ray Hiner and Joseph M. Hawes, 255–66. Urbana: University of Illinois Press, 1985.

Francis, Thomas, Jr., Robert F. Korns, Robert B. Voight, Morton Boisen, Fay M. Hemphill, John A. Napier, and Eva Tolchinsky. *An Evaluation of the 1954 Poliomyelitis Vaccine Trials: Summary Report.* Ann Arbor, MI: Edwards Brothers, 1955.

Glantz, Leonard H. "The Influence of the Nuremberg Code on U.S. Statutes and Regulations." In *The Nazi Doctors and the Nuremberg Code: Human Rights in Human Experimentation*, edited by George J. Annas and Michael A. Grodin, 183–200. New York: Oxford University Press, 1992.

Glantz, Leonard H. "The Law of Human Experimentation with Children." In *Children as Research Subjects: Science, Ethics, and Law*, edited by Michael A. Grodin and Leonard H. Glantz, 103–24. New York: Oxford University Press, 1994.

Goldberg, Robert A. *Tabloid Medicine: How the Internet Is Being Used to Hijack Medical Science for Fear and Profit.* New York: Kaplan Publishing, 2010.

Goodman, Jordan, Anthony McElligott, and Lara Marks. Editors. *Useful Bodies: Humans in the Service of Medical Science in the Twentieth Century.* Baltimore: Johns Hopkins University Press, 2003.

Gordon, Linda. "The Progressive-Era Transformation of Child Protection, 1900–1920." In *Childhood in America*, edited by Paula S. Fass and Mary Ann Mason, 543–48. New York: New York University Press, 2000.

Gordon, Linda. *Heroes of Their Own Lives: The Politics and History of Family Violence.* Urbana: University of Illinois Press, 2002.

Grant, Julia. *Raising Baby by the Book: The Education of American Mothers.* New Haven, CT: Yale University Press, 1998.

Grodin, Michael A. "Historical Origins of the Nuremberg Trial." In *The Nazi Doctors and the Nuremberg Code: Human Rights in Human Experimentation*, edited by George J. Annas and Michael A. Grodin, 121–44. New York: Oxford University Press, 1992.

Habakus, Louise Kuo. "A Human Rights Assessment." In *Vaccine Epidemic: How Corporate Greed, Biased Science, and Coercive Government Threaten Our Human Rights, Our Health, and Our Children*, edited by Louise Kuo Habakus and Mary Holland, 20–28. New York: Skyhorse Publishing, 2012.

Haggett, Peter. *The Geographical Structure of Epidemics*. New York: Oxford University Press, 2000.

Hall, Patricia. *Johnny Gruelle: Creator of Raggedy Ann and Andy*. Gretna, LA: Pelican Publishing Company, 1998.

Halpern, Sydney A. *Lesser Harms: The Morality of Risk in Medical Research*. Chicago: University of Chicago Press, 2004.

Holland, Mary. "Vaccination Choice Is a Fundamental Human Right." In *Vaccine Epidemic: How Corporate Greed, Biased Science, and Coercive Government Threaten Our Human Rights, Our Health, and Our Children*, edited by Louise Kuo Habakus and Mary Holland, 12–14. New York: Skyhorse Publishing, 2012.

Hammonds, Evelynn Maxine. *Childhood's Deadly Scourge: The Campaign to Control Diphtheria in New York City, 1880–1930*. Baltimore: Johns Hopkins University Press, 1999.

Hansen, Bert. *Picturing Medical Progress from Pasteur to Polio: A History of Mass Media Images and Popular Attitudes in America*. New Brunswick, NJ: Rutgers University Press, 2009.

Hardy, Anne. "Scarlet Fever." In *The Cambridge World History of Human Disease*, edited by Kenneth F. Kiple, 990–92. New York: Cambridge University Press, 1993.

Hatfield, Elizabeth Fish. "Celebrity Influence on Audiences' Consumption Practices as Parents." In *Building Bridges in Celebrity Studies*, edited by Jackie Raphael, Basuli Deb, and Nidhi Shrivastava, 8–19. Toronto: Waterhill Publishers, 2016.

Hays, J. N. *The Burdens of Disease: Epidemics and Human Response in Western History*. New Brunswick: Rutgers University Press, 1998.

Hilts, Philip J. *Protecting America's Health: The FDA, Business, and One-Hundred Years of Regulation*. New York: Alfred A. Knopf, 2003.

Hinman, Alan R., and Walter A. Orenstein. "A Shot at Protection: Immunizations Against Infectious Disease." In *Silent Victories: The History and Practice of Public Health in Twentieth-Century America*, edited by John W. Ward and Christian Warren, 63–80. New York: Oxford University Press, 2007.

Jacobs, Charlotte DeCroes. *Jonas Salk: A Life*. New York: Oxford University Press, 2015.

Jones, David S., and Robert L. Martenson, "Human Radiation Experiments and the Formation of Medical Physics at the University of California, San Francisco and Berkeley, 1937–1962." In *Useful Bodies: Humans in the Service of Medical Science in the Twentieth Century*, edited by Jordan Goodman, Anthony McElligott, and Lara Marks. Baltimore: Johns Hopkins University Press, 2003.

Joy, Robert J. "Tetanus." In *The Cambridge World History of Human Disease*, edited by Kenneth F. Kiple, 1043–46. New York: Cambridge University Press, 1993.

Kampmeier, R. H. "Varicella Zoster." In *The Cambridge World History of Human Disease*, edited by Kenneth F. Kiple, 1092–94. New York: Cambridge University Press, 1993.

Kang, Lydia, and Nate Pedersen. *Quackery: A Brief History of the Worst Ways to Cure Everything*. New York: Workman Publishing Company, 2017.

Kim-Farley, Robert J. "Mumps." In *The Cambridge World History of Human Disease*, edited by Kenneth F. Kiple, 1087–88. New York: Cambridge University Press, 1993.

King, Charles R. "Infant Mortality." In *Childhood in America*, edited by Paula S. Fass and Mary Ann Mason, 427–31. New York: New York University Press, 2000.

Kelley, Florence. "Children Bureau Bill Again Before Congress" (1909), In *Childhood in America*, edited by Paula S. Fass and Mary Ann Mason, 556–57. New York: New York University Press, 2000.

Kelley, Florence. "The Federal Government and the Working Children" (1906). In *Childhood in America*, edited by Paula S. Fass and Mary Ann Mason, 555–56. New York: New York University Press, 2000.

Kennedy, Robert F., Jr. Editor. *Thimerosal: Let the Science Speak.* New York: Skyhorse Publishing, 2014.

Kirschmann, Anne Taylor. *A Vital Force: Women in American Homeopathy.* New Brunswick, NJ: Rutgers University Press, 2004.

Kirschmann, Anne Taylor. "Making Friends for 'Pure' Homeopathy: Hahnemannians and the Twentieth-Century Preservation and Transformation of Homeopathy." In *The Politics of Healing: Histories of Alternative Medicine in Twentieth-Century North America*, edited by Robert D. Johnston, 29–42. New York: Routledge, 2004.

Klein, Aaron E. *Trial by Fury: The Polio Vaccine Controversy.* New York: Charles Scribner's Sons, 1972.

Kluger, Jeffrey. *Splendid Solution: Jonas Salk and the Conquest of Polio.* New York: G. P. Putnum's, 2004.

Kotelchuck, Milton. "Safe Mothers, Healthy Babies: Reproductive Health in the Twentieth Century." In *Silent Victories: The History and Practice of Public*

Health in Twentieth-Century America, edited by John W. Ward and Christian Warren, 105–34. New York: Oxford University Press, 2007.

LaFollette, Marcel C. *Making Science Our Own: Public Images of Science, 1910–1955*. Chicago: University of Chicago Press, 1990.

Leavitt, Judith Walzer. "'Be Safe. Be Sure.' New York City's Experience with Epidemic Smallpox." In *Sickness and Health in America: Readings in the History of Medicine and Public Health*, edited by Judith Walzer Leavitt and Ronald L. Numbers, 407–17. Madison: University of Wisconsin Press, 1997.

Lederer, Susan E. *Subjected to Science: Human Experimentation in America Before the Second World War*. Baltimore: Johns Hopkins University Press, 1995.

Lederer, Susan E. "Medical Ethics and the Media: Oaths, Codes, and Popular Culture." In *The American Ethics Revolution: How the AMA's Code of Ethics Has Transformed Physicians' Relationships to Patients, Professionals, and Society*, edited by Robert B. Baker, et al., 91–103. Baltimore: Johns Hopkins University Press, 1999.

Lederer, Susan E., and Michael A. Grodin. "Historical Overview: Pediatric Experimentation." In *Children as Research Subjects: Science, Ethics, and Law*, edited by Michael A. Grodin and Leonard H. Glantz, 3–20. New York: Oxford University Press, 1994.

Lederer, Susan E., and Naomi Rogers. "Media." In *Medicine in the Twentieth Century*, edited by Roger Cooter and John Pickstone, 487–502. Amsterdam: Harwood Academic Publishers, 2000.

Levine, Susan. *School Lunch Politics: The Surprising History of America's Favorite Welfare Program*. Princeton: Princeton University Press, 2008.

Levitt, Alexandra M., D. Peter Drotman, and Stephen Ostroff. "Control of Infectious Diseases: A Twentieth-Century Public Health Achievement." In *Silent Victories: The History and Practice of Public Health in Twentieth-Century America*, edited by John W. Ward and Christian Warren, 3–17. New York: Oxford University Press, 2007.

Lovett, Robert W. *Infantile Paralysis in Vermont, 1894–1922: A Memorial to Charles S. Caverly*. Burlington, VT: Vermont State Department of Health, 1924.

Lowry, Richard S. "Lewis Hine's Family Romance." In *The American Child: A Cultural Studies Reader*, edited by Caroline F. Levander and Carol J. Singley, 184–207. New Brunswick, NJ: Rutgers University Press, 2003.

Markel, Howard. *Quarantine! East European Jewish Immigrants and the New York City Epidemics of 1892*. Baltimore: Johns Hopkins University Press, 1997.

Marshall, P. David. "New Media—New Self: The Changing Power of Celebrity." In *The Celebrity Culture Reader*, edited by David P. Marshall, 634–44. New York: Routledge, 2006.

Mason, Mary Ann. "The State as Superparent." In *Childhood in America*, edited by Paula S. Fass and Mary Ann Mason, 549–54. New York: New York University Press, 2000.

Mawdsley, Stephen E. *Selling Science: Polio and the Promise of Gamma Globulin*. New Brunswick: Rutgers University Press, 2016.

McCarthy, Jenny. *Mother Warriors: A Nation of Parents Healing Autism Against All Odds*. New York: Plume Books, 2008.

Meckel, Richard A. *Classrooms and Clinics: Urban Schools and the Protection and Promotion of Child Health, 1870–1930*. New Brunswick: Rutgers University Press, 2013.

Mnookin, Seth. *The Panic Virus: The True Story Behind the Vaccine-Autism Controversy*. New York: Simon and Schuster, 2011.

Moran, Joe. "The Reign of Hype: The Contemporary (Literary) Star System." In *The Celebrity Culture Reader*, edited by David P. Marshall, 324–44. New York: Routledge, 2006.

Morantz-Sanchez, Regina. *Sympathy and Science: Women Physicians in American Medicine*. Chapel Hill: North Carolina University Press, 2000.

Moreno, Jonathan D. *Undue Risk: Secret State Experiments on Humans*. New York: Routledge, 2001.

Offit, Paul A. *The Cutter Incident: How America's First Polio Vaccine Led to the Growing Vaccine Crisis*. New Haven: Yale University Press, 2005.

Offit, Paul A. *Vaccinated: One Man's Quest to Defeat the World's Deadliest Diseases*. New York: Smithsonian Books, 2007.

Offit, Paul A. *Deadly Choices: How the Anti-Vaccine Movement Threatens Us All*. New York: Basic Books, 2011.

Offit, Paul A. *Bad Faith: When Religious Belief Undermines Modern Medicine*. New York: Basic Books, 2015.

Oliver, Wade W. *The Man Who Lived for Tomorrow: A Biography of William Hallock Park*. New York: E. P. Dutton & Co., 1941.

Oshinsky, David M. *Polio: An American Story*. New York: Oxford University Press, 2005.

Pence, Gregory E. *Classic Cases in Medical Ethics*. 4th Edition. Boston: McGraw-Hill, 2004.

Pendergrast, Mark. *Inside the Outbreaks: The Elite Medical Detectives of the Epidemic Intelligence Service*. New York: Mariner Books, 2011.

Perley, Sharon, Sev S. Fluss, Zbigniew Bankowski, and Françoise Simon. "The Nuremberg Code: An International Overview." In *Nuremberg Code: Human Rights in Human Experimentation*, edited by George J. Annas and Michael A. Grodin, 149–73. New York: Oxford University Press, 1992.

Pickstone, John. "Medicine, Society, and the State." In *The Cambridge Illustrated History of Medicine*, edited by Roy Porter, 304–41. Cambridge: Cambridge University Press, 2001.

Porter, Roy. *The Greatest Benefit to Mankind: A Medical History of Humanity.* New York: W. W. Norton Company, 1997.

Porter, Roy. "What Is Disease?" In *The Cambridge Illustrated History of Medicine*, edited by Roy Porter, 82–117. Cambridge: Cambridge University Press, 2001.

Prescott, Heather Munro. *A Doctor of Their Own: The History of Adolescent Medicine.* Cambridge: Harvard University Press, 1998.

Preston, Samuel H., and Michael R. Haines. *Fatal Years: Child Mortality in Late Nineteenth-Century America.* Princeton: University of Princeton Press, 1991.

Proctor, Robert N. "Nazi Doctors, Racial Medicine, and Human Experimentation." In *The Nazi Doctors and the Nuremberg Code: Human Rights in Human Experimentation*, edited by George J. Annas and Michael A. Grodin, 17–31. New York: Oxford University Press, 1992.

Raphael, Jackie, Basuli Deb, and Nidhi Shrivastava. Editors. *Building Bridges in Celebrity Studies.* Toronto: Waterhill Publishers, 2016.

Reagan, Leslie J., Nancy Tomes, and Paula A. Treichler. *Medicine's Moving Pictures: Medicine, Health, and Bodies in American Film and Television.* Rochester, NY: University of Rochester Press, 2007.

Reese, William J. *Power and the Promise of School Reform: Grass-Roots Movements During the Progressive Era.* Boston: Routledge and Kegan Paul, 1986.

Reilly, Gretchen Ann. "'Not a So-Called Democracy:' Anti-Fluoridationists and the Fight Over Drinking Water." In *The Politics of Healing: Histories of Alternative Medicine in Twentieth-Century North America*, edited by Robert D. Johnston, 131–50. New York: Routledge, 2004.

Reilly, Gretchen Ann. "The Task Is a Political One: The Promotion of Fluoridation." In *Silent Victories: The History and Practice of Public Health in Twentieth-Century America*, edited by John W. Ward and Christian Warren, 323–40. New York: Oxford University Press, 2007.

Riis, Jacob A. *How the Other Half Lives.* 1901; Rpt. New York: Dover Publications, 1971.

Riis, Jacob A. "The Child-Saving Movement." In *Childhood in America*, edited by Paula S. Fass and Mary Ann Mason, 539–43. New York: New York University Press, 2000.

Rivera, Geraldo. "Welcome to Willowbrook." In *Childhood in America*, edited by Paula S. Fass and Mary Ann Mason, 436–37. New York: New York University Press, 2000.

Roberts, Judith. "Commodifying Celebrity: Social Media, Sensationalism, and How the Media Plays a Role in Creating Celebrities." In *Building Bridges in Celebrity Studies*, edited by Jackie Raphael, Basuli Deb, and Nidhi Shrivastava, 45–53. Toronto: Waterhill Publishers, 2016.

Rogers, Naomi. *Dirt and Disease: Polio Before FDR*. New Brunswick, NJ: Rutgers University Press, 1996.

Rogers, Naomi. "Sister Kenny Goes to Washington: Polio, Populism, and Medical Politics in Postwar America." In *The Politics of Healing: Histories of Alternative Medicine in Twentieth-Century North America*, edited by Robert D. Johnston, 97–116. New York: Routledge, 2004.

Rogers, Naomi. "American Medicine and the Politics of Filmmaking: *Sister Kenny*." In *Medicine's Moving Pictures: Medicine, Health, and Bodies in American Film and Television*, edited by Leslie J. Reagan, Nancy Tomes, and Paula A. Treichler, 199–238. Rochester, NY: University of Rochester Press, 2007.

Rogers, Naomi. "Polio Can Be Conquered: Science and Health Propaganda in the United States from Polio Polly to Jonas Salk." In *Silent Victories: The History and Practice of Public Health in Twentieth-Century America*, edited by John W. Ward and Christian Warren, 81–101. New York: Oxford University Press, 2007.

Rogers, Naomi. *Polio Wars: Sister Kenny and the Golden Age of American Medicine*. New York: Oxford University Press, 2014.

Rose, David W. *Images of America: March of Dimes*. Charleston, SC: Arcadia Publishing, 2003.

Rosenberg, Charles E. *The Care of Strangers: The Rise of America's Hospital System*. Baltimore: Johns Hopkins University Press, 1987.

Rosenberg, Charles E. "Codes Visible and Invisible: The Twentieth-Century Fate of a Nineteenth-Century Code." In *The American Ethics Revolution: How the AMA's Code of Ethics Has Transformed Physicians' Relationships to Patients, Professionals, and Society*, edited by Robert B. Baker, et al., 207–17. Baltimore: Johns Hopkins University Press, 1999.

Ross, Lainie Friedman. *Children in Medical Research: Access Versus Protection*. New York: Oxford University Press, 2006.

Rothman, David J. *Strangers at the Bedside: A History of How Law and Bioethics Transformed Medical Decision Making*. New York: Basic Books, 1991.

Rothman, David J., and Sheila M. Rothman. *The Willowbrook Wars: Bringing the Mentally Disabled into the Community*. 1984; Rpt. New Brunswick, NJ: Transaction Press, 2005.

Schmidt, Ulf. *Justice at Nuremberg: Leo Alexander and the Nazi Doctors' Trial*. New York: Palgrave Macmillan, 2004.

Schultz, William J. *The Humane Movement in the United States, 1910–1922*. 1924; Rpt. New York: AMS Press, 1968.

Sears, Robert W. *The Vaccine Book: Making the Right Decision for Your Child*. New York: Little, Brown and Company, 2011.

Sherman, Josepha. *Johnny Gruelle and the Story of Raggedy Ann and Andy*. Hockessin, DE: Mitchell Lane Publishers, 2006.

Smith, Jane S. *Patenting the Sun: Polio and the Salk Vaccine*. New York: William Morrow and Company, 1990.

Song, Sookyung. "The International Rights Standard." In *Vaccine Epidemic: How Corporate Greed, Biased Science, and Coercive Government Threaten Our Human Rights, Our Health, and Our Children*, edited by Louise Kuo Habakus and Mary Holland, 15–19. New York: Skyhorse Publishing, 2012.

Sontag, Susan. *Illness as Metaphor and AIDS and Its Metaphors*. New York: Picador, 1989.

Starr, Paul. *The Social Transformation of American Medicine*. New York: Basic Books, 1982.

Stephens, Martha. *The Treatment: The Story of Those Who Died in the Cincinnati Radiation Tests*. Durham: Duke University Press, 2002.

Stern, Alexandra Minna. "Better Babies Contests at the Indiana State Fair: Child Health, Scientific Motherhood, and Eugenics in the Midwest, 1920–35." In *Formative Years: Children's Health in the United States, 1880–2000*, edited by Alexandra Minna Stern and Howard Markel, 121–52. Ann Arbor: University of Michigan Press, 2004.

Tomes, Nancy. *The Gospel of Germs: Men, Women, and the Microbe in American Life*. Cambridge: Harvard University Press, 1998.

Turner, Graeme. *Understanding Celebrity*. Los Angeles: Sage, 2014.

Turner, James. "Due Process and the American Government." In *Vaccine Epidemic: How Corporate Greed, Biased Science, and Coercive Government Threaten Our Human Rights, Our Health, and Our Children*, edited by Louise Kuo Habakus and Mary Holland, 29–41. New York: Skyhorse Publishing, 2012.

U.S. News and World Report. "If Polio Strikes, What to Do" (7/17/53). In *Childhood in America*, edited by Paula S. Fass and Mary Ann Mason, 432–35. New York: New York University Press, 2000.

Vaughn, Roger. *Listen to the Music: The Life of Hilary Koprowski*. Oxford, MD: Roger Vaughn, 1999.

Viner, Russell, and Janet Goldman. "Children's Experiences of Illness." In *Medicine in the Twentieth Century*, edited by Roger Cooter and John Pickstone, 575–83. Australia: Harwood Academic Publishers, 2000.

Wagner, William. "God, Government, and Parental Rights." In *Vaccine Epidemic: How Corporate Greed, Biased Science, and Coercive Government Threaten Our Human Rights, Our Health, and Our Children*, edited by Louise Kuo Habakus and Mary Holland, 53–57. New York: Skyhorse Publishing, 2012.

Wald, Lillian. "The House on Henry Street" (1915). In *Childhood in America*, edited by Paula S. Fass and Mary Ann Mason, 557–58. New York: New York University Press, 2000.

Washington, Harriet A. *Medical Apartheid: The Dark History of Medical Experimentation on Black Americans from Colonial Times to the Present.* New York: Doubleday Books, 2006.

Whipple, Guy M. Editor. *Preschool and Parental Education.* National Society for the Study of Education Yearbook, Vol. 14, No. 28. Bloomington, IL: Public School Publishing Company, 1929.

Willrich, Michael. *Pox: An American History.* New York: Penguin Books, 2011.

Williams, Gareth. *Angel of Death: The Story of Smallpox.* New York: Palgrave Macmillan, 2011.

Williams, Gareth. *Paralyzed with Fear: The Story of Polio.* New York: Palgrave Macmillan, 2013.

Wilson, Charles C. *Health Education: A Guide for Teachers and a Text for Teacher Education.* 4th Edition. Washington, DC: National Education Association, 1948.

Wilson, Daniel J. *Living with Polio: The Epidemic and Its Survivors.* Chicago: University of Chicago Press, 2005.

Wilson, Daniel J. *Polio.* Santa Barbara, CA: Greenwood Press, 2009.

Wilson, John R. *Margin of Safety: A Doctor Tells the Controversial Inside Story of the Development of the Polio Vaccine.* Garden City, NY: Doubleday and Co., 1963.

Wolf, Jacqueline H. "Saving Babies and Mothers: Pioneering Efforts to Decrease Infant and Maternal Mortality." In *Silent Victories: The History and Practice of Public Health in Twentieth-Century America*, edited by John W. Ward and Christian Warren, 135–60. New York: Oxford University Press, 2007.

Wooten, Heather Green. *The Polio Years in Texas: Battling a Terrifying Unknown.* College Station: Texas A&M University Press, 2009.

Zelizer, Viviana A. *Pricing the Priceless Child: The Changing Social Value of Children.* New York: Basic Books, 1985.

Zelizer, Viviana A. "The Changing Value of Children." In *Childhood in America*, edited by Paula S. Fass and Mary Ann Mason, 260–62. New York: New York University Press, 2000.

Ziporyn, Terra. *Disease in the Popular American Press: The Case of Diphtheria, Typhoid Fever, and Syphilis, 1870–1920.* Westport, CT: Greenwood Press, 1988.

Fiction

Lewis, Sinclair. *Arrowsmith.* New York: Harcourt, Brace and World, 1961.

Williams, Margery. *The Velveteen Rabbit or How Toys Become Real.* 1922; Rpt. New York: Henry Holt and Company, 1983.

Dissertations/Theses

Mawdsley, Stephen E. "Fighting Polio: Selling the Gamma Globulin Field Trials, 1950–1953." PhD dissertation, University of Cambridge, 2011.

Woolworth, Stephen T. "Conflict, Collaboration, and Concession: A Study of the Rise and Fall of Medical Authority in the Seattle Public Schools, 1892–1922." PhD dissertation, University of Washington, 2002.

General Reference Works

Altenbaugh, Richard J. *Historical Dictionary of American Education.* Westport, CT: Greenwood Press, 1999.

Bowman, John S. Editor. *The Cambridge Dictionary of American Biography.* Cambridge: Cambridge University Press, 1995.

Cliff, Andrew, Peter Haggett, and Matthew Smallman-Raynor. *Measles: An Historical Geography of a Major Human Viral Disease, from Global Expansion to Local Retreat, 1840–1990.* Cambridge, MA: Blackwell Publishers, 1993.

College of Physicians. *History of Vaccines.* Philadelphia: College of Physicians, 2013.

Foner, Eric, and John A. Garraty. Editors. *The Reader's Companion to American History.* Boston: Houghton Mifflin Co., 1991

Heilbron, J. L. Editor. *The Oxford Companion to the History of Modern Science.* New York: Oxford University Press, 2003.

Kiple, Kenneth F. Editor. *The Cambridge World History of Human Disease.* Cambridge: University of Cambridge Press, 1999.

Kohn, George Childs. Editor. *Encyclopedia of Plague and Pestilence: From Ancient Times to the Present.* New York: Checkmark Books, 2001.

Margotta, Roberto. *The History of Medicine.* New York: Smithmark Publishers, 1996.

Plotkin, Stanley A., and Edward A. Mortimer, Jr. Editors. *Vaccines.* Philadelphia: W.B. Saunders Company, 1994.

Porter, Roy. Editor. *The Cambridge Illustrated History of Medicine.* Cambridge: Cambridge University Press, 2001.

Media Sources: Magazines, Newspapers, Radio, Television

Deer, Brian. "Transcript: Royal Free Facilities Attack on MMR in Medical School Single Shots Videotape." (February 4, 1998). http://briandeer.com/wakefield/royal-video.htm (Accessed June 27, 2016).

"Disease Center Reports 10 Amish Polio Cases." *New York Times*, June 9, 1979: 8. http://pitt.idm.oclc.org/login?url://search.proquest.com/docview/1284141329?=14709 (Accessed June 25, 2016).

Dvorek, Peter. "Parents Who Refuse to Vaccinate Their Children Are Putting Others at Risk." *Washington Post*, January 27, 2015: No Pagination. http://search.proquest.com/docuview/1648471363? (Accessed January 31, 2015).

Ferrell, Tom, Virginia Adams, and Michael Wright. "Amish Polio Cases Cause New Alarm." *New York Times*, June 3, 1979: E9. http://pitt.oclc.org/login?url=http://search.proquest.com/docview120827897?accountid=14709 (Accessed June 25, 2016).

Freiwald, Colette. "Dr. Salk's Guinea Pig." *Pittsburgh Post-Gazette*, April 9, 2011: Section A, p. 15.

Honawar, Vaishali. "Four Amish Children Contract Polio Virus." *Education Week*, 25 (November 30, 2005): 15. http://go.galegroup.com/ps/i.do?id=GALE%7CA215123766&v=21&u=upitt_main&it=r&p=AONE&sw=w&asid=56abda1befbce83479e5a37ebeffaa62 (Accessed June 25, 2016).

Honigsbaum, Mark. "Vaccination: A Vexatious History." *Lancet*, 387 (May 14, 2016): 1988–89. http://dx.org/10.1016?S01040-6736(16)30474-3 (Accessed May 19, 2016).

Jacobs, Susan. "Watson Institute: Needs Have Changed, But Focus Is Still on Children with Disabilities." *Pittsburgh Post-Gazette*, January 16, 2002: Section N, 1–2.

Klein, Jeff. "Crosby Has Mumps, So Does Brassard." *New York Times*, December 15, 2014: Page D5. http://search.proquest.com/docview/1636086195? (Accessed February 3, 2015).

"Mailbag: The Vaccine Debate." *Los Angeles Times*, January 24, 2015: Page A13. http://search.proquest.com/docview/1647788338? (Accessed January 31, 2015).

McKay, Betsy, and Jeanne Whalen, "Vaccine Skeptics on the Rise–Latest Measles Outbreak Shines Light on New Trend Worrying Health Officials." *Wall Street Journal*, January 26, 2015: Page A3. http://search.proquest.com/docview/1648007423? (Accessed February 3, 2015). www.NVIC.org (Accessed June 22, 2016).

"Not Convinced on Vaccination." *Washington Post*, January 26, 2015: No Pagination. http://search.proquest.com/docuview/1649926257? (Accessed January 31, 2015).

"Obituary: Frank C. Robbins." *Pittsburgh Post-Gazette*, August 6, 2003: B-5.

"Obituary: Polio Pioneers Joseph Melnick, Dorothy M. Horstmann Die." *Washington Post*, January 22, 2001: B-6. http://pitt.idm.oclc.org/login?url=http://search.proquest.com/docview/409060435?accountid=14709 (Accessed November 7, 2015).

"O Pioneers!" *New Yorker*, May 1954: 24–26.

Oshinsky, David. "The Last Epidemic." *Wall Street Journal*, October 18, 2014: Page C1. http://search.proquest.com/docview/1613381472? (Accessed February 5, 2015).

Parker, Kathleen. "Seeking a Vaccine for Ignorance. How Would Salk be Treated If He Tried to Introduce His Polio Vaccine Today?" *Pittsburgh Post-Gazette*, February 4, 2015: Page B5.

Pennsylvania Coalition for Informed Consent. www.pacic.org (Accessed June 20, 2016).

Pugmire, Lance. "Mumps Is the Word on Two Ducks." *Los Angeles Times*, November 13, 2014: Page C3. http://search.proquest.com/docview/1623066791? (Accessed February 3, 2015).

Real Time with Bill Maher. Home Box Office, Inc., 2015 (Accessed April 24, 2015).

Templeton, David. "Despite Objections, Vaccinations Urged." *Pittsburgh Post-Gazette*, February 4, 2015: 1–2.

Today Show. April 13, 2016 (*NBC News*, 2016). http://www.today.com/...robert-de-niroon-anti-vaccine-film-controversy-let-s-find-out-the-truth-665031235642 (Accessed July 4, 2016).

Today Show. April 18, 2016 (*NBC News*, 2016). https://www.youtube.com/watch?v=W8c327UcjLQ (Accessed July 4, 2016).

Trace Amounts: Autism, Mercury, and the Hidden Truth. Gathr Films: Trace Amounts, LLC, 2015. https://traceamounts.vhx.tv/packages/trace.amounts (Accessed May 27, 2016).

Vaxxed: From Cover-Up to Catastrophe. Cinema Libre Studio, 2016 (Viewed at the Parkway Theater, McKees Rocks, PA. June 16, 2016).

"WHO Says Polio Virus Eliminated in Europe." *Pittsburgh Post-Gazette*, June 22, 2002: Section A, p. 2.

Autobiographies/Memoirs/Oral Histories

Daniel, Thomas M., and Frederick C. Robbins. Editors. *Polio*. Rochester, NY: University of Rochester Press, 1997.

de Kruif, Paul. *The Sweeping Wind: A Memoir*. New York: Harcourt, Brace and World, Inc., 1962.

Finger, Anne. *Elegy for a Disease: A Personal and Cultural History of Polio*. New York: St. Martin's Press, 2006.

Gould, Tony. *A Summer Plague: Polio and Its Survivors.* New Haven: Yale University Press, 1995.

Katz, Samuel L. "John F. Enders and Measles Virus Vaccine—A Reminiscence." In *Measles: History and Basic Biology*, edited by Daine E. Griffin and Michael B. A. Oldstone. Berlin: Springer-Verlag, 2009.

Paul, John R. *A History of Poliomyelitis.* New Haven: Yale University Press, 1971.

Rivers, Thomas M. *Reflections on a Life in Medicine and Science. An Oral History Prepared by Saul Benison.* Cambridge: Massachusetts Institute of Technology, 1967.

Seavey, Nina Gilden, Jane S. Smith, and Paul Wagner. *A Paralyzing Fear: The Triumph Over Polio in America.* New York: TV Books, 1998.

Serotte, Brenda. *The Fortune Teller's Kiss.* Lincoln: University of Nebraska Press, 2006.

Troan, John. *Passport to Adventure: Or, How a Typewriter from Santa Led to an Exciting Lifetime Journey.* Pittsburgh, PA: Geyer Printing, 2000.

Woods, Regina. *Tales from Inside the Iron Lung (And How I Got Out of It).* With a Forward by David E. Rogers, M. D. Philadelphia: University of Pennsylvania Press, 1994.

Zemke, Diane. *Polio: A Special Ride?* Minnetonka, MN: Diagnostic Center of Learning Patterns, Inc., 1997.

Index

R. J. Altenbaugh, *Vaccination in America*,
Palgrave Studies in the History of Science and Technology,
https://doi.org/10.1007/978-3-319-96349-5

B

V

The manufacturer's authorised representative in the EU is Springer Nature Customer Service Centre GmbH, Europaplatz 3, 69115 Heidelberg, Germany. If you have any concerns regarding our products, please contact ProductSafety@springernature.com

Printed and bound by CPI Group (UK) Ltd, Croydon, CR0 4YY

30/09/2026

02225653-0014